INTRODUCTION TO QUANTUM MECHANICS

With Applications to Chemistry

BY

LINUS PAULING

*Research Professor, Linus Pauling
Institute of Science and Medicine,
Palo Alto, California*

AND

E. BRIGHT WILSON, JR.

*Theodore William Richards
Professor of Chemistry,
Emeritus, Harvard University*

DOVER PUBLICATIONS, INC.
NEW YORK

Published in Canada by General Publishing Company, Ltd., 30 Lesmill Road, Don Mills, Toronto, Ontario.
Published in the United Kingdom by Constable and Company, Ltd., 10 Orange Street, London WC2H 7EG.

This Dover edition, first published in 1985, is an unabridged and unaltered republication of the work first published by The McGraw-Hill Book Co., New York, in 1935.

Manufactured in the United States of America
Dover Publications, Inc., 31 East 2nd Street, Mineola, N.Y. 11501

Library of Congress Cataloging in Publication Data

Pauling, Linus, 1901–
    Introduction to quantum mechanics.

    Reprint. Originally published: New York; London: McGraw-Hill, 1935.
    Bibliography: p.
    Includes index.
    1. Quantum theory. 2. Wave mechanics. 3. Chemistry, Physical and theoretical. I. Wilson, E. Bright (Edgar Bright), 1908– II. Title.
[QC174.12.P39   1985]   530.1'2   84-25919
ISBN 0-486-64871-0

PREFACE

In writing this book we have attempted to produce a textbook of practical quantum mechanics for the chemist, the experimental physicist, and the beginning student of theoretical physics. The book is not intended to provide a critical discussion of quantum mechanics, nor even to present a thorough survey of the subject. We hope that it does give a lucid and easily understandable introduction to a limited portion of quantum-mechanical theory; namely, that portion usually suggested by the name "wave mechanics," consisting of the discussion of the Schrödinger wave equation and the problems which can be treated by means of it. The effort has been made to provide for the reader a means of equipping himself with a practical grasp of this subject, so that he can apply quantum mechanics to most of the chemical and physical problems which may confront him.

The book is particularly designed for study by men without extensive previous experience with advanced mathematics, such as chemists interested in the subject because of its chemical applications. We have assumed on the part of the reader, in addition to elementary mathematics through the calculus, only some knowledge of complex quantities, ordinary differential equations, and the technique of partial differentiation. It may be desirable that a book written for the reader not adept at mathematics be richer in equations than one intended for the mathematician; for the mathematician can follow a sketchy derivation with ease, whereas if the less adept reader is to be led safely through the usually straightforward but sometimes rather complicated derivations of quantum mechanics a firm guiding hand must be kept on him. Quantum mechanics is essentially mathematical in character, and an understanding of the subject without a thorough knowledge of the mathematical methods involved and the results of their application cannot be obtained. The student not thoroughly trained in the theory of partial differential equations and orthogonal functions must

iii

learn something of these subjects as he studies quantum mechanics. In order that he may do so, and that he may follow the discussions given without danger of being deflected from the course of the argument by inability to carry through some minor step, we have avoided the temptation to condense the various discussions into shorter and perhaps more elegant forms.

After introductory chapters on classical mechanics and the old quantum theory, we have introduced the Schrödinger wave equation and its physical interpretation on a postulatory basis, and have then given in great detail the solution of the wave equation for important systems (harmonic oscillator, hydrogen atom) and the discussion of the wave functions and their properties, omitting none of the mathematical steps except the most elementary. A similarly detailed treatment has been given in the discussion of perturbation theory, the variation method, the structure of simple molecules, and, in general, in every important section of the book.

In order to limit the size of the book, we have omitted from discussion such advanced topics as transformation theory and general quantum mechanics (aside from brief mention in the last chapter), the Dirac theory of the electron, quantization of the electromagnetic field, etc. We have also omitted several subjects which are ordinarily considered as part of elementary quantum mechanics, but which are of minor importance to the chemist, such as the Zeeman effect and magnetic interactions in general, the dispersion of light and allied phenomena, and most of the theory of aperiodic processes.

The authors are severally indebted to Professor A. Sommerfeld and Professors E. U. Condon and H. P. Robertson for their own introduction to quantum mechanics. The constant advice of Professor R. C. Tolman is gratefully acknowledged, as well as the aid of Professor P. M. Morse, Dr. L. E. Sutton, Dr. G. W. Wheland, Dr. L. O. Brockway, Dr. J. Sherman, Dr. S. Weinbaum, Mrs. Emily Buckingham Wilson, and Mrs. Ava Helen Pauling.

LINUS PAULING.
E. BRIGHT WILSON, JR.

PASADENA, CALIF.,
CAMBRIDGE, MASS.,
July, 1935.

CONTENTS

v

CHAPTER III

THE SCHRÖDINGER WAVE EQUATION WITH THE
HARMONIC OSCILLATOR AS AN EXAMPLE

CONTENTS

CHAPTER XI

PERTURBATION THEORY INVOLVING THE TIME, THE
EMISSION AND ABSORPTION OF RADIATION, AND THE
RESONANCE PHENOMENON

CHAPTER XII

THE STRUCTURE OF SIMPLE MOLECULES

CONTENTS

INTRODUCTION TO QUANTUM MECHANICS

CHAPTER I

SURVEY OF CLASSICAL MECHANICS

The subject of quantum mechanics constitutes the most recent step in the very old search for the general laws governing the motion of matter. For a long time investigators confined their efforts to studying the dynamics of bodies of macroscopic dimensions, and while the science of mechanics remained in that stage it was properly considered a branch of physics. Since the development of atomic theory there has been a change of emphasis. It was recognized that the older laws are not correct when applied to atoms and electrons, without considerable modification. Moreover, the success which has been obtained in making the necessary modifications of the older laws has also had the result of depriving physics of sole claim upon them, since it is now realized that the combining power of atoms and, in fact, all the chemical properties of atoms and molecules are explicable in terms of the laws governing the motions of the electrons and nuclei composing them.

Although it is the modern theory of quantum mechanics in which we are primarily interested because of its applications to chemical problems, it is desirable for us first to discuss briefly the background of classical mechanics from which it was developed. By so doing we not only follow to a certain extent the historical development, but we also introduce in a more familiar form many concepts which are retained in the later theory. We shall also treat certain problems in the first few chapters by the methods of the older theories in preparation for their later treatment by quantum mechanics. It is for this reason that the student is advised to consider the exercises of the first few chapters carefully and to retain for later reference the results which are secured.

In the first chapter no attempt will be made to give any parts of classical dynamics but those which are useful in the treatment of atomic and molecular problems. With this restriction, we have felt justified in omitting discussion of the dynamics of rigid bodies, non-conservative systems, non-holonomic systems, systems involving impact, etc. Moreover, no use is made of Hamilton's principle or of the Hamilton-Jacobi partial differential equation. By thus limiting the subjects to be discussed, it is possible to give in a short chapter a thorough treatment of Newtonian systems of point particles.

1. NEWTON'S EQUATIONS OF MOTION IN THE LAGRANGIAN FORM

The earliest formulation of dynamical laws of wide application is that of Sir Isaac Newton. If we adopt the notation x_i, y_i, z_i for the three Cartesian coordinates of the ith particle with mass m_i, Newton's equations for n point particles are

$$\left.\begin{array}{l} m_i\ddot{x}_i = X_i, \\ m_i\ddot{y}_i = Y_i, \\ m_i\ddot{z}_i = Z_i, \end{array}\right\} \quad i = 1, 2, \cdots, n, \qquad (1\text{-}1)$$

where X_i, Y_i, Z_i are the three components of the force acting on the ith particle. There is a set of such equations for each particle. Dots refer to differentiation with respect to time, so that

$$\ddot{x}_i = \frac{d^2x_i}{dt^2}. \qquad (1\text{-}2)$$

By introducing certain familiar definitions we change Equation 1–1 into a form which will be more useful later. We define as the *kinetic energy* T (for Cartesian coordinates) the quantity

$$T = \tfrac{1}{2}m_1(\dot{x}_1^2 + \dot{y}_1^2 + \dot{z}_1^2) + \cdots + \tfrac{1}{2}m_n(\dot{x}_n^2 + \dot{y}_n^2 + \dot{z}_n^2)$$
$$= \tfrac{1}{2}\sum_{i=1}^{n}m_i(\dot{x}_i^2 + \dot{y}_i^2 + \dot{z}_i^2). \qquad (1\text{-}3)$$

If we limit ourselves to a certain class of systems, called *conservative systems*, it is possible to define another quantity, the *potential energy* V, which is a function of the coordinates $x_1y_1z_1 \cdots x_ny_nz_n$ of all the particles, such that the force components acting on each particle are equal to partial derivatives of the potential energy with respect to the coordinates of the particle (with negative sign); that is,

$$\left.\begin{aligned} X_i &= -\frac{\partial V}{\partial x_i}, \\ Y_i &= -\frac{\partial V}{\partial y_i}, \\ Z_i &= -\frac{\partial V}{\partial z_i}, \end{aligned}\right\} \qquad i = 1, 2, \cdots, n. \qquad (1\text{-}4)$$

It is possible to find a function V which will express in this manner forces of the types usually designated as mechanical, electrostatic, and gravitational. Since other types of forces (such as electromagnetic) for which such a potential-energy function cannot be set up are not important in chemical applications, we shall not consider them in detail.

With these definitions, Newton's equations become

$$\frac{d}{dt}\frac{\partial T}{\partial \dot{x}_i} + \frac{\partial V}{\partial x_i} = 0, \qquad (1\text{-}5a)$$

$$\frac{d}{dt}\frac{\partial T}{\partial \dot{y}_i} + \frac{\partial V}{\partial y_i} = 0, \qquad (1\text{-}5b)$$

$$\frac{d}{dt}\frac{\partial T}{\partial \dot{z}_i} + \frac{\partial V}{\partial z_i} = 0. \qquad (1\text{-}5c)$$

There are three such equations for every particle, as before. These results are definitely restricted to Cartesian coordinates; but by introducing a new function, the *Lagrangian function L*, defined for Newtonian systems as the difference of the kinetic and potential energy,

$$L = L(x_1, y_1, z_1, \cdots, x_n, y_n, z_n, \dot{x}_1, \cdots, \dot{z}_n) = T - V, \qquad (1\text{-}6)$$

we can throw the equations of motion into a form which we shall later prove to be valid in any system of coordinates (Sec. 1c). In Cartesian coordinates T is a function of the velocities $\dot{x}_1, \cdots, \dot{z}_n$ only, and for the systems to which our treatment is restricted V is a function of the coordinates only; hence the equations of motion given in Equation 1–5 on introduction of the function L assume the form

$$\left. \begin{aligned} \frac{d}{dt}\frac{\partial L}{\partial \dot{x}_i} - \frac{\partial L}{\partial x_i} &= 0, \\ \frac{d}{dt}\frac{\partial L}{\partial \dot{y}_i} - \frac{\partial L}{\partial y_i} &= 0, \\ \frac{d}{dt}\frac{\partial L}{\partial \dot{z}_i} - \frac{\partial L}{\partial z_i} &= 0, \end{aligned} \right\} \quad i = 1, 2, \cdots, n. \quad (1\text{--}7)$$

In the following paragraphs a simple dynamical system is discussed by the use of these equations.

1a. The Three-dimensional Isotropic Harmonic Oscillator.— As an illustration of the use of the equations of motion in this form, we choose a system which has played a very important part in the development of quantum theory. This is the *harmonic oscillator*, a particle bound to an equilibrium position by a force which increases in magnitude linearly with its distance r from the point. In the three-dimensional isotropic harmonic oscillator this corresponds to a potential function $\frac{1}{2}kr^2$, representing a force of magnitude kr acting in a negative direction; i.e., from the position of the particle to the origin. k is called the *force constant* or *Hooke's-law constant*. Using Cartesian coordinates we have

$$L = \tfrac{1}{2}m(\dot{x}^2 + \dot{y}^2 + \dot{z}^2) - \tfrac{1}{2}k(x^2 + y^2 + z^2), \quad (1\text{--}8)$$

whence

$$\left. \begin{aligned} \frac{d}{dt}(m\dot{x}) + kx = m\ddot{x} + kx &= 0, \\ m\ddot{y} + ky &= 0, \\ m\ddot{z} + kz &= 0. \end{aligned} \right\} \quad (1\text{--}9)$$

Multiplication of the first member of Equation 1–9 by $\dot{x}$ gives

$$m\dot{x}\frac{d\dot{x}}{dt} = -kx\frac{dx}{dt} \quad (1\text{--}10)$$

or

$$\frac{1}{2}m\frac{d(\dot{x})^2}{dt} = -\frac{1}{2}k\frac{d(x^2)}{dt}, \quad (1\text{--}11)$$

which integrates directly to

$$\tfrac{1}{2}m\dot{x}^2 = -\tfrac{1}{2}kx^2 + \text{constant}. \quad (1\text{--}12)$$

The constant of integration is conveniently expressed as $\frac{1}{2}kx_0^2$.

Hence

$$\frac{dx}{dt} = \sqrt{\frac{k}{m}(x_0^2 - x^2)}, \qquad (1\text{-}13)$$

or, on introducing the expression $4\pi^2 m\nu_0^2$ in place of the force constant k,

$$2\pi\nu_0 dt = \frac{dx}{(x_0^2 - x^2)^{\frac{1}{2}}},$$

which on integration becomes

$$2\pi\nu_0 t + \delta_x = \sin^{-1}\frac{x}{x_0}$$

or

$$x = x_0 \sin(2\pi\nu_0 t + \delta_x), \qquad (1\text{-}14)$$

and similarly

$$\left.\begin{aligned} y &= y_0 \sin(2\pi\nu_0 t + \delta_y), \\ z &= z_0 \sin(2\pi\nu_0 t + \delta_z). \end{aligned}\right\} \qquad (1\text{-}15)$$

In these expressions x_0, y_0, z_0, δ_x, δ_y, and δ_z are constants of integration, the values of which determine the motion in any given case. The quantity ν_0 is related to the constant of the restoring force by the equation

$$4\pi^2 m\nu_0^2 = k, \qquad (1\text{-}16)$$

so that the potential energy may be written as

$$V = 2\pi^2 m\nu_0^2 r^2. \qquad (1\text{-}17)$$

As shown by the equations for x, y, and z, ν_0 is the frequency of the motion. It is seen that the particle may be described as carrying out independent harmonic oscillations along the x, y, and z axes, with different amplitudes x_0, y_0, and z_0 and different phase angles δ_x, δ_y, and δ_z, respectively.

The energy of the system is the sum of the kinetic energy and the potential energy, and is thus equal to

$$\tfrac{1}{2}m(\dot{x}^2 + \dot{y}^2 + \dot{z}^2) + 2\pi^2 m\nu_0^2(x^2 + y^2 + z^2).$$

On evaluation, it is found to be independent of the time, with the value $2\pi^2 m\nu_0^2(x_0^2 + y_0^2 + z_0^2)$ determined by the amplitudes of oscillation.

The one-dimensional harmonic oscillator, restricted to motion along the x axis in accordance with the potential function $V = \tfrac{1}{2}kx^2 = 2\pi^2 m\nu_0^2 x^2$, is seen to carry out harmonic oscillations

along this axis as described by Equation 1–14. Its total energy
is given by the expression $2\pi^2 m \nu_0^2 x_0^2$.

1b. Generalized Coordinates.—Instead of Cartesian coordinates $x_1, y_1, z_1, \cdots, x_n, y_n, z_n$, it is frequently more convenient
to use some other set of coordinates to specify the configuration
of the system. For example, the isotropic spatial harmonic
oscillator already discussed might equally well be described using
polar coordinates; again, the treatment of a system composed of
two attracting particles in space, which will be considered
later, would be very cumbersome if it were necessary to use
rectangular coordinates.

If we choose any set of $3n$ coordinates, which we shall always
assume to be independent and at the same time sufficient in
number to specify completely the positions of the particles of
the system, then there will in general exist $3n$ equations, called
the *equations of transformation*, relating the new coordinates
q_k to the set of Cartesian coordinates x_i, y_i, z_i,

$$\left.\begin{array}{l} x_i = f_i(q_1, q_2, \cdots, q_{3n}), \\ y_i = g_i(q_1, q_2, \cdots, q_{3n}), \\ z_i = h_i(q_1, q_2, \cdots, q_{3n}). \end{array}\right\} \quad (1\text{–}18)$$

There is such a set of three equations for each particle i. The
functions f_i, g_i, h_i may be functions of any or all of the $3n$ new
coordinates q_k, so that these new variables do not necessarily
split into sets which belong to particular particles. For example,
in the case of two particles the six new coordinates may be the
three Cartesian coordinates of the center of mass together
with the polar coordinates of one particle referred to the other
particle as origin.

As is known from the theory of partial differentiation, it is
possible to transform derivatives from one set of independent
variables to another, an example of this process being

$$\frac{dx_i}{dt} = \frac{\partial x_i}{\partial q_1}\frac{dq_1}{dt} + \frac{\partial x_i}{\partial q_2}\frac{dq_2}{dt} + \cdots + \frac{\partial x_i}{\partial q_{3n}}\frac{dq_{3n}}{dt}. \quad (1\text{–}19a)$$

This same equation can be put in the much more compact form

$$\dot{x}_i = \sum_{j=1}^{3n} \frac{\partial x_i}{\partial q_j}\dot{q}_j. \quad (1\text{–}19b)$$

This gives the relation between any Cartesian component of velocity and the time derivatives of the new coordinates. Similar relations, of course, hold for $\dot{y}_i$ and $\dot{z}_i$ for any particle. The quantities $\dot{q}_j$, by analogy with $\dot{x}_i$, are called *generalized velocities*, even though they do not necessarily have the dimensions of length divided by time (for example, q_j may be an angle).

Since partial derivatives transform in just the same manner, we have

$$-\frac{\partial V}{\partial q_j} = -\frac{\partial V}{\partial x_1}\frac{\partial x_1}{\partial q_j} - \frac{\partial V}{\partial y_1}\frac{\partial y_1}{\partial q_j} - \cdots - \frac{\partial V}{\partial z_n}\frac{\partial z_n}{\partial q_j}$$

$$= -\sum_{i=1}^{n}\left(\frac{\partial V}{\partial x_i}\frac{\partial x_i}{\partial q_j} + \frac{\partial V}{\partial y_i}\frac{\partial y_i}{\partial q_j} + \frac{\partial V}{\partial z_i}\frac{\partial z_i}{\partial q_j}\right) = Q_j. \quad (1\text{--}20)$$

Since Q_j is given by an expression in terms of V and q_j which is analogous to that for the force X_i in terms of V and x_i, it is called a *generalized force*.

In exactly similar fashion, we have

$$\frac{\partial T}{\partial \dot{q}_j} = \sum_{i=1}^{n}\left(\frac{\partial T}{\partial \dot{x}_i}\frac{\partial \dot{x}_i}{\partial \dot{q}_j} + \frac{\partial T}{\partial \dot{y}_i}\frac{\partial \dot{y}_i}{\partial \dot{q}_j} + \frac{\partial T}{\partial \dot{z}_i}\frac{\partial \dot{z}_i}{\partial \dot{q}_j}\right). \quad (1\text{--}21)$$

1c. The Invariance of the Equations of Motion in the Lagrangian Form.—We are now in a position to show that when Newton's equations are written in the form given by Equation 1–7 they are valid for any choice of coordinate system. For this proof we shall apply a transformation of coordinates to Equations 1–5, using the methods of the previous section. Multiplication of Equation 1–5a by $\dfrac{\partial x_i}{\partial q_j}$, of 1–5b by $\dfrac{\partial y_i}{\partial q_j}$, etc., gives

$$\left.\begin{aligned}
\frac{\partial x_1}{\partial q_j}\frac{d}{dt}\frac{\partial T}{\partial \dot{x}_1} + \frac{\partial V}{\partial x_1}\frac{\partial x_1}{\partial q_j} &= 0, \\
\frac{\partial x_2}{\partial q_j}\frac{d}{dt}\frac{\partial T}{\partial \dot{x}_2} + \frac{\partial V}{\partial x_2}\frac{\partial x_2}{\partial q_j} &= 0, \\
\cdots\cdots\cdots\cdots\cdots\cdots&\cdots, \\
\frac{\partial x_n}{\partial q_j}\frac{d}{dt}\frac{\partial T}{\partial \dot{x}_n} + \frac{\partial V}{\partial x_n}\frac{\partial x_n}{\partial q_j} &= 0,
\end{aligned}\right\} \quad (1\text{--}22)$$

with similar equations in y and z. Adding all of these together gives

$$\sum_{i=1}^{n}\left\{\frac{\partial x_i}{\partial q_j}\frac{d}{dt}\frac{\partial T}{\partial \dot{x}_i} + \frac{\partial y_i}{\partial q_j}\frac{d}{dt}\frac{\partial T}{\partial \dot{y}_i} + \frac{\partial z_i}{\partial q_j}\frac{d}{dt}\frac{\partial T}{\partial \dot{z}_i}\right\} + \frac{\partial V}{\partial q_j} = 0, \quad (1\text{-}23)$$

where the result of Equation 1-20 has been used. In order to reduce the first sum, we note the following identity, obtained by differentiating a product,

$$\frac{\partial x_i}{\partial q_j}\frac{d}{dt}\left(\frac{\partial T}{\partial \dot{x}_i}\right) = \frac{d}{dt}\left(\frac{\partial T}{\partial \dot{x}_i}\frac{\partial x_i}{\partial q_j}\right) - \frac{\partial T}{\partial \dot{x}_i}\frac{d}{dt}\left(\frac{\partial x_i}{\partial q_j}\right). \quad (1\text{-}24)$$

From Equation 1-19b we obtain directly

$$\frac{\partial \dot{x}_i}{\partial \dot{q}_j} = \frac{\partial x_i}{\partial q_j}. \quad (1\text{-}25)$$

Furthermore, because the order of differentiation is immaterial, we see that

$$\frac{d}{dt}\left(\frac{\partial x_i}{\partial q_j}\right) = \sum_{k=1}^{3n}\frac{\partial}{\partial q_k}\left(\frac{\partial x_i}{\partial q_j}\right)\dot{q}_k = \sum_{k=1}^{3n}\frac{\partial}{\partial q_j}\left(\frac{\partial x_i}{\partial q_k}\right)\dot{q}_k$$

$$= \frac{\partial}{\partial q_j}\sum_{k=1}^{3n}\left(\frac{\partial x_i}{\partial q_k}\right)\dot{q}_k = \frac{\partial \dot{x}_i}{\partial q_j}. \quad (1\text{-}26)$$

By introducing Equations 1-26 and 1-25 in 1-24 and using the result in Equation 1-23, we get

$$\sum_{i=1}^{n}\left\{\frac{d}{dt}\left(\frac{\partial T}{\partial \dot{x}_i}\frac{\partial \dot{x}_i}{\partial \dot{q}_j} + \frac{\partial T}{\partial \dot{y}_i}\frac{\partial \dot{y}_i}{\partial \dot{q}_j} + \frac{\partial T}{\partial \dot{z}_i}\frac{\partial \dot{z}_i}{\partial \dot{q}_j}\right) - \left(\frac{\partial T}{\partial \dot{x}_i}\frac{\partial \dot{x}_i}{\partial q_j} + \frac{\partial T}{\partial \dot{y}_i}\frac{\partial \dot{y}_i}{\partial q_j}\right.\right.$$

$$\left.\left. + \frac{\partial T}{\partial \dot{z}_i}\frac{\partial \dot{z}_i}{\partial q_j}\right)\right\} + \frac{\partial V}{\partial q_j} = 0, \quad (1\text{-}27)$$

which, in view of the results of the last section, reduces to

$$\frac{d}{dt}\frac{\partial T}{\partial \dot{q}_j} - \frac{\partial T}{\partial q_j} + \frac{\partial V}{\partial q_j} = 0. \quad (1\text{-}28)$$

Finally, the introduction of the Lagrangian function $L = T - V$, with V a function of the coordinates only, gives the more compact form

$$\frac{d}{dt}\frac{\partial L}{\partial \dot{q}_j} - \frac{\partial L}{\partial q_j} = 0, \quad j = 1, 2, 3, \cdots, 3n. \quad (1\text{-}29)$$

(It is important to note that L must be expressed as a function of the coordinates and their first time-derivatives.)

Since the above derivation could be carried out for any value of j, there are $3n$ such equations, one for each coordinate q_j. They are called the *equations of motion in the Lagrangian form* and are of great importance. The method by which they were derived shows that they are independent of the coordinate system.

We have so far rather limited the types of systems considered, but Lagrange's equations are much more general than we have indicated and *by a proper choice of the function L nearly all dynamical problems can be treated with their use*. These equations are therefore frequently chosen as the fundamental postulates of classical mechanics instead of Newton's laws.

1d. An Example: The Isotropic Harmonic Oscillator in Polar Coordinates.—The example which we have treated in Section 1a can equally well be solved by the use of polar coordinates r, ϑ, and φ (Fig. 1-1). The equations of transformation corresponding to Equation 1–18 are

$$\left.\begin{aligned} x &= r \sin \vartheta \cos \varphi, \\ y &= r \sin \vartheta \sin \varphi, \\ z &= r \cos \vartheta. \end{aligned}\right\} \tag{1-30}$$

With the use of these we find for the kinetic and potential energies of the isotropic harmonic oscillator the following expressions:

$$\left.\begin{aligned} T &= \frac{1}{2} m(\dot{x}^2 + \dot{y}^2 + \dot{z}^2) = \frac{m}{2}(\dot{r}^2 + r^2\dot{\vartheta}^2 + r^2 \sin^2 \vartheta \; \dot{\varphi}^2), \\ V &= 2\pi^2 m \nu_0^2 r^2, \end{aligned}\right\} \tag{1-31}$$

and

$$L = T - V = \frac{m}{2}(\dot{r}^2 + r^2\dot{\vartheta}^2 + r^2 \sin^2 \vartheta \dot{\varphi}^2) - 2\pi^2 m \nu_0^2 r^2. \tag{1-32}$$

The equations of motion are

$$\frac{d}{dt}\frac{\partial L}{\partial \dot{\varphi}} - \frac{\partial L}{\partial \varphi} = \frac{d}{dt}(mr^2 \sin^2 \vartheta \dot{\varphi}) = 0, \tag{1-33}$$

$$\frac{d}{dt}\frac{\partial L}{\partial \dot{\vartheta}} - \frac{\partial L}{\partial \vartheta} = \frac{d}{dt}(mr^2\dot{\vartheta}) - mr^2 \sin \vartheta \cos \vartheta \dot{\varphi}^2 = 0, \tag{1-34}$$

$$\frac{d}{dt}\frac{\partial L}{\partial \dot{r}} - \frac{\partial L}{\partial r} = \frac{d}{dt}(m\dot{r}) - mr\dot{\vartheta}^2 - mr \sin^2 \vartheta \dot{\varphi}^2 + 4\pi^2 m \nu_0^2 r = 0. \tag{1-35}$$

In Appendix II it is shown that the motion takes place in a plane containing the origin. This conclusion enables us to simplify the problem by making a change of variables. Let us introduce new polar coordinates r, ϑ', χ such that at the time $t = 0$ the plane determined by the vectors $\mathbf{r}$ and $\mathbf{v}$, the position and velocity vectors of the particle at $t = 0$, is normal to the new z' axis. This transformation is known in terms of the old set of coordinates if two parameters ϑ_0 and φ_0, determining the position of the axis z' in terms of the old coordinates, are given (Fig. 1–2).

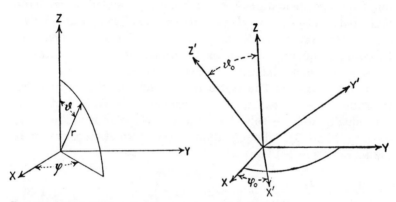

FIG. 1-1.—The relation of polar coordinates r, ϑ, and φ to Cartesian axes.   FIG. 1-2.—The rotation of axes.

In terms of the new coordinates, the Lagrangian function L and the equations of motion have the same form as previously, because the first choice of axis direction was quite arbitrary. However, since the coordinates have been chosen so that the plane of the motion is the $x'y'$ plane, the angle ϑ' is always equal to a constant, $\pi/2$. Inserting this value of ϑ' in Equation 1–33 and writing it in terms of χ instead of φ, we obtain

$$\frac{d}{dt}(mr^2\dot{\chi}) = 0, \qquad (1\text{–}36)$$

which has the solution

$$mr^2\dot{\chi} = p_\chi, \text{ a constant.} \qquad (1\text{–}37)$$

The r equation, Equation 1–35, becomes

$$\frac{d}{dt}(m\dot{r}) - mr\dot{\chi}^2 + 4\pi^2 m\nu_0^2 r = 0,$$

or, using Equation 1–37,

$$\frac{d}{dt}(m\dot{r}) - \frac{p_\chi^2}{mr^3} + 4\pi^2 m v_0^2 r = 0, \qquad (1\text{–}38)$$

an equation differing from the related one-dimensional Cartesian-coordinate equation by the additional term $-p_\chi^2/mr^3$ which represents the centrifugal force.

Multiplication by $\dot{r}$ and integration with respect to the time gives

$$\dot{r}^2 = -\frac{p_\chi^2}{m^2 r^2} - 4\pi^2 v_0^2 r^2 + b, \qquad (1\text{–}39)$$

so that $\dot{r} = \left(-\dfrac{p_\chi^2}{m^2 r^2} - 4\pi^2 v_0^2 r^2 + b\right)^{\frac{1}{2}}$.

This can be again integrated, to give

$$t - t_0 = \int \frac{r\,dr}{\left(-\dfrac{p_\chi^2}{m^2} + br^2 - 4\pi^2 v_0^2 r^4\right)^{\frac{1}{2}}}$$

$$= \frac{1}{2}\int \frac{dx}{(a + bx + cx^2)^{\frac{1}{2}}},$$

in which $x = r^2$, $a = -p_\chi^2/m^2$, b is the constant of integration in Equation 1–39, and $c = -4\pi^2 v_0^2$. This is a standard integral which yields the equation

$$r^2 = \frac{1}{8\pi^2 v_0^2}\{b + A \sin 4\pi v_0(t - t_0)\},$$

with A given by

$$A = \sqrt{b^2 - \frac{16\pi^2 v_0^2 p_\chi^2}{m^2}}.$$

We have thus obtained the dependence of r on the time, and by integrating Equation 1–37 we could obtain χ as a function of the time, completing the solution. Elimination of the time between these two results would give the equation of the orbit, which is an ellipse with center at the origin. It is seen that the constant v_0 again occurs as the frequency of the motion.

1e. The Conservation of Angular Momentum.—The example worked out in the previous section illustrates an important principle of wide applicability, the principle of the *conservation of angular momentum*.

Equation 1–37 shows that when $\dot\chi$ is the angular velocity of the particle about a fixed axis z' and r is the distance of the particle from the axis, the quantity $p_\chi = mr^2\dot\chi$ is a constant of the motion.[1] This quantity is called the *angular momentum* of the particle about the axis z'.

It is not necessary to choose an axis normal to the plane of the motion, as z' in this example, in order to apply the theorem. Thus Equation 1–33, written for arbitrary direction z, is at once integrable to

$$mr^2 \sin^2 \vartheta \dot\varphi = p_\varphi, \text{ a constant.} \qquad (1\text{–}40)$$

Here $r \sin \vartheta$ is the distance of the particle from the axis z, so that the left side of this equation is the angular momentum about the axis z.[2]   It is seen to be equal to a constant, p_φ.

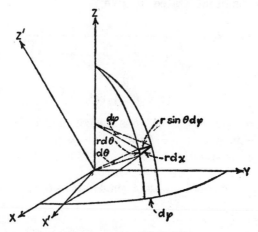

Fig. 1-3.—Figure showing the relation between $d\chi$, $d\vartheta$, and $d\varphi$.

In order to apply the principle, it is essential that the axis of reference be a fixed axis.   Thus the angle ϑ of polar coordinates has associated with it an angular momentum $p_\vartheta = mr^2\dot\vartheta$ about an axis in the xy plane, but the principle of conservation of angular momentum cannot be applied directly to this quantity because the axis is not, in general, fixed but varies with φ.   A simple relation involving p_ϑ connects the angular momenta

---

[1] The phrase *a constant of the motion* is often used in referring to a constant of integration of the equations of motion for a dynamical system.

[2] This is sometimes referred to as the component of angular momentum along the axis z.

p_χ and p_φ about different fixed axes, one of which, p_χ, relates to the axis normal to the plane of the motion. This is

$$p_\chi d\chi = p_\vartheta d\vartheta + p_\varphi d\varphi, \qquad (1\text{-}41)$$

an equation easily derived by considering Figure 1–3. The sides of the small triangle have the lengths $r \sin \vartheta d\varphi$, $rd\chi$, and $rd\vartheta$. Since they form a right triangle, these distances are connected by the relation

$$r^2(d\chi)^2 = r^2 \sin^2 \vartheta (d\varphi)^2 + r^2(d\vartheta)^2,$$

which gives, on introduction of the angular velocities $\dot\chi$, $\dot\varphi$, and $\dot\vartheta$ and multiplication by m/dt,

$$mr^2\dot\chi d\chi = mr^2 \sin^2 \vartheta \dot\varphi d\varphi + mr^2\dot\vartheta d\vartheta.$$

Equation 1–41 follows from this and the definitions of p_χ, p_ϑ, and p_φ.

Conservation of angular momentum may be applied to more general systems than the one described here. It is at once evident that we have not used the special form of the potential-energy expression except for the fact that it is independent of direction, since this function enters into the r equation only. Therefore the above results are true for a particle moving in any spherically symmetric potential field.

Furthermore, we can extend the theorem to a collection of point particles interacting with each other in any desired way but influenced by external forces only through a spherically symmetric potential function. If we describe such a system by using the polar coordinates of each particle, the Lagrangian function is

$$L = \tfrac{1}{2}\sum_{i=1}^{n} m_i(\dot r_i^2 + r_i^2\dot\vartheta_i^2 + r_i^2 \sin^2 \vartheta_i \dot\varphi_i^2) - V. \qquad (1\text{-}42)$$

Instead of φ_1, φ_2, $\cdots$, φ_n, we now introduce new angular coordinates α, β, $\cdots$, κ given by the linear equations

$$\left.\begin{aligned}
\varphi_1 &= \alpha + b_1\beta + \cdots + k_1\kappa, \\
\varphi_2 &= \alpha + b_2\beta + \cdots + k_2\kappa, \\
&\cdots\cdots\cdots\cdots\cdots\cdots\cdots, \\
\varphi_n &= \alpha + b_n\beta + \cdots + k_n\kappa.
\end{aligned}\right\} \qquad (1\text{-}43)$$

The values given the constants b_1, $\cdots$, k_n are unimportant so long as they make the above set of equations mutually independent. α is an angle about the axis z such that if α is increased by $\Delta\alpha$, holding β, $\cdots$, κ constant, the effect is to increase each φ_i by $\Delta\alpha$, or, in other words, to rotate the whole system of particles about z without changing their mutual positions. By hypothesis the value of V is not changed by such a rotation, so that V is independent of α. We therefore obtain the equation

$$\frac{d}{dt}\frac{\partial L}{\partial\dot{\alpha}} - \frac{\partial L}{\partial\alpha} = \frac{d}{dt}\frac{\partial T}{\partial\dot{\alpha}} = 0. \qquad (1\text{-}44)$$

Moreover, from Equation 1-42 we derive the relation

$$\frac{\partial T}{\partial\dot{\alpha}} = \sum_{i=1}^{n}\frac{\partial T}{\partial\dot{\varphi}_i}\frac{\partial\dot{\varphi}_i}{\partial\dot{\alpha}} = \sum_{i=1}^{n}m_i r_i^2\sin^2\vartheta_i\dot{\varphi}_i. \qquad (1\text{-}45)$$

Hence, calling the distance $r_i\sin\vartheta_i$ of the ith particle from the z axis ρ_i, we obtain the equation

$$\sum_{i=1}^{n}m_i\rho_i^2\dot{\varphi}_i = \text{constant.} \qquad (1\text{-}46)$$

This is the more general expression of the principle of the conservation of angular momentum which we were seeking. In such a system of many particles with mutual interactions, as, for example, an atom consisting of a number of electrons and a nucleus, the individual particles do not in general conserve angular momentum but the aggregate does.

The potential-energy function V need be only cylindrically symmetric about the axis z for the above proof to apply, since the essential feature was the independence of V on the angle α about z. However, in that case z is restricted to a particular direction in space, whereas if V is spherically symmetric the theorem holds for any choice of axis.

Angular momenta transform like vectors, the directions of the vectors being the directions of the axes about which the angular momenta are determined. It is customary to take the sense of the vectors such as to correspond to the right-hand screw rule.

2. THE EQUATIONS OF MOTION IN THE HAMILTONIAN FORM

2a. Generalized Momenta.—In Cartesian coordinates the momentum related to the direction x_k is $m_k\dot{x}_k$, which, since V is restricted to be a function of the coordinates only, can be written as

$$p_k = \frac{\partial T}{\partial \dot{x}_k} = \frac{\partial L}{\partial \dot{x}_k}, \quad k = 1, 2, \cdots, 3n. \quad (2\text{-}1)$$

Angular momenta can likewise be expressed in this manner. Thus, for one particle in a spherically symmetric potential field, the angular momentum about the z axis was defined in Section 1e by the expression

$$p_\varphi = m\rho^2 \dot{\varphi} = mr^2 \sin^2 \vartheta \dot{\varphi}. \quad (2\text{-}2)$$

Reference to Equation 1–31, which gives the expression for the kinetic energy in polar coordinates, shows that

$$p_\varphi = \frac{\partial T}{\partial \dot{\varphi}} = \frac{\partial L}{\partial \dot{\varphi}}. \quad (2\text{-}3)$$

Likewise, in the case of a number of particles, the angular momentum conjugate to the coordinate α is

$$p_\alpha = \frac{\partial T}{\partial \dot{\alpha}} = \frac{\partial L}{\partial \dot{\alpha}}, \quad (2\text{-}4)$$

as shown by the discussion of Equation 1–46. By extending this to other coordinate systems, the *generalized momentum* p_k conjugate to the coordinate q_k is defined as

$$p_k = \frac{\partial L}{\partial \dot{q}_k}, \quad k = 1, 2, \cdots, 3n. \quad (2\text{-}5)$$

The form taken by Lagrange's equations (Eq. 1–29) when the definition of p_k is introduced is

$$\dot{p}_k = \frac{\partial L}{\partial q_k}, \quad k = 1, 2, \cdots, 3n, \quad (2\text{-}6)$$

so that Equations 2–5 and 2–6 form a set of $6n$ first-order differential equations equivalent to the $3n$ second-order equations of Equation 1–29.

$\dfrac{\partial L}{\partial \dot{q}_k}$ being in general a function of both the q's and $\dot{q}$'s, the definition of p_k given by Equation 2–5 provides $3n$ relations between the variables q_k, $\dot{q}_k$, and p_k, permitting the elimination of the $3n$ velocities $\dot{q}_k$, so that the system can now be described in terms of the $3n$ coordinates q_k and the $3n$ conjugate momenta

p_k. Hamilton in 1834 showed that the equations of motion can in this way be thrown into an especially simple form, involving a function H of the p_k's and q_k's called the *Hamiltonian function*.

2b. The Hamiltonian Function and Equations.—For conservative systems[1] we shall show that the function H is the total energy (kinetic plus potential) of the system, expressed in terms of the p_k's and q_k's. In order to have a definition which holds for more general systems, we introduce H by the relation

$$H = \sum_{k=1}^{3n} p_k \dot{q}_k - L(q_k, \dot{q}_k). \qquad (2\text{--}7)$$

Although this definition involves the velocities $\dot{q}_k$, H may be made a function of the coordinates and momenta only, by eliminating the velocities through the use of Equation 2–5. From the definition we obtain for the total differential of H the equation

$$dH = \sum_{k=1}^{3n} p_k d\dot{q}_k + \sum_{k=1}^{3n} \dot{q}_k dp_k - \sum_{k=1}^{3n} \frac{\partial L}{\partial q_k} dq_k - \sum_{k=1}^{3n} \frac{\partial L}{\partial \dot{q}_k} d\dot{q}_k, \qquad (2\text{--}8)$$

or, using the expressions for p_k and $\dot{p}_k$ given in Equations 2–5 and 2–6 (equivalent to Lagrange's equations),

$$dH = \sum_{k=1}^{3n} (\dot{q}_k dp_k - \dot{p}_k dq_k), \qquad (2\text{--}9)$$

whence, if H is regarded as a function of the q_k's and p_k's, we obtain the equations

$$\left. \begin{aligned} \frac{\partial H}{\partial p_k} &= \dot{q}_k, \\ \frac{\partial H}{\partial q_k} &= -\dot{p}_k, \end{aligned} \right\} \qquad k = 1, 2, \cdots, 3n. \qquad (2\text{--}10)$$

These are the *equations of motion in the Hamiltonian or canonical form*.

2c. The Hamiltonian Function and the Energy.—Let us consider the time dependence of H for a conservative system. We have

---

[1] A conservative system is a system for which H does not depend explicitly on the time t. We have restricted our discussion to conservative systems by assuming that the potential function V does not depend on t.

$$\frac{dH}{dt} = \frac{d}{dt}\left(\sum_{k=1}^{3n} p_k \dot{q}_k - L\right) = \sum_{k=1}^{3n}\left(p_k \ddot{q}_k + \dot{p}_k \dot{q}_k - \frac{\partial L}{\partial q_k}\dot{q}_k - \frac{\partial L}{\partial \dot{q}_k}\ddot{q}_k\right)$$

$$= \sum_{k=1}^{3n}(\dot{p}_k \dot{q}_k - \dot{p}_k \dot{q}_k) = 0, \tag{2-11}$$

using the same substitutions for p_k and $\dot{p}_k$ (Eqs. 2–5 and 2–6) as before. H is hence a constant of the motion, which is called the *energy* of the system. For Newtonian systems, in which we shall be chiefly interested, the Hamiltonian function is the sum of the kinetic energy and the potential energy,

$$H = T + V, \tag{2-12}$$

expressed as a function of the coordinates and momenta. This is proved by considering the expression for T for such systems. For any set of coordinates, T will be a homogeneous quadratic function of the velocities

$$T = \sum_{i,j=1}^{3n} a_{ij}\dot{q}_i\dot{q}_j, \tag{2-13}$$

where the a_{ij}'s may be functions of the coordinates. Hence

$$\sum_{k=1}^{3n} p_k \dot{q}_k = \sum_{k=1}^{3n}\frac{\partial L}{\partial \dot{q}_k}\dot{q}_k = \sum_{k=1}^{3n}\frac{\partial T}{\partial \dot{q}_k}\dot{q}_k$$

$$= 2\sum_{i,k=1}^{3n} a_{ik}\dot{q}_i\dot{q}_k = 2T \tag{2-14}$$

so that

$$H = 2T - L = T + V.$$

2d. A General Example.—The use of the Hamiltonian equations may be illustrated by the example of two point particles with masses m_1 and m_2, respectively, moving under the influence of a mutual attraction given by the potential energy function $V(r)$, in which r is the distance between the two particles. The hydrogen atom is a special case of such a system, so that the results obtained below will be used in Chapter II. If the coordinates of the first particle are x_1, y_1, z_1 and those of the second x_2, y_2, z_2, the Lagrangian function L is

$$L = \frac{m_1}{2}(\dot{x}_1^2 + \dot{y}_1^2 + \dot{z}_1^2) + \frac{m_2}{2}(\dot{x}_2^2 + \dot{y}_2^2 + \dot{z}_2^2) - V(r). \quad (2\text{-}15)$$

The solution of this problem is facilitated by the introduction, in place of x_1, y_1, z_1, x_2, y_2, z_2, of the Cartesian coordinates x, y, z of the center of mass of the system and the polar coordinates r, ϑ, φ of one particle referred to the other as origin.

The coordinates of the center of gravity are determined by the equations

$$m_1 x_1 + m_2 x_2 = (m_1 + m_2)x \quad (2\text{-}16)$$

with similar equations for y_1, y_2, y and z_1, z_2, z. The polar coordinates r, ϑ, φ are given by

$$\left.\begin{array}{l} x_2 - x_1 = r \sin \vartheta \cos \varphi, \\ y_2 - y_1 = r \sin \vartheta \sin \varphi, \\ z_2 - z_1 = r \cos \vartheta. \end{array}\right\} \quad (2\text{-}17)$$

Elimination of x_2, y_2, and z_2 between these two sets of equations leads to the relations

$$\left.\begin{array}{l} x_1 = x - \dfrac{m_2}{m_1 + m_2} r \sin \vartheta \cos \varphi, \\[2mm] y_1 = y - \dfrac{m_2}{m_1 + m_2} r \sin \vartheta \sin \varphi, \\[2mm] z_1 = z - \dfrac{m_2}{m_1 + m_2} r \cos \vartheta, \end{array}\right\} \quad (2\text{-}18)$$

while elimination of x_1, y_1, and z_1 gives the set

$$\left.\begin{array}{l} x_2 = x + \dfrac{m_1}{m_1 + m_2} r \sin \vartheta \cos \varphi, \\[2mm] y_2 = y + \dfrac{m_1}{m_1 + m_2} r \sin \vartheta \sin \varphi, \\[2mm] z_2 = z + \dfrac{m_1}{m_1 + m_2} r \cos \vartheta. \end{array}\right\} \quad (2\text{-}19)$$

The substitution of the time derivatives of these quantities in the Lagrangian function L results in the expression

$$L = \tfrac{1}{2}(m_1 + m_2)(\dot{x}^2 + \dot{y}^2 + \dot{z}^2) + \\ \tfrac{1}{2}\mu(\dot{r}^2 + r^2\dot{\vartheta}^2 + r^2 \sin^2 \vartheta \dot{\varphi}^2) - V(r), \quad (2\text{-}20)$$

in which μ, called the *reduced mass*, is given by

$$\mu = \frac{m_1 m_2}{m_1 + m_2}. \quad (2\text{-}21)$$

The momenta conjugate to x, y, z and r, ϑ, φ are

$$
\left.\begin{aligned}
p_x &= \frac{\partial L}{\partial \dot{x}} = (m_1 + m_2)\dot{x}, \\
p_y &= (m_1 + m_2)\dot{y}, \\
p_z &= (m_1 + m_2)\dot{z}, \\
p_r &= \mu\dot{r}, \\
p_\vartheta &= \mu r^2 \dot{\vartheta}, \\
p_\varphi &= \mu r^2 \sin^2 \vartheta \dot{\varphi}.
\end{aligned}\right\} \tag{2-22}
$$

The Hamiltonian function is therefore

$$
\begin{aligned}
H &= (m_1 + m_2)(\dot{x}^2 + \dot{y}^2 + \dot{z}^2) + \mu(\dot{r}^2 + r^2\dot{\vartheta}^2 + r^2 \sin^2 \vartheta \dot{\varphi}^2) - L \\
&= \tfrac{1}{2}(m_1 + m_2)(\dot{x}^2 + \dot{y}^2 + \dot{z}^2) + \tfrac{1}{2}\mu(\dot{r}^2 + r^2\dot{\vartheta}^2 + r^2 \sin^2 \vartheta \dot{\varphi}^2) \\
&\qquad\qquad\qquad\qquad\qquad\qquad\qquad\qquad\qquad + V(r) \\
&= \frac{1}{2(m_1 + m_2)}(p_x^2 + p_y^2 + p_z^2) + \frac{1}{2\mu}\left(p_r^2 + \frac{p_\vartheta^2}{r^2} + \frac{p_\varphi^2}{r^2 \sin^2 \vartheta}\right) + \\
&\qquad\qquad\qquad\qquad\qquad\qquad\qquad\qquad\qquad V(r). \quad (2\text{-}23)
\end{aligned}
$$

The equations of motion become

$$
\left.\begin{aligned}
\dot{p}_x &= -\frac{\partial H}{\partial x} = 0, \\
\dot{p}_y &= 0, \\
\dot{p}_z &= 0;
\end{aligned}\right\} \tag{2-24}
$$

$$
\left.\begin{aligned}
\dot{p}_r &= \frac{1}{\mu}\left(\frac{p_\vartheta^2}{r^3} + \frac{p_\varphi^2}{r^3 \sin^2 \vartheta}\right) - \frac{\partial V}{\partial r}, \\
\dot{p}_\vartheta &= \frac{1}{\mu}\frac{p_\varphi^2 \cos \vartheta}{r^2 \sin^3 \vartheta}, \\
\dot{p}_\varphi &= 0;
\end{aligned}\right\} \tag{2-25}
$$

$$
\left.\begin{aligned}
\dot{x} &= \frac{\partial H}{\partial p_x} = \frac{p_x}{m_1 + m_2}, \\
\dot{y} &= \frac{p_y}{m_1 + m_2}, \\
\dot{z} &= \frac{p_z}{m_1 + m_2};
\end{aligned}\right\} \tag{2-26}
$$

$$
\left.\begin{aligned}
\dot{r} &= \frac{p_r}{\mu}, \\
\dot{\vartheta} &= \frac{p_\vartheta}{\mu r^2}, \\
\dot{\varphi} &= \frac{p_\varphi}{\mu r^2 \sin^2 \vartheta}.
\end{aligned}\right\} \tag{2-27}
$$

It is noticed that the last six of these equations (2–26, 2–27) are identical with the equations which define the momenta involved. An inspection of Equations 2–25 indicates that they are closely related to Equations 1–33, 1–34, and 1–35. If in these equations m is replaced by μ and if in Equation 1–35 $4\pi^2 m\nu_0^2 r$ is replaced by $\frac{\partial V}{\partial r}$, we obtain just the equations which result from substituting for p_r, p_ϑ, p_φ their expressions in terms of $\dot{r}$, $\dot{\vartheta}$, and $\dot{\varphi}$ in Equation 2–25. The first three, 2–24, show that the center of gravity of the system moves with a constant velocity, while the next three are the equations of motion of a particle of mass μ bound to a fixed center by a force whose potential energy function is $V(r)$.

This problem illustrates the fact that in most actual problems the Lagrangian equations are reached in the process of solution of the equations of motion in the Hamiltonian form. The great value of the Hamiltonian equations lies in their particular suitability for general considerations, such as, for example, Liouville's theorem in statistical mechanics, the rules of quantization in the old quantum theory, and the formulation of the Schrödinger wave equation. This usefulness is in part due to the symmetrical or conjugate form of the equations in p and q.

Problem 2–1. Discuss the motion of a charged particle in a uniform electric field.

Problem 2–2. Solve the equation of motion for a charged particle of mass m constrained to move on the x axis in a uniform electric field (the potential energy due to the field being $-eFx$, where e is the electric charge constant) and connected to the origin by a spring of force constant k. Assuming a fixed charge $-e$ at the origin, obtain an expression for the average electric moment of the system as a function of the quantities e, m, F, k, and the energy of the system. See Equation 3–5.

Problem 2–3. Derive an expression for the kinetic energy of a particle in terms of cylindrical coordinates and then treat the equations of motion for a cylindrically symmetrical potential function.

Problem 2–4. Using spherical polar coordinates, solve the equations of motion for a free particle and discuss the results.

Problem 2–5. Obtain the solution for χ in Section 1d.

Problem 2–6. By eliminating the time in the result of Problem 2–5 and the equation for r in Section 1d, show that the orbit of the particle is an ellipse.

Problem 2–7. Prove the identity of the motion of the plane isotropic harmonic oscillator found by solution in Cartesian and polar coordinates.

Problem 2–8. Show how to obtain an immediate integral of one equation of motion, if the Lagrangian function does not involve the corresponding

The average rate of emission of radiant energy by such a system is consequently

$$-\frac{dE}{dt} = \frac{16\pi^4 \nu^4 e^2 x_0^2}{3c^3},\qquad(3\text{-}4)$$

inasmuch as the average value $\cos^2 2\pi\nu t$ over a cycle is one-half. As a result of the emission of energy, the amplitude x_0 of the motion will decrease with time; if the fractional change in energy during a cycle of the motion is small, however, this equation retains its validity.

The radiation emitted by such a system has the frequency ν of the emitting system. It is plane-polarized, the plane of the electric vector being the plane which includes the x axis and the direction of propagation of the light.

In case that the particle carries out harmonic oscillations along all three axes x, y, and z, with frequencies ν_x, ν_y, and ν_z and amplitudes (at a given time) x_0, y_0, and z_0, respectively, the total rate of emission of radiant energy will be given as the sum of three terms similar to the right side of Equation 3–4, one giving the rate of emission of energy as light of frequency ν_x, one of ν_y, and one of ν_z.

If the motion of the particle is not simple harmonic, it can be represented by a Fourier series or Fourier integral as a sum or integral of harmonic terms similar to that of Equation 3–2; light of frequency characteristic of each of these terms will then be emitted at a rate given by Equation 3–4, the coefficient of the Fourier term being introduced in place of x_0.

The emission of light by a system composed of several interacting electrically charged particles is conveniently discussed in the following way. A Fourier analysis is first made of the motion of the system in a given state to resolve it into harmonic terms. For a given term, corresponding to a given frequency of motion ν, the coefficient resulting from the analysis (which is a function of the coordinates of the particles) is expanded as a power series in the quantities x_1/λ, $\cdots$, z_n/λ, in which x_1, $\cdots$, z_n are the coordinates of the particles relative to some origin (such as the center of mass) and $\lambda = c/\nu$ is the wave length of the radiation with frequency ν. The term of zero degree in this expansion is zero, inasmuch as the electric charge of the system does not change with time. The term of first degree involves, in addition to the harmonic function of the time, only

coordinate but only its derivative. Such a coordinate is cal
coordinate.

3. THE EMISSION AND ABSORPTION OF RADIAT

The classical laws of mechanical and electromagne
permit the complete discussion of the emission and abs
electromagnetic radiation by a system of electricall
particles. In the following paragraphs we shall ou
results of this discussion. It is found that these resul
in agreement with experiments involving atoms and 1
it was, indeed, just this disagreement which was the
factor in leading to the development of the Bohr thec
atom and later of the quantum mechanics. Even at tl
time, when an apparently satisfactory theoretical tre£
dynamical systems composed of electrons and nuclei is
by the quantum mechanics, the problem of the emis
absorption of radiation still lacks a satisfactory solutioi
the concentration of attention on it by the most able tl
physicists. It will be shown in a subsequent chapter,
that, despite our lack of a satisfactory conception of tl
of electromagnetic radiation, equations similar to the
equations of this section can be formulated which 1
correctly the emission and absorption of radiation b
systems to within the limits of error of experiment.

According to the classical theory the rate of emission o
energy by an accelerated particle of electric charge e is

$$-\frac{dE}{dt} = \frac{2e^2\dot{v}^2}{3c^3},$$

in which $-\dfrac{dE}{dt}$ is the rate at which the energy E of the
is converted into radiant energy, $\dot{v}$ is the acceleratioi
particle, and c the velocity of light.

Let us first consider a system of a special type, in
particle of charge e carries out simple harmonic os
with frequency ν along the x axis, according to the equa

$$x = x_0 \cos 2\pi\nu t.$$

Differentiating this expression, assuming that x_0 is indej
of the time, we obtain for the acceleration the value

$$\dot{v} = \ddot{x} = -4\pi^2\nu^2 x_0 \cos 2\pi\nu t.$$

a function of the coordinates. The aggregate of these first-degree terms in the coordinates with their associated time factors, summed over all frequency values occurring in the original Fourier analysis, represents a dynamical quantity known as the *electric moment* of the system, a vector quantity $\mathbf{P}$ defined as

$$\mathbf{P} = \sum_i e_i \mathbf{r}_i, \qquad (3\text{-}5)$$

in which $\mathbf{r}_i$ denotes the vector from the origin to the position of the ith particle, with charge e_i. Consequently to this degree of approximation the radiation emitted by a system of several particles can be discussed by making a Fourier analysis of the electric moment $\mathbf{P}$. Corresponding to each term of frequency ν in this representation of $\mathbf{P}$, there will be emitted radiation of frequency ν at a rate given by an equation similar to Equation 3-4, with ex_0 replaced by the Fourier coefficient in the electric-moment expansion. The emission of radiation by this mechanism is usually called *dipole emission*, the radiation itself sometimes being described as *dipole radiation*.

The quadratic terms in the expansions in powers of x_1/λ, $\cdots$, z_n/λ form a quantity Q called the *quadrupole moment* of the system, and higher powers form higher moments. The rate of emission of radiant energy as a result of the change of quadrupole and higher moments of an atom or molecule is usually negligibly small in comparison with the rate of dipole emission, and in consequence dipole radiation alone is ordinarily discussed. Under some circumstances, however, as when the intensity of dipole radiation is zero and the presence of very weak radiation can be detected, the process of quadrupole emission is important.

4. SUMMARY OF CHAPTER I

The purpose of this survey of classical mechanics is twofold: first, to indicate the path whereby the more general formulations of classical dynamics, such as the equations of motion of Lagrange and of Hamilton, have been developed from the original equations of Newton; and second, to illustrate the application of these methods to problems which are later discussed by quantum-mechanical methods.

In carrying out the first purpose, we have discussed Newton's equations in Cartesian coordinates and then altered their form by the introduction of the kinetic and potential energies. By defining the Lagrangian function for the special case of Newtonian systems and introducing it into the equations of motion, Newton's equations were then thrown into the Lagrangian form. Following an introductory discussion of generalized coordinates, the proof of the general validity of the equations of motion in the Lagrangian form for any system of coordinates has been given; and it has also been pointed out that the Lagrangian form of the equations of motion, although we have derived it from the equations of Newton, is really more widely applicable than Newton's postulates, because by making a suitable choice of the Lagrangian function a very wide range of problems can be treated in this way.

In the second section there has been derived a third form for the equations of motion, the Hamiltonian form, following the introduction of the concept of generalized momenta, and the relation between the Hamiltonian function and the energy has been discussed.

In Section 3 a very brief discussion of the classical theory of the radiation of energy from accelerated charged particles has been given, in order to have a foundation for later discussions of this topic. Mention is made of both dipole and quadrupole radiation.

Finally, several examples (which are later solved by the use of quantum mechanics), including the three-dimensional harmonic oscillator in Cartesian and in polar coordinates, have been treated by the methods discussed in this chapter.

General References on Classical Mechanics

W. D. MacMillan: "Theoretical Mechanics. Statics and the Dynamics of a Particle," McGraw-Hill Book Company, Inc., New York, 1932.

S. L. Loney: "Dynamics of a Particle and of Rigid Bodies," Cambridge University Press, Cambridge, 1923.

J. H. Jeans: "Theoretical Mechanics," Ginn and Company, Boston, 1907.

E. T. Whittaker: "Analytical Dynamics," Cambridge University Press, Cambridge, 1928.

R. C. Tolman: "Statistical Mechanics with Applications to Physics and Chemistry," Chemical Catalog Company, Inc., New York, 1927, Chap. II, The Elements of Classical Mechanics.

W. E. Byerly: "Generalized Coordinates," Ginn and Company, Boston, 1916.

CHAPTER II

THE OLD QUANTUM THEORY

5. THE ORIGIN OF THE OLD QUANTUM THEORY

The old quantum theory was born in 1900, when Max Planck[1] announced his theoretical derivation of the distribution law for black-body radiation which he had previously formulated from empirical considerations. He showed that the results of experiment on the distribution of energy with frequency of radiation in equilibrium with matter at a given temperature can be accounted for by postulating that the vibrating particles of matter (considered to act as harmonic oscillators) do not emit or absorb light continuously but instead only in discrete quantities of magnitude $h\nu$ proportional to the frequency ν of the light. The constant of proportionality, h, is a new constant of nature; it is called *Planck's constant* and has the magnitude 6.547×10^{-27} erg sec. Its dimensions (energy $\times$ time) are those of the old dynamical quantity called *action;* they are such that the product of h and frequency ν (with dimensions sec^{-1}) has the dimensions of energy. The dimensions of h are also those of angular momentum, and we shall see later that just as $h\nu$ is a *quantum* of radiant energy of frequency ν, so is $h/2\pi$ a natural unit or quantum of angular momentum.

The development of the quantum theory was at first slow. It was not until 1905 that Einstein[2] suggested that the quantity of radiant energy $h\nu$ was sent out in the process of emission of light not in all directions but instead unidirectionally, like a particle. The name *light quantum* or *photon* is applied to such a portion of radiant energy. Einstein also discussed the photoelectric effect, the fundamental processes of photochemistry, and the heat capacities of solid bodies in terms of the quantum theory. When light falls on a metal plate, electrons are emitted from it. The maximum speed of these photoelectrons, however,

[1] M. PLANCK, *Ann. d. Phys.* (4) **4**, 553 (1901).
[2] A. EINSTEIN, *Ann. d. Phys.* (4) **17**, 132 (1905).

is not dependent on the intensity of the light, as would be expected from classical electromagnetic theory, but only on its frequency; Einstein pointed out that this is to be expected from the quantum theory, the process of photoelectric emission involving the conversion of the energy $h\nu$ of one photon into the kinetic energy of a photoelectron (plus the energy required to remove the electron from the metal). Similarly, Einstein's law of photochemical equivalence states that one molecule may be activated to chemical reaction by the absorption of one photon.

The third application, to the heat capacities of solid bodies, marked the beginning of the quantum theory of material systems. Planck's postulate regarding the emission and absorption of radiation in quanta $h\nu$ suggested that a dynamical system such as an atom oscillating about an equilibrium position with frequency ν_0 might not be able to oscillate with arbitrary energy, but only with energy values which differ from one another by integral multiples of $h\nu_0$. From this assumption and a simple extension of the principles of statistical mechanics it can be shown that the heat capacity of a solid aggregate of particles should not remain constant with decreasing temperature, but should at some low temperature fall off rapidly toward zero. This prediction of Einstein, supported by the earlier experimental work of Dewar on diamond, was immediately verified by the experiments of Nernst and Eucken on various substances; and quantitative agreement between theory and experiment for simple crystals was achieved through Debye's brilliant refinement of the theory.[1]

5a. The Postulates of Bohr.—The quantum theory had developed to this stage before it became possible to apply it to the hydrogen atom; for it was not until 1911 that there occurred the discovery by Rutherford of the nuclear constitution of the atom—its composition from a small heavy positively charged nucleus and one or more extranuclear electrons. Attempts were made immediately to apply the quantum theory to the hydrogen atom. The successful effort of Bohr[2] in 1913, despite its simplicity, may well be considered the greatest single step in the development of the theory of atomic structure.

[1] P. DEBYE, *Ann. d. Phys.* (4) **39**, 789 (1912); see also M. BORN and T. VON KÁRMÁN, *Phys. Z.* **13**, 297 (1912); **14**, 15 (1913).

[2] N. BOHR, *Phil. Mag.* **26**, 1 (1913).

It was clearly evident that the laws of classical mechanical and electromagnetic theory could not apply to the Rutherford hydrogen atom. According to classical theory the electron in a hydrogen atom, attracted toward the nucleus by an inverse-square Coulomb force, would describe an elliptical or circular orbit about it, similar to that of the earth about the sun. The acceleration of the charged particles would lead to the emission of light, with frequencies equal to the mechanical frequency of the electron in its orbit, and to multiples of this as overtones. With the emission of energy, the radius of the orbit would diminish and the mechanical frequency would change. Hence the emitted light should show a wide range of frequencies. This is not at all what is observed—the radiation emitted by hydrogen atoms is confined to spectral lines of sharply defined frequencies, and, moreover, these frequencies are not related to one another by integral factors, as overtones, but instead show an interesting additive relation, expressed in the *Ritz combination principle*, and in addition a still more striking relation involving the squares of integers, discovered by Balmer. Furthermore, the existence of stable non-radiating atoms was not to be understood on the basis of classical theory, for a system consisting of electrons revolving about atomic nuclei would be expected to emit radiant energy until the electrons had fallen into the nuclei.

Bohr, no doubt inspired by the work of Einstein mentioned above, formulated the two following postulates, which to a great extent retain their validity in the quantum mechanics.

I. *The Existence of Stationary States.* An atomic system can exist in certain *stationary states*, each one corresponding to a definite value of the energy W of the system; and transition from one stationary state to another is accompanied by the emission or absorption as radiant energy, or the transfer to or from another system, of an amount of energy equal to the difference in energy of the two states.

II. *The Bohr Frequency Rule.* The frequency of the radiation emitted by a system on transition from an initial state of energy W_2 to a final state of lower energy W_1 (or absorbed on transition from the state of energy W_1 to that of energy W_2) is given by the equation[1]

---

[1] This relation was suggested by the *Ritz combination principle*, which it closely resembles. It was found empirically by Ritz and others that if

$$\nu = \frac{W_2 - W_1}{h}. \qquad (5\text{-}1)$$

Bohr in addition gave a method of determining the quantized states of motion—the stationary states—of the hydrogen atom. His method of quantization, involving the restriction of the angular momentum of circular orbits to integral multiples of the quantum $h/2\pi$, though leading to satisfactory energy levels, was soon superseded by a more powerful method, described in the next section.

Problem 5-1. Consider an electron moving in a circular orbit about a nucleus of charge Ze. Show that when the centrifugal force is just balanced by the centripetal force Ze^2/r^2, the total energy is equal to one-half the potential energy $-Ze^2/r$. Evaluate the energy of the stationary states for which the angular momentum equals $nh/2\pi$, with $n = 1, 2, 3, \cdots$.

5b. The Wilson-Sommerfeld Rules of Quantization.—In

1915 W. Wilson and A. Sommerfeld discovered independently[1] a powerful method of quantization, which was soon applied, especially by Sommerfeld and his coworkers, in the discussion

---

lines of frequencies ν_1 and ν_2 occur in the spectrum of a given atom it is frequently possible to find also a line with frequency $\nu_1 + \nu_2$ or $\nu_1 - \nu_2$. This led directly to the idea that a set of numbers, called *term values*, can be assigned to an atom, such that the frequencies of all the spectral lines can be expressed as differences of pairs of term values. Term values are usually given in wave numbers, since this unit, which is the reciprocal of the wave length expressed in centimeters, is a convenient one for spectroscopic use. We shall use the symbol $\tilde{\nu}$ for term values in wave numbers, reserving the simpler symbol ν for frequencies in $\sec^{-1}$. The normal state of the ionized atom is usually chosen as the arbitrary zero, and the term values which represent states of the atom with lower energy than the ion are given the positive sign, so that the relation between W and $\tilde{\nu}$ is

$$\tilde{\nu} = -\frac{W}{hc}.$$

The modern student, to whom the Bohr frequency rule has become commonplace, might consider that this rule is clearly evident in the work of Planck and Einstein. This is not so, however; the confusing identity of the mechanical frequencies of the harmonic oscillator (the only system discussed) and the frequency of the radiation absorbed and emitted by this quantized system delayed recognition of the fact that a fundamental violation of electromagnetic theory was imperative.

[1] W. WILSON, *Phil. Mag.* **29**, 795 (1915); A. SOMMERFELD, *Ann. d. Phys.* **51**, 1 (1916).

of the fine structure of the spectra of hydrogen and ionized helium, their Zeeman and Stark effects, and many other phenomena. The first step of their method consists in solving the classical equations of motion in the Hamiltonian form (Sec. 2), therefore making use of the coordinates $q_1, \cdots, q_{3n}$ and the canonically conjugate momenta $p_1, \cdots, p_{3n}$ as the independent variables. The assumption is then introduced that only those classical orbits are allowed as stationary states for which the following conditions are satisfied:

$$\oint p_k dq_k = n_k h, \quad k = 1, 2, \cdots, 3n; \quad n_k = \text{an integer.} \quad (5\text{-}2)$$

These integrals, which are called *action integrals*, can be calculated only for conditionally periodic systems; that is, for systems for which coordinates can be found each of which goes through a cycle as a function of the time, independently of the others. The definite integral indicated by the symbol $\oint$ is taken over one cycle of the motion. Sometimes the coordinates can be chosen in several different ways, in which case the shapes of the quantized orbits depend on the choice of coordinate systems, but the energy values do not.

We shall illustrate the application of this postulate to the determination of the energy levels of certain specific problems in Sections 6 and 7.

5c. Selection Rules. The Correspondence Principle.—The old quantum theory did not provide a satisfactory method of calculating the intensities of spectral lines emitted or absorbed by a system, that is, the probabilities of transition from one stationary state to another with the emission or absorption of a photon. Qualitative information was provided, however, by an auxiliary postulate, known as *Bohr's correspondence principle*, which correlated the quantum-theory transition probabilities with the intensity of the light of various frequencies which would have been radiated by the system according to classical electromagnetic theory. In particular, if no light of frequency corresponding to a given transition would have been emitted classically, it was assumed that the transition would not take place. The results of such considerations were expressed in selection rules.

For example, the energy values $nh\nu_0$ of a harmonic oscillator (as given in the following section) are such as apparently to permit the emission or absorption of light of frequencies which are arbitrary multiples $(n_2 - n_1)\nu_0$ of the fundamental frequency ν_0. But a classical harmonic oscillator would emit only the fundamental frequency ν_0, with no overtones, as discussed in Section 3; consequently, in accordance with the correspondence principle, it was assumed that the selection rule $\Delta n = \pm 1$ was valid, the quantized oscillator being thus restricted to transitions to the adjacent stationary states.

6. THE QUANTIZATION OF SIMPLE SYSTEMS

6a. The Harmonic Oscillator. Degenerate States.—It was shown in the previous chapter that for a system consisting of a particle of mass m bound to the equilibrium position $x = 0$ by a restoring force $-kx = -4\pi^2 m \nu_0^2 x$ and constrained to move along the x axis the classical motion consists in a harmonic oscillation with frequency ν_0, as described by the equation

$$x = x_0 \sin 2\pi\nu_0 t. \tag{6-1}$$

The momentum $p_x = m\dot{x}$ has the value

$$p_x = 2\pi m\nu_0 x_0 \cos 2\pi\nu_0 t, \tag{6-2}$$

so that the quantum integral can be evaluated at once:

$$\oint p_x dx = \int_0^{1/\nu_0} m(2\pi\nu_0 x_0 \cos 2\pi\nu_0 t)^2 dt = 2\pi^2 \nu_0 m x_0^2 = nh. \tag{6-3}$$

The amplitude x_0 is hence restricted to the quantized values $x_{0_n} = \{nh/2\pi^2\nu_0 m\}^{1/2}$. The corresponding energy values are

$$W_n = T + V = 2\pi^2 m\nu_0^2 x_{0_n}^2 (\sin^2 2\pi\nu_0 t + \cos^2 2\pi\nu_0 t) = 2\pi^2 m\nu_0^2 x_{0_n}^2,$$

or

$$W_n = nh\nu_0, \qquad n = 0, 1, 2, \cdots . \tag{6-4}$$

Thus we see that the energy levels allowed by the old quantum theory are integral multiples of $h\nu_0$, as indicated in Figure 6-1. The selection rule $\Delta n = \pm 1$ permits the emission and absorption of light of frequency ν_0 only.

A particle bound to an equilibrium position in a plane by restoring forces with different force constants in the x and y directions, corresponding to the potential function

$$V = 2\pi^2 m(\nu_x^2 x^2 + \nu_y^2 y^2), \tag{6-5}$$

is similarly found to carry out independent harmonic oscillations along the two axes. The quantization restricts the energy to the values

$$W_{n_x n_y} = n_x h \nu_x + n_y h \nu_y, \qquad n_x, n_y = 0, 1, 2, \cdots , \qquad (6-6)$$

determined by the two quantum numbers n_x and n_y. The amplitudes of motion x_0 and y_0 are given by two equations similar to Equation 6-3.

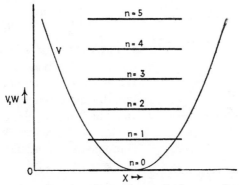

FIG. 6-1.—Potential-energy function and quantized energy levels for the harmonic oscillator according to the old quantum theory.

In case that $\nu_x = \nu_y = \nu_0$, the oscillator is said to be isotropic. The energy levels are then given by the equation

$$W_n = (n_x + n_y)h\nu_0 = nh\nu_0. \qquad (6-7)$$

Different states of motion, corresponding to different sets of values of the two quantum numbers n_x and n_y, may then correspond to the same energy level. Such an energy level is said to be *degenerate*, the degree of degeneracy being given by the number of independent sets of quantum numbers. In this case the nth level shows $(n + 1)$-fold degeneracy. The nth level of the three-dimensional isotropic harmonic oscillator shows

$$\frac{(n + 1)(n + 2)}{2}\text{-fold degeneracy.}$$

6b. The Rigid Rotator.—The configuration of the system of a rigid rotator restricted to a plane is determined by a single angular coordinate, say χ. The canonically conjugate angular momentum, $p_\chi = I\dot{\chi}$, where I is the moment of inertia,[1] is a

---

[1] See Section 36a, footnote, for a definition of moment of inertia.

constant of the motion.[1] Hence the quantum rule is

$$\int_0^{2\pi} p_\chi d\chi = 2\pi p_\chi = Kh$$

or

$$p_\chi = \frac{Kh}{2\pi}, \qquad K = 0, 1, 2, \cdots . \qquad (6\text{–}8)$$

Thus the angular momentum is an integral multiple of $h/2\pi$, as originally assumed by Bohr. The allowed energy values are

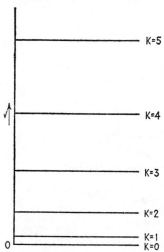

$$W_K = \frac{p_\chi^2}{2I} = \frac{K^2 h^2}{8\pi^2 I}. \qquad (6\text{–}9)$$

The rigid rotator in space can be described by polar coordinates of the figure axis, φ and ϑ. On applying the quantum rules it is found that the total angular momentum is given by Equation 6–8, and the component of angular momentum along the z axis by

$$p_\varphi = \frac{Mh}{2\pi}, \qquad M = -K, -K+1,$$
$$\cdots, 0, \cdots, +K. \qquad (6\text{–}10)$$

The energy levels are given by Equation 6–9, each level being $(2K + 1)$-fold degenerate, inasmuch as the quantum number M

FIG. 6–2.—Energy levels for the rotator according to the old quantum theory.

does not affect the energy (Fig. 6–2).

6c. The Oscillating and Rotating Diatomic Molecule.—A molecule consisting of two atoms bonded together by forces which hold them near to the distance r_0 apart may be approximately considered as a harmonic oscillator joined with a rigid rotator of moment of inertia $I = \mu r_0^2$, μ being the reduced mass. The quantized energy levels are then given by the equation

$$W_{vK} = vh\nu_0 + \frac{K^2 h^2}{8\pi^2 I}, \qquad (6\text{–}11)$$

v being the oscillational or *vibrational quantum number*[2] and K

---

[1] Section 1e, footnote.

[2] The symbol v is now used by band spectroscopists rather than n for this quantum number.

the *rotational quantum number*. The selection rules for such a molecule involving two unlike atoms are $\Delta K = \pm 1$, $\Delta v = \pm 1$. Actual molecules show larger values of Δv, resulting from deviation of the potential function from that corresponding to harmonic oscillation.

The frequency of light absorbed in a transition from the state with quantum numbers v'', K'' to that with quantum numbers v', K' is

$$\nu_{v''K'',v'K'} = (v' - v'')\nu_0 + (K'^2 - K''^2)\frac{h}{8\pi^2 I},$$

or, introducing the selection rule $\Delta K = \pm 1$,

$$\nu_{v''K'',v'K''\pm 1} = (v' - v'')\nu_0 + (\pm 2K'' + 1)\frac{h}{8\pi^2 I}. \quad (6\text{-}12)$$

The lines corresponding to this equation are shown in Figure 6–3 for the fundamental oscillational band $v = 0 \rightarrow v = 1$, together

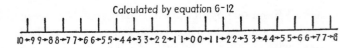

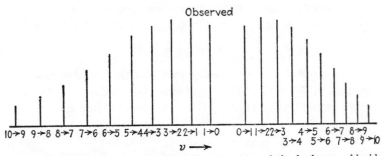

Fig. 6–3.—The observed rotational fine structure of the hydrogen chloride fundamental oscillational band $v = 0 \rightarrow v = 1$, showing deviation from the equidistant spacing of Equation 6–12.

with the experimentally observed absorption band for hydrogen chloride. It is seen that there is rough agreement; the observed lines are not equally spaced, however, indicating that our theoretical treatment, with its assumption of constancy of the moment of inertia I, is too strongly idealized.

6d. The Particle in a Box.—Let us consider a particle of mass m in a box in the shape of a rectangular parallelepiped with edges a, b, and c, the particle being under the influence of no forces except during collision with the walls of the box, from which it rebounds elastically. The linear momenta p_x, p_y, and p_z will then be constants of the motion, except that they will change sign on collision of the particle with the corresponding walls. Their values are restricted by the rule for quantization as follows:

$$\oint p_x dx = 2ap_x = n_x h, \qquad p_x = \frac{n_x h}{2a}, \quad n_x = 0, 1, 2, \cdots ,$$

$$p_y = \frac{n_y h}{2b}, \quad n_y = 0, 1, 2, \cdots , \qquad (6\text{-}13)$$

$$p_z = \frac{n_z h}{2c}, \quad n_z = 0, 1, 2, \cdots .$$

Consequently the total energy is restricted to the values

$$W_{n_x n_y n_z} = \frac{1}{2m}(p_x^2 + p_y^2 + p_z^2) = \frac{h^2}{8m}\left(\frac{n_x^2}{a^2} + \frac{n_y^2}{b^2} + \frac{n_z^2}{c^2}\right). \quad (6\text{-}14)$$

6e. Diffraction by a Crystal Lattice.—Let us consider an infinite crystal lattice, involving a sequence of identical planes spaced with the regular interval d. The allowed states of motion of this crystal along the z axis we assume, in accordance with the rules of the old quantum theory, to be those for which

$$\oint p_z dz = n_z h.$$

For this crystal it is seen that a cycle for the coordinate z is the identity distance d, so that (p_z being constant in the absence of forces acting on the crystal) the quantum rule becomes

$$\int_0^d p_z dz = n_z h, \qquad \text{or} \qquad p_z = \frac{n_z h}{d}. \quad (6\text{-}15)$$

Any interaction with another system must be such as to leave p_z quantized; that is, to change it by the amount $\Delta p_z = \Delta n_z h/d$ or nh/d, in which $n = \Delta n_z$ is an integer. One such type of interaction is collision with a photon of frequency ν, represented in Figure 6–4 as impinging at the angle ϑ and being specularly reflected. Since the momentum of a photon is $h\nu/c$, and its component along the z axis $\frac{h\nu}{c} \sin \vartheta$, the momentum transferred to the crystal is $\frac{2h\nu}{c} \sin \vartheta = \frac{2h}{\lambda} \sin \vartheta$. Equating this with the allowed momentum change of the crystal nh/d, we obtain the expression

$$n\lambda = 2d \sin \vartheta. \tag{6-16}$$

This is, however, just the Bragg equation for the diffraction of x-rays by a crystal. This derivation from the corpuscular view of the nature of light was given by Duane and Compton[1] in 1923.

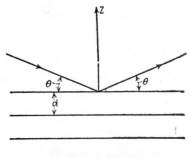

Let us now consider a particle, say an electron, of mass m similarly reflected by the crystal. The momentum transferred to the crystal will be $2mv \sin \vartheta$, which is equal to a quantum for the crystal when

$$n\frac{h}{mv} = 2d \sin \vartheta. \tag{6-17}$$

FIG. 6-4.—The reflection of a photon by a crystal.

Thus we see that a particle would be scattered by a crystal only when a diffraction equation similar to the Bragg equation for x-rays is satisfied. The wave length of light is replaced by the expression

$$\lambda = \frac{h}{mv}, \tag{6-18}$$

which is indeed the de Broglie expression for the wave length associated with an electron moving with the speed v. This simple consideration, which might have led to the discovery of the wave character of material particles in the days when the old quantum theory had not yet been discarded, was overlooked at that time.

In the above treatment, which is analogous to the Bragg treatment of x-ray diffraction, the assumption of specular reflection is made. This can be avoided by a treatment similar to Laue's derivation of his diffraction equations.

The foregoing considerations provide a simple though perhaps somewhat extreme illustration of the power of the old quantum theory as well as of its indefinite character. That a formal argument of this type leading to diffraction equations usually derived

[1] W. DUANE, *Proc. Nat. Acad. Sci.* **9**, 158 (1923); A. H. COMPTON, *ibid.* **9**, 359 (1923).

by the discussion of interference and reinforcement of waves could be carried through from the corpuscular viewpoint with the old quantum theory, and that a similar treatment could be given the scattering of electrons by a crystal, with the introduction of the de Broglie wave length for the electron, indicates that the gap between the old quantum theory and the new wave mechanics is not so wide as has been customarily assumed. The indefiniteness of the old quantum theory arose from its incompleteness—its inability to deal with any systems except multiply-periodic ones. Thus in this diffraction problem we are able to derive only the simple diffraction equation for an infinite crystal, the interesting questions of the width of the diffracted beam, the distribution of intensity in different diffraction maxima, the effect of finite size of the crystal, etc., being left unanswered.[1]

7. THE HYDROGEN ATOM

The system composed of a nucleus and one electron, whose treatment underlies any theoretical discussion of the electronic structure of atoms and molecules, was the subject of Bohr's first paper on the quantum theory.[2] In this paper he discussed circular orbits of the planetary electron about a fixed nucleus. Later[3] he took account of the motion of the nucleus as well as the electron about their center of mass and showed that with the consequent introduction of the reduced mass of the two particles a small numerical deviation from a simple relation between the spectral frequencies of hydrogen and ionized helium is satisfactorily explained. Sommerfeld[4] then applied his more general rules for quantization, leading to quantized elliptical orbits with definite spatial orientations, and showed that the relativistic change in mass of the electron causes a splitting of energy levels correlated with the observed fine structure of hydrogenlike spectra. In this section we shall reproduce the Sommerfeld treatment, except for the consideration of the relativistic correction.

7a. Solution of the Equations of Motion.—The system consists of two particles, the heavy nucleus, with mass m_1 and

---

[1] The application of the correspondence principle to this problem was made by P. S. Epstein and P. Ehrenfest, *Proc. Nat. Acad. Sci.* **10**, 133 (1924).

[2] N. Bohr, *Phil. Mag.* **26**, 1 (1913).

[3] N. Bohr, *ibid.* **27**, 506 (1914).

[4] A. Sommerfeld, *Ann. d. Phys.* **51**, 1 (1916).

electric charge $+Ze$, and the electron, with mass m_2 and charge
$-e$, between which there is operative an inverse-square attractive force corresponding to the potential-energy function

$$V(r) = -Ze^2/r,$$

r being the distance between the two particles. (The gravitational attraction is negligibly small relative to the electrostatic attraction.) The system is similar to that of the sun and a
planet, or the earth and moon. It was solved by Sir Isaac
Newton in his "Philosophiae Naturalis Principia Mathematica,"
wherein he showed that the orbits of one particle relative to the
other are conic sections. Of these we shall discuss only the
closed orbits, elliptical or circular, inasmuch as the old quantum
theory was incapable of dealing with the hyperbolic orbits of the
ionized hydrogen atom.

The system may be described by means of Cartesian coordi-
nates x_1, y_1, z_1 and x_2, y_2, z_2 of the two particles. As shown in
Section 2d by the introduction of coordinates x, y, z of the center
of mass and of polar coordinates r, ϑ, φ of the electron relative
to the nucleus, the center of mass of the system undergoes
translational motion in a fixed direction with constant speed,
like a single particle in field-free space, and the relative motion
of electron and nucleus is that of a particle of mass $\mu = \dfrac{m_1 m_2}{m_1 + m_2}$,
the *reduced mass* of the two particles, about a fixed center to
which it is attracted by the same force as that between the
electron and nucleus. Moreover, the orbit representing any
state of motion lies in a plane (Sec. 1d).

In terms of variables r and χ in the plane of motion, the
Lagrangian equations of motion are

$$\mu \ddot{r} = \mu r \dot{\chi}^2 - \frac{Ze^2}{r^2} \qquad (7\text{-}1)$$

and

$$\frac{d}{dt}(\mu r^2 \dot{\chi}) = 0. \qquad (7\text{-}2)$$

The second of these can be integrated at once (as in Sec. 1d), to
give

$$\mu r^2 \dot{\chi} = p, \text{ a constant.} \qquad (7\text{-}3)$$

This first result expresses Kepler's area law: The radius vector

from sun to planet sweeps out equal areas in equal times. The
constant p is the total angular momentum of the system.

Eliminating $\dot{\chi}$ from Equations 7–1 and 7–3, we obtain

$$\mu\ddot{r} = \frac{p^2}{\mu r^3} - \frac{Ze^2}{r^2}, \tag{7-4}$$

which on multiplication by $\dot{r}$ and integration leads to

$$\frac{\mu\dot{r}^2}{2} = -\frac{p^2}{2\mu r^2} + \frac{Ze^2}{r} + W. \tag{7-5}$$

The constant of integration W is the total energy of the system
(aside from the translational energy of the system as a whole).
Instead of solving this directly, let us eliminate t to obtain an
equation involving r and χ.   Since

$$\dot{r} = \frac{dr}{dt} = \frac{dr}{d\chi}\frac{d\chi}{dt} = \frac{dr}{d\chi}\frac{p}{\mu r^2}, \tag{7-6}$$

Equation 7–5 reduces to

$$\left(\frac{1}{r^2}\frac{dr}{d\chi}\right)^2 = -\frac{1}{r^2} + \frac{2Ze^2\mu}{p^2 r} + \frac{2\mu W}{p^2},$$

or, introducing the new variable

$$u = \frac{1}{r}, \tag{7-7}$$

$$\pm d\chi = \frac{du}{\sqrt{\dfrac{2\mu W}{p^2} + \dfrac{2Ze^2\mu}{p^2}u - u^2}}. \tag{7-8}$$

This can be integrated at once, for W either positive or negative.
In the latter case (closed orbits) there is obtained

$$u = \frac{1}{r} = \frac{Ze^2\mu}{p^2} + \frac{1}{2}\sqrt{\frac{4\mu^2 Z^2 e^4}{p^4} + \frac{8\mu W}{p^2}}\sin{(\chi - \chi_0)}. \tag{7-9}$$

This is the equation of an ellipse with the origin at one focus, as
in Figure 7–1.   In terms of the eccentricity ϵ and the semimajor
and semiminor axes a and b, the equation of such an ellipse is

$$u = \frac{1}{r} = \frac{1 + \epsilon\sin{(\chi - \chi_0)}}{a(1 - \epsilon^2)} = \frac{a}{b^2} + \frac{\sqrt{a^2 - b^2}}{b^2}\sin{(\chi - \chi_0)}, \tag{7-10}$$

with
$$b = a\sqrt{1 - \epsilon^2}.$$

Thus it is found that the elements of the elliptical orbit are given by the equations

$$a = -\frac{Ze^2}{2W}, \qquad b = \frac{p}{\sqrt{-2\mu W}}, \qquad 1 - \epsilon^2 = -\frac{2Wp^2}{\mu Z^2 e^4}. \quad (7\text{–}11)$$

The energy W is determined by the major axis of the ellipse alone.

As shown in Problem 5–1, the total energy for a circular orbit is equal to one-half the potential energy and to the kinetic energy with changed sign. It can be shown also that similar relations

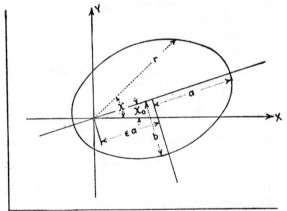

Fig. 7–1.—An elliptical electron-orbit for the hydrogen atom according to the old quantum theory.

hold for the time-average values of these quantities for elliptic orbits, that is, that

$$W = \tfrac{1}{2}\bar{V} = -\bar{T}, \qquad (7\text{–}12)$$

in which the barred symbols indicate the time-average values of the dynamical quantities.

7b. Application of the Quantum Rules. The Energy Levels.— The Wilson-Sommerfeld quantum rules, in terms of the polar coordinates r, ϑ, and φ, are expressed by the three equations

$$\oint p_r dr = n_r h, \qquad (7\text{–}13a)$$
$$\oint p_\vartheta d\vartheta = n_\vartheta h, \qquad (7\text{–}13b)$$
$$\oint p_\varphi d\varphi = mh. \qquad (7\text{–}13c)$$

Since p_φ is a constant (Sec. 1e), the third of these can be integrated at once, giving

$$2\pi p_\varphi = mh, \quad \text{or} \quad p_\varphi = \frac{mh}{2\pi}, \quad m = \pm 1, \pm 2, \cdots .$$

$$(7\text{--}14)$$

Hence the component of angular momentum of the orbit along the z axis can assume only the quantized values which are integral multiples of $h/2\pi$. The quantum number m is called the *magnetic quantum number*, because it serves to distinguish the various slightly separated levels into which the field-free energy levels are split upon the application of a magnetic field to the atom. This quantum number is closely connected with the orientation of the old-quantum-theory orbit in space, a question discussed in Section $7d$.

The second integral is easily discussed by the introduction of the angle χ and its conjugate momentum $p_\chi = p$, the total angular momentum of the system, by means of the relation, given in Equation 1–41, Section $1e$,

$$p_\chi d\chi = p_\vartheta d\vartheta + p_\varphi d\varphi. \tag{7--15}$$

In this way we obtain the equation

$$\oint p_\chi d\chi = kh, \tag{7--16}$$

in which p_χ is a constant of the motion and k is the sum of n_ϑ and m. This integrates at once to

$$2\pi p = kh, \quad \text{or} \quad p = \frac{kh}{2\pi}, \quad k = 1, 2, \cdots . \tag{7--17}$$

Hence the total angular momentum of the orbit was restricted by the old quantum theory to values which are integral multiples of the quantum unit of angular momentum $h/2\pi$. The quantum number k is called the *azimuthal quantum number*.

To evaluate the first integral it is convenient to transform it in the following way, involving the introduction of the angle χ and the variable $u = 1/r$ with the use of Equation 7–6:

$$p_r dr = \mu \dot{r} dr = \frac{p}{r^2}\left(\frac{dr}{d\chi}\right)^2 d\chi = p \cdot \frac{1}{u^2}\left(\frac{du}{d\chi}\right)^2 d\chi. \tag{7--18}$$

From Equation 7–10 we find on differentiation

$$\frac{du}{d\chi} = \frac{\epsilon \cos (\chi - \chi_0)}{a(1 - \epsilon^2)}, \tag{7--19}$$

with the use of which the r quantum condition reduces to the form

$$p\epsilon^2 \int_0^{2\pi} \frac{\cos^2 (\chi - \chi_0)}{\{1 + \epsilon \sin (\chi - \chi_0)\}^2} d\chi = n_r h. \qquad (7\text{-}20)$$

The definite integral was evaluated by Sommerfeld.[1] The resultant equation is

$$2\pi p \left(\frac{1}{\sqrt{1 - \epsilon^2}} - 1 \right) = n_r h. \qquad (7\text{-}21)$$

This, with the value of p of Equation 7–17 and the relation $b = a\sqrt{1 - \epsilon^2}$, leads to the equation

$$\frac{a}{b} = \frac{n_r + k}{k} = \frac{n}{k}. \qquad (7\text{-}22)$$

In this equation we have introduced a new quantum number n, called the *total quantum number*, as the sum of the azimuthal quantum number k and the radial quantum number n_r:

$$n = n_r + k. \qquad (7\text{-}23)$$

With these equations and Equation 7–11, the energy values of the quantized orbits and the values of the major and minor semiaxes can be expressed in terms of the quantum numbers and the physical constants involved. The energy is seen to have the value

$$W_n = -\frac{Z^2 2\pi^2 \mu e^4}{n^2 h^2} = -\frac{Z^2}{n^2} Rhc, \qquad (7\text{-}24)$$

being a function of the total quantum number alone. The value of R, the *Rydberg constant*, which is given by the equation

$$R = \frac{2\pi^2 \mu e^4}{h^3 c}, \qquad (7\text{-}25)$$

depends on the reduced mass μ of the electron and the nucleus. It is known very accurately, being obtained directly from spectroscopic data, the values as reported by Birge for hydrogen, ionized helium, and infinite nuclear mass being

$$R_H = 109,677.759 \pm 0.05 \text{ cm}^{-1},$$
$$R_{He} = 109,722.403 \pm 0.05 \text{ cm}^{-1},$$
$$R_\infty = 109,737.42 \pm 0.06 \text{ cm}^{-1}.$$

[1] A. SOMMERFELD, *Ann. d. Phys.* **51,** 1 (1916).

The major and minor semiaxes have the values

$$a = \frac{n^2 a_0}{Z}, \qquad b = \frac{nka_0}{Z}, \qquad (7\text{-}26)$$

in which the constant a_0 has the value

$$a_0 = \frac{h^2}{4\pi^2 \mu e^2}. \qquad (7\text{-}27)$$

The value of this quantity, which for hydrogen is the distance of the electron from the nucleus in the circular orbit with $n = 1$, $k = 1$, also depends on the reduced mass, but within the experimental error in the determination of e the three cases mentioned above lead to the same value[1]

$$a_0 = 0.5285\text{Å},$$

in which $1\text{Å} = 1 \times 10^{-8}$ cm. The energy may also be expressed in terms of a_0 as

$$W_n = -\frac{Ze^2}{2a} = -\frac{Z^2 e^2}{2n^2 a_0}. \qquad (7\text{-}28)$$

The total energy required to remove the electron from the normal hydrogen atom to infinity is hence

$$W_H = \frac{2\pi^2 \mu_H e^4}{h^2} = R_H hc = \frac{e^2}{2a_0}. \qquad (7\text{-}29)$$

This quantity, $W_H = 2.1528 \times 10^{-11}$ ergs, is often expressed in volt electrons, $W_H = 13.530$ v.e., or in reciprocal centimeters or wave numbers, $W_H = 109,677.76$ cm^{-1} (the factor hc being omitted), or in calories per mole, $W_H = 311,934$ cal/mole.

The energy levels of hydrogen are shown in Figure 7-2. It is seen that the first excitation energy, the energy required to raise the hydrogen atom from the normal state, with $n = 1$, to the first excited state, with $n = 2$, is very large, amounting to 10.15 v.e. or 234,000 cal/mole. The spectral lines emitted by an excited hydrogen atom as it falls from one stationary state to another would have wave numbers or reciprocal wave lengths $\tilde{\nu}$ given by the equation

$$\tilde{\nu} = R_H \left(\frac{1}{n''^2} - \frac{1}{n'^2} \right), \qquad (7\text{-}30)$$

The value given by Birge for infinite mass is

$$0.5281_{69} \pm 0.0004 \times 10^{-8} \text{ cm},$$

that for hydrogen being 0.0003 larger (Appendix I).

in which n'' and n' are the values of the total quantum number for the lower and the upper state, respectively. The series of lines corresponding to $n'' = 1$, that is, to transitions to the normal state, is called the *Lyman series*, and those corresponding to $n'' = 2, 3,$ and 4 are called the *Balmer, Paschen,* and *Brackett series*, respectively. The Lyman series lies in the ultraviolet region, the lower members of the Balmer series are in the visible region, and the other series all lie in the infrared.

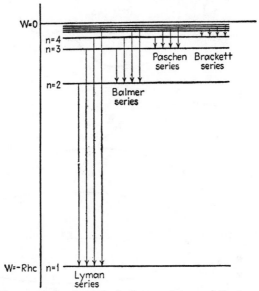

Fig. 7–2.—The energy levels of the hydrogen atom, and the transitions giving rise to the Lyman, Balmer, Paschen, and Brackett series.

7c. Description of the Orbits.—Although the allowed orbits given by the treatment of Section 7b are not retained in the quantum-mechanical model of hydrogen, they nevertheless serve as a valuable starting point for the study of the more subtle concepts of the newer theories. The old-quantum-theory orbits are unsatisfactory chiefly because they restrict the motion too rigidly, a criticism which is generally applicable to the results of this theory.

For the simple non-relativistic model of the hydrogen atom in field-free space the allowed orbits are certain ellipses whose common focus is the center of mass of the nucleus and the electron, and whose dimensions are certain functions of the quantum numbers, as we have seen. For a given energy level of the atom there is in general more than one allowed ellipse, since the energy depends only on the major axis of the ellipse and not on its eccentricity or orientation in space. These different ellipses are distinguished by having different values of the azimuthal

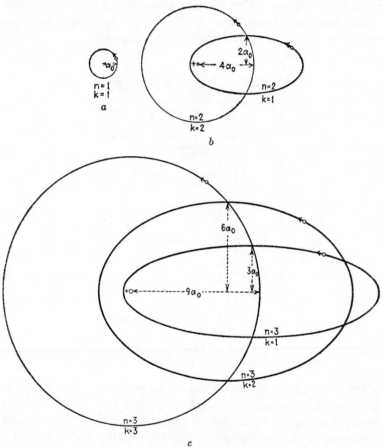

Fig. 7–3a, b, c.—Bohr-Sommerfeld electron-orbits for $n = 1$, 2, and 3, drawn to the same scale.

quantum number k, which may be any integer from 1 to n. When k equals n, the orbit is a circle, as is seen from Equation 7–26. For k less than n, the minor semiaxis b is less than the major semiaxis a, the eccentricity ϵ of the orbit increasing as k decreases relative to n. The value zero for k was somewhat arbitrarily excluded, on the basis of the argument that the corresponding orbit is a degenerate line ellipse which would cause the electron to strike the nucleus.

Figure 7-3 shows the orbits for $n = 1$, 2, and 3 and for the allowed values of k. The three different ellipses with $n = 3$ have major axes of the same length and minor axes which decrease with decreasing k. Figure 7-3 also illustrates the expansion of the orbits with increasing quantum number, the radii of the circular orbits increasing as the square of n.

A property of these orbits which is of particular importance in dealing with heavier atoms is the distance of closest approach of the electron to the nucleus. Using the expressions for a and b given in Equation 7-26 and the properties of the ellipse, we obtain for this distance the value $\dfrac{n(n - \sqrt{n^2 - k^2})a_0}{Z}$. This formula and the orbits drawn in Figure 7-3 show that the most eccentric orbit for a given n, i.e., that with the smallest value of k, comes the nearest to the nucleus. In many-electron atoms, this causes a separation of the energies corresponding to these different elliptical orbits with the same n, since the presence of the other electrons, especially the inner or core electrons, causes a modification of the field acting on the electron when it enters the region near the nucleus.

Since the charge on the nucleus enters the expression for the radius of the orbit given by Equations 7-26 and 7-27, the orbits for He^+ are smaller than the corresponding ones for hydrogen, the major semiaxis being reduced one-half by the greater charge on the helium-ion nucleus.

7d. Spatial Quantization.—So far we have said nothing of the orientation of the orbits in space. If a weak field, either electric or magnetic, is applied to the atom, so that the z direction in space can be distinguished but no appreciable change in energy occurs, the z component of the angular momentum of the atom must be an integral multiple of $h/2\pi$, as mentioned in Section 7b following Equation 7-14. This condition, which restricts the orientation of the plane of the orbit to certain definite directions, is called *spatial quantization*. The vector representing the total angular momentum **p** is a line perpendicular to the plane of the orbit (see Sec. 1e) and from Equation 7-17 has the length $kh/2\pi$. The z component of the angular momentum is of length $k \cos \omega(h/2\pi)$, if ω is the angle between the vector **p** and the z axis. This results in the following expression for cos ω:

$$\cos \omega = \frac{m}{k}.$$

The value zero for m was excluded for reasons related to those used in barring $k = 0$, so that m may be ± 1, ± 2, $\cdot \cdot \cdot$, $\pm k$.

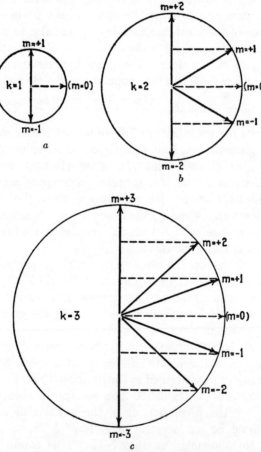

Fig. 7–4a, b, c.—Spatial quantization of Bohr-Sommerfeld orbits with $k = 1, 2$ and 3.

For the lowest state of hydrogen, in which $k = 1$ (and for all orbits for which $k = 1$), there are only two values of m, $+1$ and -1, which correspond to motion in the xy plane in a counterclockwise or in a clockwise sense. For $k = 2$ four orientations are permitted, as shown in Figure 7–4. Values $\pm k$ for m always correspond to orbits lying in the xy plane.

It can be shown by the methods of classical electromagnetic theory that the motion of an electron with charge $-e$ and mass m_0 in an orbit with angular momentum $\dfrac{kh}{2\pi}$ gives rise to a magnetic field corresponding to a magnetic dipole of magnitude $\dfrac{kh}{2\pi}\dfrac{e}{2m_0c}$ oriented in the same direction as the angular momentum vector. The component of magnetic moment in the direction of the z axis is $m\dfrac{he}{4\pi m_0c}$. The energy of magnetic interaction of the atom with a magnetic field of strength H parallel to the z axis is $m\dfrac{he}{4\pi m_0c}H$. It was this interaction energy which was considered to give rise to the *Zeeman effect* (the splitting of spectral lines by a magnetic field) and the phenomenon of paramagnetism. It is now known that this explanation is only partially satisfactory, inasmuch as the magnetic moment associated with the spin of the electron, discussed in Chapter VIII, also makes an important contribution. The magnetic moment $\dfrac{he}{4\pi m_0c}$ is called a *Bohr magneton*.

Problem 7–1. Calculate the frequencies and wave lengths of the first five members of the Balmer series for the isotopic hydrogen atom whose mass is approximately 2.0136 on the atomic weight scale, and compare with those for ordinary hydrogen.

Problem 7–2. Quantize the system consisting of two neutral particles of masses equal to those of the electron and proton held together by gravitational attraction, obtaining expressions for the axes of the orbits and the energy levels.

8. THE DECLINE OF THE OLD QUANTUM THEORY

The historical development of atomic and molecular mechanics up to the present may be summarized by the following division into periods (which, of course, are not so sharply demarcated as indicated):

1913–1920. The origin and extensive application of the old quantum theory of the atom.

1920–1925. The decline of the old quantum theory.

1925– The origin of the new quantum mechanics and its application to physical problems.

1927–   The application of the new quantum mechanics to chemical problems.

The present time may well be also the first part of the era of the development of a more fundamental quantum mechanics, including the theory of relativity and of the electromagnetic field, and dealing with the mechanics of the atomic nucleus as well as of the extranuclear structure.

The decline of the old quantum theory began with the introduction of half-integral values for quantum numbers in place of integral values for certain systems, in order to obtain agreement with experiment. It was discovered that the pure rotation spectra of the hydrogen halide molecules are not in accordance with Equation 6–9 with $K = 0, 1, 2, \cdots$, but instead require $K = \frac{1}{2}, \frac{3}{2}, \cdots$. Similarly, half-integral values of the oscillational quantum number v in Equation 6–11 were found to be required in order to account for the observed isotope displacements for diatomic molecules. Half-integral values for the azimuthal quantum number k were also indicated by observations on both polarization and penetration of the atom core by a valence electron. Still more serious were cases in which agreement with the observed energy levels could not be obtained by the methods of the old quantum theory by any such subterfuge or arbitrary procedure (such as the normal state of the helium atom, excited states of the helium atom, the normal state of the hydrogen molecule ion, etc.), and cases where the methods of the old quantum theory led to definite qualitative disagreement with experiment (the influence of a magnetic field on the dielectric constant of a gas, etc.). Moreover, the failure of the old quantum theory to provide a method of calculating transition probabilities and the intensities of spectral lines was recognized more and more clearly as a fundamental flaw. Closely related to this was the lack of a treatment of the phenomenon of the dispersion of light, a problem which attracted a great amount of attention.

This dissatisfaction with the old quantum theory culminated in the formulation by Heisenberg[1] in 1925 of his quantum mechanics, as a method of treatment of atomic systems leading to values of the intensities as well as frequencies of spectral lines. The quantum mechanics of Heisenberg was rapidly

[1] W. Heisenberg, Z. f. Phys. 33, 879 (1925).

developed by Heisenberg, Born, and Jordan[1] by the introduction of matrix methods.   In the meantime Schrödinger had independently discovered and developed his wave mechanics,[2] stimulated by the earlier attribution of a wave character to the electron by de Broglie[3] in 1924.   The mathematical identity of matrix mechanics and wave mechanics was then shown by Schrödinger[4] and by Eckart.[5]   The further development of the quantum mechanics was rapid, especially because of the contributions of Dirac, who formulated[6] a relativistic theory of the electron and contributed to the generalization of the quantum mechanics (Chap. XV).

General References on the Old Quantum Theory

A. SOMMERFELD: "Atomic Structure and Spectral Lines," E. P. Dutton & Co., Inc., New York, 1923.

A. E. RUARK and H. C. UREY: "Atoms, Molecules and Quanta," McGraw-Hill Book Company, Inc., New York, 1930.

[1] M. BORN and P. JORDAN, *ibid.* **34**, 858 (1925); M. BORN, W. HEISENBERG, and P. JORDAN, *ibid.* **35**, 557 (1926).

[2] E. SCHRÖDINGER, *Ann. d. Phys.* **79**, 361, 489; **80**, 437; **81**, 109 (1926).

[3] L. DE BROGLIE, Thesis, Paris, 1924; *Ann. de phys.* (10) **3**, 22 (1925).

[4] E. SCHRÖDINGER, *Ann. d. Phys.* **79**, 734 (1926).

[5] C. ECKART, *Phys. Rev.* **28**, 711 (1926).

[6] P. A. M. DIRAC, *Proc. Roy. Soc.* A **113**, 621; **114**, 243 (1927); **117**, 610 (1928).

CHAPTER III

THE SCHRÖDINGER WAVE EQUATION WITH THE HARMONIC OSCILLATOR AS AN EXAMPLE

In the preceding chapters we have given a brief discussion of the development of the theory of mechanics before the discovery of the quantum mechanics. Now we begin the study of the quantum mechanics itself, starting in this chapter with the Schrödinger wave equation for a system with only one degree of freedom, the general principles of the theory being illustrated by the special example of the harmonic oscillator, which is treated in great detail because of its importance in many physical problems. The theory will then be generalized in the succeeding chapter to systems of point particles in three-dimensional space.

9. THE SCHRÖDINGER WAVE EQUATION

In the first paragraph of his paper[1] Quantisierung als Eigenwertproblem, communicated to the *Annalen der Physik* on January 27, 1926, Erwin Schrödinger stated essentially:

In this communication I wish to show, first for the simplest case of the non-relativistic and unperturbed hydrogen atom, that the usual rules of quantization can be replaced by another postulate, in which there occurs no mention of whole numbers. Instead, the introduction of integers arises in the same natural way as, for example, in a vibrating string, for which the number of nodes is integral. The new conception can be generalized, and I believe that it penetrates deeply into the true nature of the quantum rules.

In this and four other papers, published during the first half of 1926, Schrödinger communicated his wave equation and applied it to a number of problems, including the hydrogen atom, the harmonic oscillator, the rigid rotator, the diatomic molecule, and

---

[1] E. SCHRÖDINGER, *Ann. d. Phys.* **79**, 361 (1926), and later papers referred to on the preceding page. An English translation of these papers has appeared under the title E. Schrödinger, "Collected Papers on Wave Mechanics," Blackie and Son, London and Glasgow, 1928.

the hydrogen atom in an electric field (Stark effect). For the
last problem he developed his perturbation theory, and for
the discussion of dispersion he also developed the theory of a
perturbation varying with the time. His methods were rapidly
adopted by other investigators, and applied with such success
that there is hardly a field of physics or chemistry that has
remained untouched by Schrödinger's work.

Schrödinger's system of dynamics differs from that of Newton,
Lagrange, and Hamilton in its aim as well as its method. Instead
of attempting to find equations, such as Newton's equations,
which enable a prediction to be made of the exact positions and
velocities of the particles of a system in a given state of motion,
he devised a method of calculating a function of the coordinates
of the system and the time (and not the momenta or velocities),
with the aid of which, in accordance with the interpretation
developed by Born,[1] probable values of the coordinates and
of other dynamical quantities can be predicted for the system.
It was later recognized that the acceptance of dynamical equations of this type involves the renunciation of the hope of describing in exact detail the behavior of a system. The degree of
accuracy with which the behavior of a system can be discussed
by quantum-mechanical methods forms the subject of *Heisenberg's uncertainty principle*,[2] to which we shall recur in Chapter
XV.

The Schrödinger wave equation and its auxiliary postulates
enable us to determine certain functions Ψ of the coordinates of a
system and the time. These functions are called the *Schrödinger
wave functions* or *probability amplitude functions*. The square
of the absolute value of a given wave function is interpreted as
a *probability distribution function* for the coordinates of the
system in the state represented by this wave function, as will
be discussed in Section 10a. The wave equation has been
given this name because it is a differential equation of the second
order in the coordinates of the system, somewhat similar to the
wave equation of classical theory. The similarity is not close,
however, and we shall not utilize the analogy in our exposition.

Besides yielding the probability amplitude or wave function Ψ,
the Schrödinger equation provides a method of calculating **values**

[1] M. BORN, *Z. f. Phys.* **37**, 863; **38**, 803 (1926).
[2] W. HEISENBERG, *Z. f. Phys.* **43**, 172 (1927).

of the energy of the stationary states of a system, the existence
of which we have discussed in connection with the old quantum
theory. No arbitrary postulates concerning quantum numbers
are required in this calculation; instead, integers enter automatically in the process of finding satisfactory solutions of the
wave equation.

For our purposes, the Schrödinger equation, the auxiliary
restrictions upon the wave function Ψ, and the interpretation of
the wave function are conveniently taken as fundamental
postulates, with no derivation from other principles necessary.

This idea may be clarified by a comparison with other branches
of physics. Every department of deductive science must
necessarily be founded on certain postulates which are regarded
as fundamental. Frequently these fundamental postulates are
so closely related to experiment that their acceptance follows
directly upon the acceptance of the experiments upon which
they are based, as, for example, the inverse-square law of electrical
attraction. In other cases the primary postulates are not so
directly obvious from experiment, but owe their acceptance to the
fact that conclusions drawn from them, often by long chains of
reasoning, agree with experiment in all of the tests which have
been made. The second law of thermodynamics is representative
of this type of postulate. It is not customary to attempt to
derive the second law for general systems from anything more
fundamental, nor is it obvious that it follows directly from
some simple experiment; nevertheless, it is accepted as correct
because deductions made from it agree with experiment. It is
an assumption, justified only by the success achieved by its
consequences.

The wave equation of Schrödinger belongs to this latter class
of primary assumption. It is not derived from other physical
laws nor obtained as a necessary consequence of any experiment;
instead, it is assumed to be correct, and then results predicted
by it are compared with data from the laboratory.

A clear distinction must frequently be made between the way
in which a discoverer arrives at a given hypothesis and the
logical position which this hypothesis occupies in the theory when
it has been completed and made orderly and deductive. In
the process of discovery, analogy often plays a very important
part. Thus the analogies between geometrical optics and

classical mechanics on the one hand and undulatory optics and wave mechanics on the other may have assisted Schrödinger to formulate his now famous equation; but these analogies by no means provide a logical derivation of the equation. In many cases there is more than one way of stating the fundamental postulates. Thus either Lagrange's or Hamilton's form of the equations of motion may be regarded as fundamental for classical mechanics, and if one is so chosen, the other can be derived from it. Similarly, there are other ways of expressing the basic assumptions of quantum mechanics, and if they are used, the wave equation can be derived from them, but, no matter which mode of presenting the theory is adopted, some starting point must be chosen, consisting of a set of assumptions not deduced from any deeper principles.

It often happens that principles which have served as the basis for whole branches of theory are superseded by other principles of wider applicability. Newton's laws of motion, adopted because they were successful in predicting the motions of the planets and in correlating celestial and terrestrial phenomena, were replaced by Lagrange's and Hamilton's equations because these are more general. They include Newton's laws as a special case and in addition serve for the treatment of motions involving electric, magnetic, and relativistic phenomena. Likewise, quantum mechanics includes Newton's laws for the special case of heavy bodies and in addition is successful in problems involving atoms and electrons. A still more general theory than that of Schrödinger has been developed (we shall discuss it in Chap. XV), but for nearly all purposes the wave equation is a convenient and sufficient starting point.

9a. The Wave Equation Including the Time.—Let us first consider a Newtonian system with one degree of freedom, consisting of a particle of mass m restricted to motion along a fixed straight line, which we take as the x axis, and let us assume that the system is further described by a potential-energy function $V(x)$ throughout the region $-\infty < x < +\infty$. For this system the Schrödinger wave equation is assumed to be

$$-\frac{h^2}{8\pi^2 m}\frac{\partial^2 \Psi(x, t)}{\partial x^2} + V(x)\Psi(x, t) = -\frac{h}{2\pi i}\frac{\partial \Psi(x, t)}{\partial t}. \quad (9\text{-}1)$$

In this equation the function $\Psi(x, t)$ is called the *Schrödinger wave function including the time,* or the probability amplitude function. It will be noticed that the equation is somewhat similar in form to the wave equations occurring in other branches of theoretical physics, as in the discussion of the motion of a vibrating string. The student facile in mathematical physics may well profit from investigating this similarity and also the analogy between classical mechanics and geometrical optics on the one hand, and wave mechanics and undulatory optics on the other.[1] However, it is not necessary to do this. An extensive previous knowledge of partial differential equations and their usual applications in mathematical physics is not a necessary prerequisite for the study of wave mechanics, and indeed the study of wave mechanics may provide a satisfactory introduction to the subject for the more physically minded or chemically minded student.

The Schrödinger time equation is closely related to the equation of classical Newtonian mechanics

$$H(p_x, x) = T(p_x) + V(x) = W, \qquad (9\text{--}2)$$

which states that the total energy W is equal to the sum of the kinetic energy T and the potential energy V and hence to the Hamiltonian function $H(p_x, x)$. Introducing the coordinate x and momentum p_x, this equation becomes

$$H(p_x, x) = \frac{1}{2m}p_x^2 + V(x) = W. \qquad (9\text{--}3)$$

If we now arbitrarily replace p_x by the differential operator $\dfrac{h}{2\pi i}\dfrac{\partial}{\partial x}$ and W by $-\dfrac{h}{2\pi i}\dfrac{\partial}{\partial t}$, and introduce the function $\Psi(x, t)$ on which these operators can operate, this equation becomes

$$H\!\left(\frac{h}{2\pi i}\frac{\partial}{\partial x}, x\right)\!\Psi(x, t) = -\frac{h^2}{8\pi^2 m}\frac{\partial^2\Psi}{\partial x^2} + V\Psi = -\frac{h}{2\pi i}\frac{\partial\Psi}{\partial t}, \qquad (9\text{--}4)$$

which is identical with Equation 9–1. The wave equation is

---

[1] See, for example, Condon and Morse, "Quantum Mechanics," p. 10, McGraw-Hill Book Company, Inc., New York, 1929; Ruark and Urey, "Atoms, Molecules and Quanta," Chap. XV, McGraw-Hill Book Company, Inc., New York, 1930; E. Schrödinger, *Ann. d. Phys.* **79**, 489 (1926); K. K. Darrow, *Rev. Mod. Phys.* **6**, 23 (1934); or other treatises on wave mechanics, listed at the end of this chapter.

consequently often conveniently written as

$$H\Psi = W\Psi, \qquad (9\text{-}5)$$

in which it is understood that the operators $\dfrac{h}{2\pi i} \dfrac{\partial}{\partial x}$ and $-\dfrac{h}{2\pi i} \dfrac{\partial}{\partial t}$ are to be introduced.

In replacing p_x by the operator $\dfrac{h}{2\pi i} \dfrac{\partial}{\partial x}$, p_x^2 is to be replaced by $\left(\dfrac{h}{2\pi i}\right)^2 \dfrac{\partial^2}{\partial x^2}$, and so on. (In some cases, which, however, do not arise in the simpler problems which we are discussing in this book, there may be ambiguity regarding the formulation of the operator.[1]) It might be desirable to distinguish between the classical Hamiltonian function $H = H(p_x, x)$ and the Hamiltonian operator

$$H = H\left(\frac{h}{2\pi i} \frac{\partial}{\partial x}, x\right),$$

as by writing H_{operator} for the latter. We shall not do this, however, since the danger of confusion is small. Whenever H is followed by Ψ (or by ψ, representing the wave functions not including the time, discussed in the following sections), it is understood to be the Hamiltonian operator. Similarly, whenever W is followed by Ψ it represents the operator $-\dfrac{h}{2\pi i} \dfrac{\partial}{\partial t}$. The symbol W will also be used to represent the energy constant (Secs. 9b, 9c). We shall, indeed, usually restrict the symbol W to this use, and write $-\dfrac{h}{2\pi i} \dfrac{\partial}{\partial t}$ for the operator.

It must be recognized that this correlation of the wave equation and the classical energy equation, as well as the utilization which we shall subsequently make of many other classical dynamical expressions, has only formal significance. It provides a convenient way of describing the system for which we are setting up a wave equation by making use of the terminology developed over a long period of years by the workers in classical dynamics. Thus our store of direct knowledge regarding the nature of the system known as the hydrogen atom consists in the results of a large number of experiments—spectroscopic, chemical, etc. It is found that all of the known facts about this system can be correlated and systematized (and, we say, explained) by associating with this system a certain wave equation. Our confidence in the significance of this association increases when predictions regarding previously uninvestigated properties of

[1] B. Podolsky, *Phys. Rev.* **32**, 812 (1928).

the hydrogen atom are subsequently verified by experiment. We might then describe the hydrogen atom by giving its wave equation; this description would be complete. It is unsatisfactory, however, because it is unwieldy. On observing that there is a formal relation between this wave equation and the classical energy equation for a system of two particles of different masses and electrical charges, we seize on this as providing a simple, easy, and familiar way of describing the system, and we say that the hydrogen atom consists of two particles, the electron and proton, which attract each other according to Coulomb's inverse-square law. Actually we do not know that the electron and proton attract each other in the same way that two macroscopic electrically charged bodies do, inasmuch as the force between the two particles in a hydrogen atom has never been directly measured. All that we do know is that the wave equation for the hydrogen atom bears a certain formal relation to the classical dynamical equations for a system of two particles attracting each other in this way.

Having emphasized the formal nature of this correlation and of the usual description of wave-mechanical systems in terms of classical concepts, let us now point out the extreme practical importance of this procedure. It is found that satisfactory wave equations can be formulated for nearly all atomic and molecular systems by accepting the descriptions of them developed during the days of the classical and old quantum theory and translating them into quantum-mechanical language by the methods discussed above. Indeed, in many cases the wave-mechanical expressions for values of experimentally observable properties of systems are identical with those given by the old quantum theory, and in other cases only small changes are necessary. Throughout the following chapters we shall make use of such locutions as "a system of two particles with inverse-square attraction" instead of "a system whose wave equation involves six coordinates and a function e^2/r_{12}," etc.

9b. The Amplitude Equation.—In order to solve Equation 9–1, let us (as is usual in the solution of a partial differential equation of this type) first study the solutions Ψ (if any exist) which can be expressed as the product of two functions, one involving the time alone and the other the coordinate alone:

$$\Psi(x, t) = \psi(x)\varphi(t).$$

On introducing this in Equation 9–1 and dividing through by $\psi(x)\varphi(t)$, it becomes

$$\frac{1}{\psi(x)}\left\{-\frac{h^2}{8\pi^2 m}\frac{d^2\psi(x)}{dx^2} + V(x)\psi(x)\right\} = -\frac{h}{2\pi i}\cdot\frac{1}{\varphi(t)}\frac{d\varphi(t)}{dt}. \quad (9\text{--}6)$$

The right side of this equation is a function of the time t alone and the left side a function of the coordinate x alone. It is consequently necessary that the value of the quantity to which each side is equal be dependent on neither x nor t; that is, that it be a constant. Let us call it W. Equation 9–6 can then be written as two equations, namely,

$$\left.\begin{array}{c}\dfrac{d\varphi(t)}{dt} = -\dfrac{2\pi i}{h}W\varphi(t) \\[2ex] -\dfrac{h^2}{8\pi^2 m}\dfrac{d^2\psi(x)}{dx^2} + V(x)\psi(x) = W\psi(x).\end{array}\right\} \quad (9\text{--}7)$$

and

The second of these is customarily written in the form

$$\frac{d^2\psi}{dx^2} + \frac{8\pi^2 m}{h^2}\{W - V(x)\}\psi = 0, \quad (9\text{--}8)$$

obtained on multiplying by $-8\pi^2 m/h^2$ and transposing the term in W.

Equation 9–8 is often itself called the Schrödinger wave equation, or sometimes the amplitude equation, inasmuch as $\psi(x)$ determines the amplitude of the function $\Psi(x, t)$. It is found that the equation possesses various satisfactory solutions, corresponding to various values of the constant W. Let us indicate these values of W by attaching the subscript n, and similarly represent the amplitude function corresponding to W_n as $\psi_n(x)$. The corresponding equation for $\varphi(t)$ can be integrated at once to give

$$\varphi_n(t) = e^{-2\pi i \frac{W_n}{h} t}. \quad (9\text{--}9)$$

The general solution of Equation 9–1 is the sum of all the particular solutions with arbitrary coefficients. We consequently write as the general expression for the wave function for this system

$$\Psi(x, t) = \sum_n a_n \Psi_n(x, t) = \sum_n a_n \psi_n(x) e^{-2\pi i \frac{W_n}{h} t}, \quad (9\text{--}10)$$

in which the quantities a_n are constants. The symbol $\sum_n$ is to be considered as representing the process of summation over discrete values of W_n or integration over a continuous range or both, according to the requirements of the particular case. It will be shown later that the general postulates which we shall make regarding the physical interpretation of the wave function require that the constant W_n represent the energy of the system in its various stationary states.

9c. Wave Functions. Discrete and Continuous Sets of Characteristic Energy Values.—The functions $\psi_n(x)$ which satisfy Equation 9–8 and also certain auxiliary conditions, discussed below, are variously called *wave functions* or eigenfunctions (Eigenfunktionen), or sometimes amplitude functions, characteristic functions, or proper functions. It is found that satisfactory solutions ψ_n of the wave equation exist only for certain values of the parameter W_n (which is interpreted as the energy of the system). These values W_n are *characteristic energy values* or eigenvalues (Eigenwerte) of the wave equation. A wave equation of this type is called a *characteristic value equation*.

Inasmuch as we are going to interpret the square of the absolute value of a wave function as having the physical significance of a probability distribution function, it is not unreasonable that the wave function be required to possess certain properties, such as single-valuedness, necessary in order that this interpretation be possible and unambiguous. It has been found that a satisfactory wave mechanics can be constructed on the basis of the following auxiliary postulates regarding the nature of wave functions:

To be a satisfactory wave function, a solution of the Schrödinger wave equation must be continuous, single-valued, and finite[1] through-

---

[1] The assumption that the wave function be finite at all points in configuration space may be more rigorous than necessary. Several alternative postulates have been suggested by various investigators. Perhaps the most satisfying of these is due to W. Pauli ("Handbuch der Physik," 2d ed., Vol. XXVI, Part 1, p. 123). In Section 10 we shall interpret the function $\Psi^*\Psi$ as a probability distribution function. In order that this interpretation may be made, it is necessary that the integral of $\Psi^*\Psi$ over configuration space be a constant with changing time. Pauli has shown that this condition is satisfied provided that Ψ is finite throughout configuration space, but that it is also satisfied in certain cases by functions which are not finite everywhere. The exceptional cases are rare and do not occur in the problems treated in this book.

out the configuration space of the system (that is, for all values of
the coordinate x which the system can assume).

These conditions are those usually applied in mathematical
physics to functions representing physical quantities. For
example, the function representing the displacement of a vibrating string from its equilibrium configuration would have to
satisfy them.

For a given system the characteristic energy values W_n may
occur only as a set of discrete values, or as a set of values covering
a continuous range, or as both. From analogy with spectroscopy
it is often said that in these three cases the energy values comprise
a discrete spectrum, a continuous spectrum, or both. The way

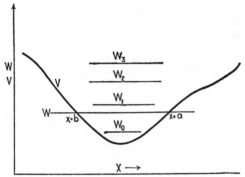

Fig. 9–1.—Potential-energy function for a general system with one degree of
freedom.

in which the above postulates regarding the wave equation and
its acceptable solutions lead to the selection of definite energy
values may be understood by the qualitative consideration of a
simple example. Let us consider, for our system of one degree
of freedom, that the potential-energy function $V(x)$ has the form
given in Figure 9–1, such that for very large positive or negative
values of x, $V(x)$ increases without limit. For a given value of
the energy parameter W, the wave equation is

$$\frac{d^2\psi}{dx^2} = \frac{8\pi^2 m}{h^2}\{V(x) - W\}\psi. \tag{9–11}$$

In the region of large x $(x > a)$ the quantity $V(x) - W$ will be
positive. Hence in this region the curvature $\dfrac{d^2\psi}{dx^2}$ will be positive
if ψ is positive, and negative if ψ is negative. Now let us assume

that at an arbitrary point $x = c$ the function ψ has a certain value (which may be chosen arbitrarily, inasmuch as the wave equation is a homogeneous equation[1]) and a certain slope $\frac{d\psi}{dx}$, as indicated for Curve 1 in Figure 9–2. The behavior of the function, as it is continued both to the right and to the left, is completely determined by the values assigned to two quantities; to wit, the slope $\frac{d\psi}{dx}$ at the point $x = c$, and the energy parameter W in the wave equation, which determines the value of the second deriva-

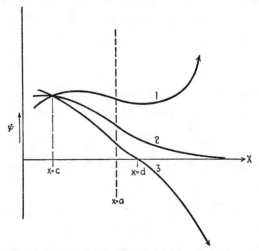

Fig. 9–2.—The behavior of ψ for $x > a$.

tive. As we have drawn Curve 1, the curvature is determined by the wave equation to be negative in the region $x < a$, where $V(x) - W$ is negative, ψ being positive, and hence the curve can be continued to the right as shown. At the point $x = a$, the function remaining positive, the curvature becomes positive, the curve then being concave upward. If the slope becomes positive, as indicated, then the curve will increase without limit for increasing x, and as a result of this "infinity catastrophe" the function will not be an acceptable wave function.

[1] An equation is homogeneous in ψ, if the same power of ψ (in our case the first power) occurs in every term. The function obtained by multiplying any solution of a homogeneous equation by a constant is also a solution of the equation.

We can now make a second attempt, choosing the slope at
$x = c$ as indicated for Curve 3. In this case the curve as drawn
intersects the x axis at a point $x = d$ to the right of a. For
values of x larger than d the function ψ is negative, and the curvature is negative. The function decreases in value more and more
rapidly with increasing x, again suffering the infinity catastrophe,
and hence it too is not an acceptable wave function in this
region.

Thus we see that, for a given value of W, only by a very careful
selection of the slope of the function at the point $x = c$ can the
function be made to behave properly for large values of x. This
selection, indicated by Curve 2, is such as to cause the wave
function to approach the value zero asymptotically with increasing x.

Supposing that we have in this way determined, for a given
value of W, a value of the slope at $x = c$ which causes the
function to behave properly for large positive values of x, we
extend the function to the left and consider its behavior for large
negative values of x. In view of our experience on the right,
it will not be surprising if our curve on extension to the left
behaves as Curve 1 or Curve 3 on the right, eliminating the
function from consideration; in fact, it is this behavior which
is expected for an arbitrarily chosen value of W. We can now
select another value of W for trial, and determine for it the value
of the slope at $x = c$ necessary to cause the function to behave
properly on the right, and then see if, for it, the curve behaves
properly on the left also. Finally, by a very careful choice of
the value of the energy parameter W, we are able to choose a
slope at $x = c$ which causes the function to behave properly
both for very large and for very small values of x. This value
of W is one of the characteristic values of the energy of the
system. In view of the sensitiveness of the curve to the parameter W, an infinitesimal change from this satisfactory value will
cause the function to behave improperly.

We conclude that the parameter W and the slope at the point
$x = c$ (for a given value of the function itself at this point) can
have only certain values if ψ is to be an acceptable wave function.
For each satisfactory value of W there is one (or, in certain
cases discussed later, more than one) satisfactory value of the
slope, by the use of which the corresponding wave function can

be built up. For this system the characteristic values W_n of the energy form a discrete set, and only a discrete set, inasmuch as for every value of W, no matter how large, $V(x) - W$ is positive for sufficiently large positive or negative values of x.

It is customary to number the characteristic energy values for such a system as indicated in Figure 9–1, W_0 being the lowest, W_1 the next, and so on, corresponding to the wave functions $\psi_0(x)$, $\psi_1(x)$, etc. The integer n, which is written as a subscript in W_n and $\psi_n(x)$, is called the *quantum number*. For such a one-dimensional system it is equal to the number of zeros[1] possessed by ψ_n. A slight extension of the argument given above

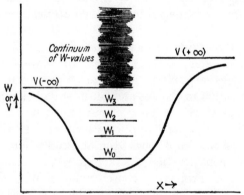

Fig. 9–3.—The energy levels for a system with $V(-\infty)$ or $V(+\infty)$ finite.

shows that all of the zeros lie in the region between the points $x = b$ and $x = a$, outside of which $V(x) - W_n$ remains positive. The natural and simple way in which integral quantum numbers are introduced and in which the energy is restricted to definite values contrasts sharply with the arbitrary and uncertain procedure of the old quantum theory.

Let us now consider a system in which the potential-energy function remains finite at $x \to +\infty$ or at $x \to -\infty$ or at both limits, as shown in Figure 9–3. For a value of W smaller than both $V(+\infty)$ and $V(-\infty)$ the argument presented above is valid. Consequently the energy levels will form a discrete set for this region. If W is greater than $V(+\infty)$, however, a similar argument shows that the curvature will be such as always to return the wave function to the x axis, about which it will

---

[1] A zero of $\psi_n(x)$ is a point $(x = x_1)$ at which ψ_n is equal to zero.

oscillate. Hence any value of W greater than $V(+\infty)$ or $V(-\infty)$ will be an allowed value, corresponding to an acceptable wave function, and the system will have a continuous spectrum of energy values in this region.

9d. The Complex Conjugate Wave Function $\Psi^*(x, t)$.—In the physical interpretation of the wave equation and its solutions, as discussed in the following section, the quantity $\Psi^*(x, t)$, the complex conjugate of $\Psi(x, t)$, enters on an equivalent basis with $\Psi(x, t)$. The wave equation satisfied by Ψ^* is the complex conjugate of Equation 9–1, namely,

$$-\frac{h^2}{8\pi^2 m}\frac{\partial^2 \Psi^*(x, t)}{\partial x^2} + V^*(x)\Psi^*(x, t) = \frac{h}{2\pi i}\frac{\partial \Psi^*(x, t)}{\partial t}. \quad (9\text{–}12)$$

The general solution of this conjugate wave equation is the following, the conjugate of 9–10:

$$\Psi^*(x, t) = \sum_n a_n^* \Psi_n^*(x, t) = \sum_n a_n^* \psi_n^*(x) e^{2\pi i \frac{W_n}{h} t}. \quad (9\text{–}13)$$

(Some authors have adopted the convention of representing by the symbol Ψ the wave function which is the solution of Equation 9–12 and by Ψ^* that of 9–1. This is only a matter of nomenclature.)

It will be noticed that in the complex conjugate wave function the exponential terms containing the time are necessarily different from the corresponding terms in Ψ itself, the minus sign being removed to form the complex conjugate. The amplitude functions $\psi_n(x)$, on the other hand, are frequently real, in which case $\psi_n^*(x) = \psi_n(x)$.

10. THE PHYSICAL INTERPRETATION OF THE WAVE FUNCTIONS

10a. $\Psi^*(x, t)\Psi(x, t)$ as a Probability Distribution Function.—Let us consider a given general solution $\Psi(x, t)$ of the wave equation. For a given value of the time t, the function $\Psi^*(x, t)\Psi(x, t)$, the product of Ψ and its complex conjugate, is a function defined for all values of x between $-\infty$ and $+\infty$; that is, throughout the configuration space of this one-dimensional system. We now make the following postulate regarding the physical significance of Ψ:

The quantity $\Psi^*(x, t)\Psi(x, t)dx$ is the probability that the system in the physical situation represented by the wave function $\Psi(x, t)$

have at the time t the configuration represented by a point in the region dx of configuration space. In other words, $\Psi^*(x, t)\Psi(x, t)$ is a *probability distribution function* for the configuration of the system. In the simple system under discussion, $\Psi^*(x, t)\Psi(x, t)dx$ is the probability that the particle lie in the region between x and $x + dx$ at the time t.

In order that this postulate may be made, the wave function $\Psi(x, t)$ must be *normalized to unity* (or, briefly, *normalized*); that is, the constants a_n of Equation 9–10 must be so chosen as to satisfy the relation

$$\int_{-\infty}^{+\infty} \Psi^*(x, t)\Psi(x, t)dx = 1, \tag{10–1}$$

inasmuch as the probability that the coordinate x of the particle lie somewhere between $-\infty$ and $+\infty$ is necessarily unity. It is also convenient to normalize the individual amplitude functions $\psi_r(x)$ to unity, so that each satisfies the equation

$$\int_{-\infty}^{+\infty} \psi_n^*(x)\psi_n(x)dx = 1. \tag{10–2}$$

Moreover, as proved in Appendix III, it is found that the independent solutions of any amplitude equation can always be chosen in such a way that for any two of them, $\psi_m(x)$ and $\psi_n(x)$, the integral $\int \psi_m^*(x)\psi_n(x)dx$ over all of configuration space vanishes; that is,

$$\int_{-\infty}^{+\infty} \psi_m^*(x)\psi_n(x)dx = 0, \qquad m \neq n. \tag{10–3}$$

The functions are then said to be *mutually orthogonal.* Using these relations and Equations 9–10 and 9–13, it is found that a wave function $\Psi(x, t) = \sum_n a_n \Psi_n(x, t)$ is normalized when the coefficients a_n satisfy the relation

$$\sum_n a_n^* a_n = 1. \tag{10–4}$$

10b. Stationary States.—Let us consider the probability distribution function $\Psi^*\Psi$ for a system in the state represented by the wave function $\Psi(x, t) = \sum_n a_n \psi_n(x)c^{-2\pi i\frac{W_n}{h}t}$ and its conjugate $\Psi^*(x, t) = \sum_m a_m^* \psi_m^*(x)e^{2\pi i\frac{W_m}{h}t}$. On multiplying these series together, $\Psi^*\Psi$ is seen to have the form

$$\Psi^*(x, t)\Psi(x, t) = \sum_n a_n^* a_n \psi_n^*(x)\psi_n(x) +$$

$$\sum_m \sum_n{}' a_m^* a_n \psi_m^*(x)\psi_n(x) e^{2\pi i \frac{(W_m - W_n)}{h} t},$$

in which the prime on the double-summation symbol indicates that only terms with $m \neq n$ are included. In general, then, the probability function and hence the properties of the system depend on the time, inasmuch as the time enters in the exponential factors of the double sum. Only if the coefficients a_n are zero for all except one value of W_n is $\Psi^*\Psi$ independent of t. In such a case the wave function will contain only a single term (with $n = n'$, say) $\Psi_{n'}(x, t) = \psi_{n'}(x) e^{-2\pi i \frac{W_{n'}}{h} t}$, the amplitude function $\psi_{n'}(x)$ being a particular solution of the amplitude equation. For such a state the properties of the system as given by the probability function $\Psi^*\Psi$ are independent of the time, and the state is called a *stationary state*.

10c. Further Physical Interpretation. Average Values of Dynamical Quantities.—If we inquire as to what average value would be expected on measurement at a given time t of the coordinate x of the system in a physical situation represented by the wave function Ψ, the above interpretation of $\Psi^*\Psi$ leads to the answer

$$\bar{x} = \int_{-\infty}^{+\infty} \Psi^*(x, t)\Psi(x, t)x\,dx;$$

that is, the value of x is averaged over all configurations, using the function $\Psi^*\Psi$ as a weight or probability function. A similar integral gives the average value predicted for x^2, or x^3, or any function $F(x)$ of the coordinate x:

$$\bar{F} = \int_{-\infty}^{+\infty} \Psi^*(x, t)\Psi(x, t)F(x)\,dx. \tag{10-5}$$

In order that the same question can be answered for a more general dynamical function $G(p_x, x)$ involving the momentum p_x as well as the coordinate x, we now make the following more general postulate:

The average value of the dynamical function $G(p_x, x)$ predicted for a system in the physical situation represented by the wave function $\Psi(x, t)$ is given by the integral

$$\bar{G} = \int_{-\infty}^{+\infty} \Psi^*(x, t) G\left(\frac{h}{2\pi i} \frac{\partial}{\partial x}, x\right) \Psi(x, t) dx, \qquad (10\text{-}6)$$

in which the operator G, obtained from $G(p_x, x)$ by replacing p_x by $\frac{h}{2\pi i} \frac{\partial}{\partial x}$, operates on the function $\Psi(x, t)$ and the integration is extended throughout the configuration space of the system.[1]

In general, the result of a measurement of G will not be given by this expression for $\bar{G}$. $\bar{G}$ rather is the average of a very large number of measurements made on a large number of identical systems in the physical situation represented by Ψ, or repeated on the same system, which before each measurement must be in the same physical situation. For example, if Ψ is

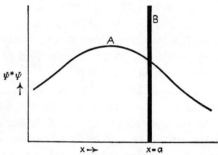

Fig. 10–1.—Two types of probability distribution function $\Psi^*\Psi$.

finite for a range of values of x (Curve A, Figure 10–1), then a measurement of x might lead to any value within this range, the probability being given by $\Psi^*\Psi$. Only if $\Psi^*\Psi$ were zero for all values of x except $x = a$, as indicated by Curve B in Figure 10–1, would the probability of obtaining a particular value $x = a$ on measurement of x be unity. In this case the value a^r would be predicted with probability unity to be obtained on measurement of the rth power of x; so that for such a probability distribution function $\overline{x^r}$ is equal to $(\bar{x})^r$. It has also been shown by mathematicians that the existence of this identity of $\overline{G^r}$ and $(\bar{G})^r$ for all values of r is sufficient to establish that the probability distribution function for the dynamical quantity G is of type B; that is, that the value of G can be predicted accurately.

---

[1] In some cases further considerations are necessary in order to determine the exact form of the operator, but we shall not encounter such difficulties.

Even if the system is in a stationary state, represented by the wave function $\Psi_n(x,\ t) = \psi_n(x)e^{-2\pi i\frac{W_n}{h}t}$, only an average value can be predicted for an arbitrary dynamical quantity. The energy of the system, corresponding to the Hamiltonian function $H(p_x,\ x)$, has, however, a definite value for a stationary state of the system, equal to the characteristic value W_n found on solution of the wave equation, so that the result of a measurement of the energy of the system in a given stationary state can be predicted accurately. To prove this, we evaluate $\overline{H^r}$ and $(\overline{H})^r$. $\overline{H}$ is given by the integral

$$\overline{H} = \int_{-\infty}^{+\infty} \psi_n^*(x)\left\{-\frac{h^2}{8\pi^2 m}\frac{d^2\psi_n(x)}{dx^2} + V(x)\psi_n(x)\right\}dx,$$

the factor involving the time being equal to unity. This transforms with the use of Equation 9–8 into

$$\overline{H} = \int_{-\infty}^{+\infty} \psi_n^*(x)W_n\psi_n(x)dx,$$

or, since W_n is a constant and $\int_{-\infty}^{+\infty}\psi_n^*(x)\psi_n(x)dx = 1$,

$$\overline{H} = W_n, \quad \text{and} \quad (\overline{H})^r = W_n^r. \tag{10–7}$$

By a similar procedure, involving repeated use of Equation 9–8, it is seen that $\overline{H^r}$ is equal to W_n^r. We have thus shown $\overline{H^r}$ to be equal to $(\overline{H})^r$, in consequence of which, in accordance with the argument set forth above, the energy of the system has the definite value W_n.

Further discussion of the physical significance of wave functions will be given in connection with the treatment of the harmonic oscillator in this chapter and of other systems in succeeding chapters, and especially in Chapter XV, in which the question of deciding which wave function to associate with a given system under given circumstances will be treated. In the earlier sections we shall restrict the discussion mainly to the properties of stationary states.

11. THE HARMONIC OSCILLATOR IN WAVE MECHANICS

11a. Solution of the Wave Equation.—As our first example of the solution of the Schrödinger wave equation for a dynamical system we choose the one-dimensional harmonic oscillator, not only because this provides a good illustration of the methods employed in applying the wave equation, but also because this system is of considerable importance in applications which we shall discuss later, such as the calculation of the vibrational energies of molecules. The more difficult problem of the three-dimensional oscillator was treated by the methods of classical mechanics in Section 1a, while the simple one-dimensional case was discussed according to the old quantum theory in Section 6a.

The potential energy may be written, as before, in the form $V(x) = 2\pi^2 m \nu_0^2 x^2$, in which x is the displacement of the particle of mass m from its equilibrium position $x = 0$. Insertion of this in the general wave equation for a one-dimensional system (Eq. 9–8) gives the equation

$$\frac{d^2\psi}{dx^2} + \frac{8\pi^2 m}{h^2}(W - 2\pi^2 m \nu_0^2 x^2)\psi = 0, \tag{11-1}$$

or, introducing for convenience the quantities $\lambda = 8\pi^2 m W/h^2$ and $\alpha = 4\pi^2 m \nu_0/h$,

$$\frac{d^2\psi}{dx^2} + (\lambda - \alpha^2 x^2)\psi = 0. \tag{11-2}$$

We desire functions $\psi(x)$ which satisfy this equation throughout the region of values $-\infty$ to $+\infty$ for x, and which are acceptable wave functions, i.e., functions which are continuous, single-valued, and finite throughout the region. A straightforward method of solution which suggests itself is the use of a power-series expansion for ψ, the coefficients of the successive powers of x being determined by substitution of the series for ψ in the wave equation. There is, however, a very useful procedure which we may make use of in this and succeeding problems, consisting of the determination of the form of ψ in the regions of large positive or negative values of x, and the subsequent discussion, by the introduction of a factor in the form of a power series (which later reduces to a polynomial), of the behavior of ψ for $|x|$ small. This procedure may be called the *polynomial method*.[1]

The first step is the asymptotic solution of the wave equation when $|x|$ is very large. For any value of the energy constant W, a value of $|x|$ can be found such that for it and all larger values

[1] A. SOMMERFELD, "Wave Mechanics," p. 11.

of $|x|$, λ is negligibly small relative to α^2x^2, the asymptotic form of the wave equation thus becoming

$$\frac{d^2\psi}{dx^2} = \alpha^2x^2\psi. \tag{11-3}$$

This equation is satisfied asymptotically by the exponential functions

$$\psi = e^{\pm\frac{\alpha}{2}x^2},$$

inasmuch as the derivatives of ψ have the values

$$\frac{d\psi}{dx} = \pm\alpha x e^{\pm\frac{\alpha}{2}x^2}$$

and

$$\frac{d^2\psi}{dx^2} = \alpha^2x^2e^{\pm\frac{\alpha}{2}x^2} \pm \alpha e^{\pm\frac{\alpha}{2}x^2}$$

and the second term in $\frac{d^2\psi}{dx^2}$ is negligible in the region considered.

Of the two asymptotic solutions $e^{-\frac{\alpha}{2}x^2}$ and $e^{+\frac{\alpha}{2}x^2}$, the second is unsatisfactory as a wave function since it tends rapidly to infinity with increasing values of $|x|$; the first, however, leads to a satisfactory treatment of the problem.

We now proceed to obtain an accurate solution of the wave equation throughout configuration space $(-\infty < x < +\infty)$, based upon the asymptotic solution, by introducing as a factor a power series in x and determining its coefficients by substitution in the wave equation.

Let $\psi = e^{-\frac{\alpha}{2}x^2}f(x)$. Then

$$\frac{d^2\psi}{dx^2} = e^{-\frac{\alpha}{2}x^2}\{\alpha^2x^2f - \alpha f - 2\alpha xf' + f''\},$$

in which f' and f'' represent $\frac{df}{dx}$ and $\frac{d^2f}{dx^2}$, respectively. Equation 11-2 then becomes, on division by $e^{-\frac{\alpha}{2}x^2}$,

$$f'' - 2\alpha xf' + (\lambda - \alpha)f = 0, \tag{11-4}$$

the terms in α^2x^2f cancelling.

It is now convenient to introduce a new variable ξ, related to x by the equation

$$\xi = \sqrt{\alpha}\,x, \tag{11-5}$$

and to replace the function $f(x)$ by $H(\xi)$, to which it is equal. The differential equation 11-4 then becomes

$$\frac{d^2H}{d\xi^2} - 2\xi\frac{dH}{d\xi} + \left(\frac{\lambda}{\alpha}-1\right)H = 0. \tag{11-6}$$

We now represent $H(\xi)$ as a power series, which we differentiate to obtain its derivatives,

$$H(\xi) = \sum_\nu a_\nu\xi^\nu = a_0 + a_1\xi + a_2\xi^2 + a_3\xi^3 + \cdots,$$

$$\frac{dH}{d\xi} = \sum_\nu \nu a_\nu\xi^{\nu-1} = a_1 + 2a_2\xi + 3a_3\xi^2 + \cdots,$$

$$\frac{d^2H}{d\xi^2} = \sum_\nu \nu(\nu-1)a_\nu\xi^{\nu-2} = 1\cdot2a_2 + 2\cdot3a_3\xi + \cdots.$$

On substitution of these expressions, Equation 11-6 assumes the following form:

$$1\cdot2a_2 + 2\cdot3a_3\xi + 3\cdot4a_4\xi^2 + 4\cdot5a_5\xi^3 + \cdots$$
$$- 2a_1\xi - 2\cdot2a_2\xi^2 - 2\cdot3a_3\xi^3 - \cdots$$
$$+ \left(\frac{\lambda}{\alpha}-1\right)a_0 + \left(\frac{\lambda}{\alpha}-1\right)a_1\xi + \left(\frac{\lambda}{\alpha}-1\right)a_2\xi^2 +$$
$$\left(\frac{\lambda}{\alpha}-1\right)a_3\xi^3 + \cdots = 0.$$

In order for this series to vanish for all values of ξ (i.e., for $H(\xi)$ to be a solution of 11-6), the coefficients of individual powers of ξ must vanish separately[1]:

$$1\cdot2a_2 + \left(\frac{\lambda}{\alpha}-1\right)a_0 = 0,$$

$$2\cdot3a_3 + \left(\frac{\lambda}{\alpha}-1-2\right)a_1 = 0,$$

$$3\cdot4a_4 + \left(\frac{\lambda}{\alpha}-1-2\cdot2\right)a_2 = 0,$$

$$4\cdot5a_5 + \left(\frac{\lambda}{\alpha}-1-2\cdot3\right)a_3 = 0,$$

[1] See footnote, Sec. 23.

or, in general, for the coefficient of ξ^ν,

$$(\nu + 1)(\nu + 2)a_{\nu+2} + \left(\frac{\lambda}{\alpha} - 1 - 2\nu\right)a_\nu = 0$$

or

$$a_{\nu+2} = -\frac{\left(\frac{\lambda}{\alpha} - 2\nu - 1\right)}{(\nu + 1)(\nu + 2)}a_\nu \tag{11-7}$$

This expression is called a *recursion formula*. It enables the coefficients a_2, a_3, a_4, $\cdots$ to be calculated successively in terms of a_0 and a_1, which are arbitrary. If a_0 is set equal to zero, only odd powers appear; with a_1 zero, the series contains even powers only.

For arbitrary values of the energy parameter λ, the above given series consists of an infinite number of terms and does not correspond to a satisfactory wave function, because, as we shall show, the value of the series increases too rapidly as x increases, with the result that the total function, even though it includes the negative exponential factor, increases without limit as x increases. To prove this we compare the series for H and that for e^{ξ^2},

$$e^{\xi^2} = 1 + \xi^2 + \frac{\xi^4}{2!} + \frac{\xi^6}{3!} + \cdots + \frac{\xi^\nu}{\left(\frac{\nu}{2}\right)!} + \frac{\xi^{\nu+2}}{\left(\frac{\nu}{2} + 1\right)!} + \cdots.$$

For large values of ξ the first terms of these series will be unimportant. Suppose that the ratio of the coefficients of the νth terms in the expansion of $H(\xi)$ and e^{ξ^2} is called c, which may be small or large, i.e., $a_\nu/b_\nu = c$, if b_ν is the coefficient of ξ^ν in the expansion of e^{ξ^2}. For large enough values of ν, we have the asymptotic relations

$$a_{\nu+2} = \frac{2}{\nu}a_\nu \quad \text{and} \quad b_{\nu+2} = \frac{2}{\nu}b_\nu,$$

so that

$$\frac{a_{\nu+2}}{b_{\nu+2}} = \frac{a_\nu}{b_\nu} = c,$$

if ν is large enough. Therefore, the higher terms of the series for H differ from those for e^{ξ^2} only by a multiplicative constant, so that for large values of $|\xi|$, for which the lower terms are unimportant, H will behave like e^{ξ^2} and the product $e^{-\frac{\xi^2}{2}}H$ will behave like $e^{+\frac{\xi^2}{2}}$ in this region, thus making it unacceptable as a wave function.

We must therefore choose the values of the energy parameter which will cause the series for H to break off after a finite number of terms, leaving a polynomial. This yields a satisfactory wave function, because the negative exponential factor $e^{-\frac{\xi^2}{2}}$ will cause the function to approach zero for large values of $|\xi|$. The value of λ which causes the series to break off after the nth term is seen from Equation 11-7 to be

$$\lambda = (2n + 1)\alpha. \tag{11-8}$$

It is, moreover, also necessary that the value either of a_0 or of a_1 be put equal to zero, according as n is odd or even, inasmuch as a suitably chosen value of λ can cause either the even or the odd series to break off, but not both. The solutions are thus either odd or even functions of ξ. This condition is a sufficient condition to insure that the wave equation 11-2 have satisfactory solutions, and it is furthermore a necessary condition; no other values of λ lead to satisfactory solutions. For each integral value 0, 1, 2, 3, $\cdots$ of n, which we may call the quantum number of the corresponding state of the oscillator, a satisfactory solution of the wave equation will exist. The straightforward way in which the quantum number enters in the treatment of the wave equation, as the degree of the polynomial $H(\xi)$, is especially satisfying when compared with the arbitrary assumption of integral or half-integral multiples of h for the phase integral of the old quantum theory.

The condition expressed in Equation 11-8 for the existence of the nth wave function becomes

$$W = W_n = (n + \tfrac{1}{2})h\nu_0, \qquad n = 0, 1, 2, \cdots, \tag{11-9}$$

when λ and α are replaced by the quantities they represent. A comparison with the result $W = nh\nu_0$ obtained in Section 6a by the old quantum theory shows that the only difference is that all the energy levels are shifted upward, as shown in Figure 11-1, by an amount equal to half the separation of the energy levels, the so-called *zero-point energy* $\tfrac{1}{2}h\nu_0$. From this we see that even in its lowest state the system has an energy greater than that which it would have if it were at rest in its equilibrium position. The existence of a zero-point energy, which leads to an improved agreement with experiment, is an important feature of the quantum mechanics and recurs in many problems.[1] Just as in the old-quantum-theory treatment, the frequency emitted or absorbed by a transition between adjacent energy levels is equal to the classical vibration frequency ν_0 (Sec. 40c).

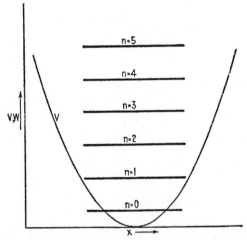

Fig. 11–1.—Energy levels for the harmonic oscillator according to wave mechanics (see Fig. 6–1).

11b. The Wave Functions for the Harmonic Oscillator and Their Physical Interpretation.—For each of the characteristic values W_n of the energy, a satisfactory solution of the wave equation 11–1 can be constructed by the use of the recursion formula 11–7. Energy levels such as these, to each of which there corresponds only one independent wave function, are said to be *non-degenerate* to distinguish them from *degenerate* energy levels (examples of which we shall consider later), to which several

---

[1] The name *zero-point energy* is used for the energy of a system in its lowest stationary state because the system in thermodynamic equilibrium with its environment at a temperature approaching the absolute zero would be in this stationary state. The zero-point energy is of considerable importance in many statistical-mechanical and thermodynamic discussions. The existence of zero-point energy is correlated with the uncertainty principle (Chap. XV).

independent wave functions correspond. The solutions of 11–1 may be written in the form

$$\psi_n(x) = N_n e^{-\frac{\xi^2}{2}} H_n(\xi), \qquad (11\text{--}10)$$

in which $\xi = \sqrt{\alpha}x$. $H_n(\xi)$ is a polynomial of the nth degree in ξ, and N_n is a constant which is adjusted so that ψ_n is normalized, i.e., so that ψ_n satisfies the relation

$$\int_{-\infty}^{+\infty} \psi_n^*(x)\psi_n(x)dx = 1, \qquad (11\text{--}11)$$

in which ψ_n^*, the complex conjugate of ψ_n, is in this case equal to ψ_n. In the next section we shall discuss the nature and properties

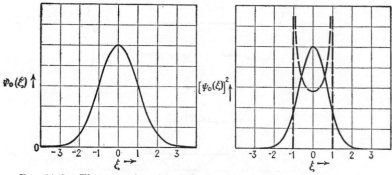

Fig. 11–2.—The wave function $\psi_0(\xi)$ for the normal state of the harmonic oscillator (left), and the corresponding probability distribution function $[\psi_0(\xi)]^2$ (right). The classical distribution function for an oscillator with the same total energy is shown by the dashed curve.

of these solutions ψ_n in great detail. The first of them, which corresponds to the state of lowest energy for the system, is

$$\psi_0(x) = \left(\frac{\alpha}{\pi}\right)^{\frac{1}{4}} e^{-\frac{\xi^2}{2}} = \left(\frac{\alpha}{\pi}\right)^{\frac{1}{4}} e^{-\frac{\alpha}{2}x^2}. \qquad (11\text{--}12)$$

Figure 11–2 shows this function. From the postulate discussed in Section 10a, $\psi_0^*\psi_0 = \psi_0^2$, which is also plotted in Figure 11–2, represents the probability distribution function for the coordinate x. In other words, the quantity $\psi_0^2(x)dx$ at any point x gives the probability of finding the particle in the range dx at that point. We see from the figure that the result of quantum mechanics for this case does not agree at all with the probability function which is computed classically for a harmonic oscillator with the same energy. Classically the particle is most likely to be found at the ends of its motion, which are clearly defined points (the classical probability distribution is shown by the dotted curve in Figure 11-2), whereas ψ_0^2 has its maximum at the origin of x and, furthermore, shows a rapidly decreasing but nevertheless finite probability of finding the particle outside the region allowed classically. This surprising result, that it is possible for a particle to penetrate into a region in which its total energy is less than its potential energy, is closely connected with Heisenberg's

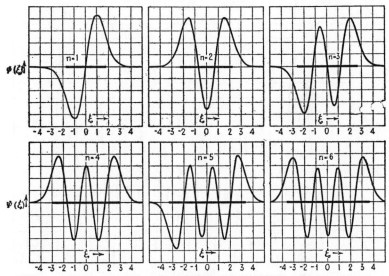

Fig. 11-3.—The wave functions $\psi_n(\xi)$, $n = 1$ to 6, for the harmonic oscillator. For each case the heavy horizontal line indicates the region traversed by the classical harmonic oscillator with the same total energy.

uncertainty principle, which leads to the conclusion that it is not possible to measure exactly both the position and the velocity of a particle at the same time. We shall discuss this phenomenon further in Chapter XV. It may be mentioned at this point, however, that the extension of the probability distribution function into the region of negative kinetic energy will not require that the law of the conservation of energy be abandoned.

The form of ψ_n for larger values of n is shown in Figure 11-3. Since H_n is a polynomial of degree n, ψ_n will have n zeros or points where ψ_n crosses the zero line. The probability of finding the particle at these points is zero. Inspection of Figure 11-3 shows that all the solutions plotted show a general behavior in agreement with that obtained by the general arguments of Section 9c; that is, inside the classically permitted region of motion of the particle (in which $V(x)$ is less than W_n) the wave function oscillates, having n zeros, while outside that region the wave function falls rapidly to zero in an exponential manner and has no zeros. Furthermore, we see in this example an illustration of still another general principle: The larger the value of n, the

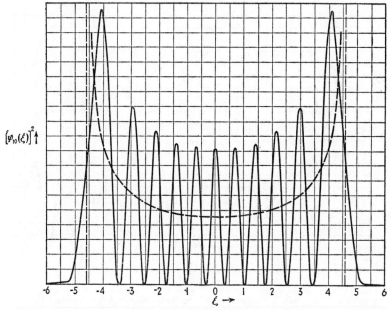

$[\psi_{10}(\xi)]^2 \uparrow$

$\xi \rightarrow$

Fig. 11–4.—The probability distribution function $[\psi_{10}(\xi)]^2$ for the state $n = 10$ of the harmonic oscillator. Note how closely the function approximates in its average value the probability distribution function for the classical harmonic oscillator with the same total energy, represented by the dashed curve.

more nearly does the wave-mechanical probability distribution function approximate to the classical expression for a particle with the same energy. Figure 11–4 shows $\psi^2(x)$ for the state with $n = 10$ compared with the classical probability curve for the harmonic oscillator with the same value $\frac{21}{2}h\nu_0$ for the energy.

It is seen that, aside from the rapid fluctuation of the wave-mechanical curve, the general agreement of the two functions is good. This agreement permits us to visualize the motion of the particle in a wave-mechanical harmonic oscillator as being similar to its classical to-and-fro motion, the particle speeding up in the center of its orbit and slowing down as it approaches its maximum displacement from its equilibrium position. The amplitude of the oscillation cannot be considered to be constant, as for the classical oscillator; instead, we may picture the particle as oscillating sometimes with very large amplitude, and sometimes with very small amplitude, but usually with an amplitude in the neighborhood of the classical value for the same energy. Other properties of the oscillator also are compatible with this picture; thus the wave-mechanical root-mean-square value of the momentum is equal to the classical value (Prob. 11-4).

A picture of this type, while useful in developing an intuitive feeling for the wave-mechanical equations, must not be taken too seriously, for it is not completely satisfactory. Thus it cannot be reconciled with the existence of zeros in the wave functions for the stationary states, corresponding to points where the probability distribution function becomes vanishingly small.

11c. Mathematical Properties of the Harmonic Oscillator Wave Functions.—The polynomials $H_n(\xi)$ and the functions $e^{-\frac{\xi^2}{2}} H_n(\xi)$ obtained in the solution of the wave equation for the harmonic oscillator did not originate with Schrödinger's work but were well known to mathematicians in connection with other problems. Their properties have been intensively studied.

For the present purpose, instead of developing the theory of the polynomials $H_n(\xi)$, called the *Hermite polynomials*, from the relation between successive coefficients given in Equation 11–7, it is more convenient to introduce them by means of another definition:

$$H_n(\xi) = (-1)^n e^{\xi^2} \frac{d^n e^{-\xi^2}}{d\xi^n}. \qquad (11\text{--}13)$$

We shall show later that this leads to the same functions as Equation 11–7. A third definition involves the use of a *generating function*, a method which is useful in many calculations and which is also applicable to other functions. The generating function for the Hermite polynomials is

$$S(\xi, s) \equiv e^{\xi^2 - (s-\xi)^2} \equiv \sum_{n=0}^{\infty} \frac{H_n(\xi)}{n!} s^n. \qquad (11\text{--}14)$$

This identity in the auxiliary variable s means that the function $e^{\xi^2-(s-\xi)^2}$ has the property that, if it is expanded in a power series in s, the coefficients of successive powers of s are just the Hermite polynomials $H_n(\xi)$, multiplied by $1/n!$. To show the equivalence of the two definitions 11-13 and 11-14, we differentiate S n times with respect to s and then let s tend to zero, using first one and then the other expression for S; the terms with $\nu < n$ vanish on differentiation, and those with $\nu > n$ vanish for $s \to 0$, leaving only the term with $\nu = n$:

$$\left(\frac{\partial^n S}{\partial s^n}\right)_{s\to 0} = \left(\frac{\partial^n}{\partial s^n}\sum_\nu \frac{H_\nu(\xi)s^\nu}{\nu!}\right)_{s\to 0} = H_n(\xi);$$

and

$$\left(\frac{\partial^n S}{\partial s^n}\right)_{s\to 0} = \left(\frac{\partial^n e^{\xi^2-(s-\xi)^2}}{\partial s^n}\right)_{s\to 0} = e^{\xi^2}\left(\frac{\partial^n e^{-(s-\xi)^2}}{\partial(s-\xi)^n}\right)_{s\to 0}$$

$$= e^{\xi^2}(-1)^n\left(\frac{\partial^n e^{-(s-\xi)^2}}{\partial \xi^n}\right)_{s\to 0} = (-1)^n e^{\xi^2}\frac{d^n e^{-\xi^2}}{d\xi^n}.$$

Comparing these two equations, we see that we obtain Equation 11-13, so that the two definitions of $H_n(\xi)$ are equivalent. Equation 11-13 is useful for obtaining the individual functions, while Equation 11-14 is frequently convenient for deriving their properties, such as in the case we shall now discuss.

To show that the functions we have defined above are the same as those used in the solution of the harmonic oscillator problem, we look for the differential equation satisfied by $H_n(\xi)$. It is first convenient to derive certain relations between successive Hermite polynomials and their derivatives. We note that since $S = e^{\xi^2-(s-\xi)^2}$, its partial derivative with respect to s is given by the equation

$$\frac{\partial S}{\partial s} = -2(s-\xi)S.$$

Similarly differentiating the series $S = \sum \frac{H_n(\xi)}{n!}s^n$, and equating the two different expressions for $\partial S/\partial s$, we obtain the equation

$$\sum_n \frac{H_n(\xi)}{(n-1)!}s^{n-1} = -2(s-\xi)\sum_n \frac{H_n(\xi)}{n!}s^n,$$

or, collecting terms corresponding to the same power of s,

$$\sum_n \left\{ \frac{H_{n+1}(\xi)}{n!} + 2\frac{H_{n-1}(\xi)}{(n-1)!} - 2\xi\frac{H_n(\xi)}{n!} \right\} s^n = 0.$$

Since this equation is true for all values of s, the coefficients of individual powers of s must vanish, giving as the recursion formula for the Hermite polynomials the expression

$$H_{n+1}(\xi) - 2\xi H_n(\xi) + 2nH_{n-1}(\xi) = 0. \qquad (11\text{-}15)$$

Similarly, by differentiation with respect to ξ, we derive the equation

$$\frac{\partial S}{\partial \xi} = 2sS,$$

which gives, in just the same manner as above, the equation

$$\sum_n \left\{ \frac{H_n'(\xi)}{n!} s^n - 2\frac{H_n(\xi)}{n!} s^{n+1} \right\} = 0$$

or

$$H_n'(\xi) = \frac{dH_n(\xi)}{d\xi} = 2nH_{n-1}(\xi), \qquad (11\text{-}16)$$

involving the first derivatives of the Hermite polynomials. This can be further differentiated with respect to ξ to obtain expressions involving higher derivatives.

Equations 11–15 and 11–16 lead to the differential equation for $H_n(\xi)$, for from 11–16 we obtain

$$H_n''(\xi) = 2nH_{n-1}'(\xi) = 4n(n-1)H_{n-2}(\xi), \qquad (11\text{-}17)$$

while Equation 11–15 may be rewritten as

$$H_n(\xi) - 2\xi H_{n-1}(\xi) + 2(n-1)H_{n-2}(\xi) = 0, \qquad (11\text{-}18)$$

which becomes, with the use of Equations 11–16 and 11–17,

$$H_n(\xi) - \frac{2\xi}{2n}H_n'(\xi) + \frac{1}{2n}H_n''(\xi) = 0$$

or

$$H_n''(\xi) - 2\xi H_n'(\xi) + 2nH_n(\xi) = 0. \qquad (11\text{-}19)$$

This is just the equation, 11–6, which we obtained from the harmonic oscillator problem, if we put $2n$ in place of $\frac{\lambda}{\alpha} - 1$, as required by Equation 11–8. Since for each integral value of n this equation has only one solution with the proper behavior at infinity, the polynomials $H_n(\xi)$ introduced in Section 11a are the Hermite polynomials.

The functions

$$\psi_n(x) = N_n \cdot e^{-\frac{\xi^2}{2}} \cdot H_n(\xi), \qquad \xi = \sqrt{\alpha}x, \qquad (11\text{-}20)$$

are called the *Hermite orthogonal functions;* they are, as we have seen, the wave functions for the harmonic oscillator. The value of N_n which makes $\int_{-\infty}^{+\infty} \psi_n^2(x)dx = 1$, i.e., which normalizes ψ_n, is

$$N_n = \left\{ \left(\frac{\alpha}{\pi}\right)^{\frac{1}{2}} \frac{1}{2^n n!} \right\}^{\frac{1}{2}}. \qquad (11\text{-}21)$$

The functions are mutually orthogonal if the integral over configuration space of the product of any two of them vanishes:

$$\int_{-\infty}^{+\infty} \psi_n(x)\psi_m(x)dx = 0, \qquad n \neq m. \qquad (11\text{-}22)$$

To prove the orthogonality of the functions and to evaluate the normalization constant given in Equation 11-21, it is convenient to consider two generating functions:

$$S(\xi, s) = \sum_n \frac{H_n(\xi)}{n!} s^n = e^{\xi^2-(s-\xi)^2}$$

and

$$T(\xi, t) = \sum_m \frac{H_m(\xi)}{m!} t^m = e^{\xi^2-(t-\xi)^2}.$$

Using these, we obtain the relations

$$\int_{-\infty}^{+\infty} STe^{-\xi^2}d\xi = \sum_n \sum_m s^n t^m \int_{-\infty}^{+\infty} \frac{H_n(\xi)H_m(\xi)}{n!m!} e^{-\xi^2}d\xi$$

$$= \int_{-\infty}^{+\infty} e^{-s^2-t^2+2s\xi+2t\xi-\xi^2}d\xi = e^{2st} \int_{-\infty}^{+\infty} e^{-(\xi-s-t)^2}d(\xi - s - t)$$

$$= \sqrt{\pi}e^{2st} = \sqrt{\pi}\left(1 + \frac{2st}{1!} + \frac{2^2 s^2 t^2}{2!} + \cdots + \frac{2^n s^n t^n}{n!} + \cdots \right).$$

Considering coefficients of $s^n t^m$ in the two equal series expansions, we see that $\int_{-\infty}^{+\infty} H_n(\xi)H_m(\xi)e^{-\xi^2}d\xi$ vanishes for $m \neq n$, and has the value $2^n n!\sqrt{\pi}$ for $m = n$, in consequence of which the functions are orthogonal and the normalization constant has the value given above.

The first few Hermite polynomials are

$$
\begin{aligned}
H_0(\xi) &= 1 \\
H_1(\xi) &= 2\xi \\
H_2(\xi) &= 4\xi^2 - 2 \\
H_3(\xi) &= 8\xi^3 - 12\xi \\
H_4(\xi) &= 16\xi^4 - 48\xi^2 + 12 \\
H_5(\xi) &= 32\xi^5 - 160\xi^3 + 120\xi \\
H_6(\xi) &= 64\xi^6 - 480\xi^4 + 720\xi^2 - 120 \\
H_7(\xi) &= 128\xi^7 - 1344\xi^5 + 3360\xi^3 - 1680\xi \\
H_8(\xi) &= 256\xi^8 - 3584\xi^6 + 13440\xi^4 - 13440\xi^2 + 1680 \\
H_9(\xi) &= 512\xi^9 - 9216\xi^7 + 48384\xi^5 - 80640\xi^3 + 30240\xi \\
H_{10}(\xi) &= 1024\xi^{10} - 23040\xi^8 + 161280\xi^6 - 403200\xi^4 + 302400\xi^2 \\
&\qquad - 30240.
\end{aligned}
$$

(11-23)

The list may easily be extended by the use of the recursion formula, Equation 11–15. Figure 11–3 shows curves for the first few wave functions, i.e., the functions given by Equation 11–20.

By using the generating functions S and T we can evaluate certain integrals involving ψ_n which are of importance. For example, we may study the integral which, as we shall later show (Sec. 40c), determines the probability of transition from the state n to the state m. This is

$$
x_{nm} = \int_{-\infty}^{+\infty} \psi_n \psi_m x\, dx = \frac{N_n N_m}{\alpha} \int_{-\infty}^{+\infty} H_n H_m e^{-\xi^2} \xi\, d\xi. \quad (11\text{–}24)
$$

Using S and T we obtain the relation

$$
\int_{-\infty}^{+\infty} ST e^{-\xi^2} \xi\, d\xi = \sum_n \sum_m \frac{1}{n!\,m!} s^n t^m \int_{-\infty}^{+\infty} H_n H_m e^{-\xi^2} \xi\, d\xi
$$

$$
= e^{2st} \int_{-\infty}^{+\infty} e^{-(\xi - s - t)^2} \xi\, d\xi = e^{2st} \int_{-\infty}^{+\infty} e^{-(\xi - s - t)^2}
$$

$$
(\xi - s - t)\, d(\xi - s - t)
$$

$$
+ e^{2st}(s + t) \int_{-\infty}^{+\infty} e^{-(\xi - s - t)^2} d(\xi - s - t).
$$

The first integral vanishes, and the second gives $\sqrt{\pi}$. On expanding the exponential, we obtain

$$\sqrt{\pi}\left(s + 2s^2t + \frac{2^2s^3t^2}{2!} + \cdots + \frac{2^ns^{n+1}t^n}{n!} + \cdots\right.$$

$$\left. + t + 2st^2 + \frac{2^2s^2t^3}{2!} + \cdots + \frac{2^ns^nt^{n+1}}{n!} + \cdots\right).$$

Hence, comparing coefficients of s^nt^m, we see that x_{nm} is zero except for $m = n \pm 1$, its values then being

$$x_{n,n+1} = \sqrt{\frac{n+1}{2\alpha}} \qquad (11\text{-}25a)$$

and

$$x_{n,n-1} = \sqrt{\frac{n}{2\alpha}}. \qquad (11\text{-}25b)$$

It will be shown later that this result requires that transitions occur only between adjacent energy levels of the harmonic oscillator, in agreement with the conclusion drawn from the correspondence principle in Section 5c.

Problem 11-1. Show that if $V(-x) = V(x)$, with V real, the solutions $\psi_n(x)$ of the amplitude equation 9–8 have the property that $\psi_n(-x) = \pm\psi_n(x)$.

Problem 11-2. Evaluate the integrals

$$(x^2)_{nm} = \int\psi_n\psi_mx^2dx, \qquad (x^3)_{nm} = \int\psi_n\psi_mx^3dx, \qquad (x^4)_{nm} = \int\psi_n\psi_mx^4dx,$$

where ψ_n is a solution of the wave equation for the harmonic oscillator.

Problem 11-3. Calculate the average values of x, x^2, x^3, and x^4 for a harmonic oscillator in the nth stationary state. Is it true that $\overline{x^2} = (\bar{x})^2$ or that $\overline{x^4} = (\overline{x^2})^2$? What conclusions can be drawn from these results concerning the results of a measurement of x?

Problem 11-4. Calculate the average values of p_x and p_x^2 for a harmonic oscillator in the nth stationary state and compare with the classical values for the same total energy. From the results of this and of the last problem, compute the average value of the energy $W = T + V$ for the nth stationary state.

Problem 11-5. *a.* Calculate the zero-point energy of a system consisting of a mass of 1 g. connected to a fixed point by a spring which is stretched 1 cm. by a force of 10,000 dynes. The particle is constrained to move only in the x direction.

b. Calculate the quantum number of the system when its energy is about equal to kT, where k is Boltzmann's constant and $T = 298°$ A. This corresponds to thermodynamic equilibrium at room temperature (Sec. 49).

General References on Quantum Mechanics

A. SOMMERFELD: "Wave Mechanics," E. P. Dutton & Co., Inc., New York, 1930.

E. U. CONDON and P. M. MORSE: "Quantum Mechanics," McGraw-Hill Book Company, Inc., New York, 1929.

A. E. RUARK and H. C. UREY: "Atoms, Molecules and Quanta," McGraw-Hill Book Company, Inc., New York, 1930.

N. F. MOTT: "An Outline of Wave Mechanics," Cambridge University Press, Cambridge, 1930.

J. FRENKEL: "Wave Mechanics," Oxford University Press, 1933–1934.

K. K. DARROW: Elementary Notions of Quantum Mechanics, *Rev. Mod. Phys.* 6, 23 (1934).

E. C. KEMBLE: General Principles of Quantum Mechanics, Part I, *Rev. Mod. Phys.* 1, 157 (1929).

E. C. KEMBLE: "Fundamental Principles of Quantum Mechanics," McGraw-Hill Book Company, Inc., 1937.

E. C. KEMBLE and E. L. HILL: General Principles of Quantum Mechanics, Part II, *Rev. Mod. Phys.* 2, 1 (1930).

S. DUSHMAN: "Elements of Quantum Mechanics," John Wiley & Sons Inc., 1938.

CHAPTER IV

THE WAVE EQUATION FOR A SYSTEM OF POINT PARTICLES IN THREE DIMENSIONS

12. THE WAVE EQUATION FOR A SYSTEM OF POINT PARTICLES

The Schrödinger equation for a system of N interacting point particles in three-dimensional space is closely similar to that for the simple one-dimensional system treated in the preceding chapter. The time equation is a partial differential equation in $3N + 1$ independent variables (the $3N$ Cartesian coordinates, say, of the N particles, and the time) instead of only two independent variables, and the wave function is a function of these $3N + 1$ variables. The same substitution as that used for the simpler system leads to the separation of the time equation into an equation involving the time alone and an amplitude equation involving the $3N$ coordinates. The equation involving the time alone is found to be the same as for the simpler system, so that the time dependency of the wave functions for the stationary states of a general system of point particles is the same as for the one-dimensional system. The amplitude equation, however, instead of being a total differential equation in one independent variable, is a partial differential equation in $3N$ independent variables, the $3N$ coordinates. It is convenient to say that this is an equation in a $3N$-dimensional configuration space, meaning by this that solutions are to be found for all values of the $3N$ Cartesian coordinates $x_1 \cdots z_N$ from $-\infty$ to $+\infty$. The amplitude function, dependent on these $3N$ coordinates, is said to be a function in configuration space. A point in configuration space corresponds to a definite value of each of the $3N$ coordinates $x_1 \cdots z_N$, and hence to definite positions of the N particles in ordinary space, that is, to a definite configuration of the system.

The wave equation, the auxiliary conditions imposed on the wave functions, and the physical interpretation of the wave functions for the general system are closely similar to those for the one-dimensional system, the only changes being those consequent to the increase in the number of dimensions of configuration space. A detailed account of the postulates made regarding the wave equation and its solutions for a general system of point particles is given in the following sections, together with a discussion of various simple systems for illustration.

12a. The Wave Equation Including the Time.—Let us consider a system consisting of N point particles of masses m_1, $m_2, \cdots, m_N$ moving in three-dimensional space under the influence of forces expressed by the potential function $V(x_1 \ y_1 \cdots z_N, t)$, $x_1 \cdots z_N$ being the $3N$ Cartesian coordinates of the N particles. The potential function V, representing the interaction of the particles with one another or with an external field or both, may be a function of the $3N$ coordinates alone or may depend on the time also. The former case, with $V = V(x_1 \cdots z_N)$, corresponds to a conservative system. Our main interest lies in systems of this type, and we shall soon restrict our discussion to them.

We assume with Schrödinger that the wave equation for this system is

$$-\frac{h^2}{8\pi^2}\sum_{i=1}^{N}\frac{1}{m_i}\left(\frac{\partial^2\Psi}{\partial x_i^2} + \frac{\partial^2\Psi}{\partial y_i^2} + \frac{\partial^2\Psi}{\partial z_i^2}\right) + V\Psi = -\frac{h}{2\pi i}\frac{\partial\Psi}{\partial t}. \quad (12\text{--}1)$$

This equation is often written as

$$-\frac{h^2}{8\pi^2}\sum_{i=1}^{N}\frac{1}{m_i}\nabla_i^2\Psi + V\Psi = -\frac{h}{2\pi i}\frac{\partial\Psi}{\partial t},$$

in which ∇_i^2 is the *Laplace operator* or *Laplacian* for the ith particle.[1] In Cartesian coordinates, it is given by the expression

$$\nabla_i^2 \equiv \frac{\partial^2}{\partial x_i^2} + \frac{\partial^2}{\partial y_i^2} + \frac{\partial^2}{\partial z_i^2}.$$

The wave function $\Psi = \Psi(x_1 \cdots z_N, t)$ is a function of the $3N$ coordinates of the system and the time.

It will be noted that the Schrödinger time equation for this general system is formally related to the classical energy equation in the same way as for the one-dimensional system of the preced-

---

[1] The symbol Δ is sometimes used in place of ∇^2. The symbol ∇^2 is commonly read as *del squared*.

ing chapter. The energy equation for a Newtonian system of point particles is

$$H(p_{x_1} \cdots p_{z_N}, x_1 \cdots z_N, t) = T(p_{x_1} \cdots p_{z_N}) + V(x_1 \cdots z_N, t) = W, \quad (12\text{-}2)$$

which on explicit introduction of the momenta $p_{x_1} \ldots p_{z_N}$ becomes

$$H(p_{x_1} \cdots p_{z_N}, x_1 \cdots z_N, t) = \sum_i \frac{1}{2m_i}(p_{x_i}^2 + p_{y_i}^2 + p_{z_i}^2) + V(x_1 \cdots z_N, t) = W. \quad (12\text{-}3)$$

We now arbitrarily replace the momenta $p_{x_1} \cdots p_{z_N}$ by the differential operators $\dfrac{h}{2\pi i}\dfrac{\partial}{\partial x_1} \cdots \dfrac{h}{2\pi i}\dfrac{\partial}{\partial z_N}$, respectively, and W by the operator $-\dfrac{h}{2\pi i}\dfrac{\partial}{\partial t}$, and introduce the function $\Psi(x_1 \cdots z_N, t)$ on which these operators can operate. The equation then becomes

$$H\left(\frac{h}{2\pi i}\frac{\partial}{\partial x_1} \cdots \frac{h}{2\pi i}\frac{\partial}{\partial z_N}, x_1 \cdots z_N, t\right)\Psi$$

$$= -\frac{h^2}{8\pi^2}\sum_{i=1}^{N}\frac{1}{m_i}\nabla_i^2\Psi + V\Psi = -\frac{h}{2\pi i}\frac{\partial\Psi}{\partial t}, \quad (12\text{-}4)$$

which is identical with Equation 12-1. Just as for the one-dimensional case, the wave equation is often symbolically written

$$H\Psi = W\Psi. \quad (12\text{-}5)$$

The discussion in Section 9a of the significance of this formal relation is also appropriate to this more general case.

12b. The Amplitude Equation.—Let us now restrict our attention to conservative systems, for which V is a function of the $3N$ coordinates only. To solve the wave equation for this case, we proceed exactly as in the simpler problem of Section 9b, investigating the solutions Ψ of the wave equation which can be expressed as the product of two functions, one of which involves only the time and the other only the $3N$ coordinates:

$$\Psi(x_1 \cdots z_N, t) = \psi(x_1 \cdots z_N)\varphi(t). \quad (12\text{-}6)$$

On introducing this expression in Equation 12-1, the wave equation can be separated into two equations, one for $\varphi(t)$ and one for $\psi(x_1 \cdots z_N)$. These equations are

and
$$
\left.
\begin{aligned}
\frac{d\varphi(t)}{dt} &= -\frac{2\pi i}{h} W \varphi(t) \\[2mm]
-\frac{h^2}{8\pi^2}\sum_{i=1}^{N}\frac{1}{m_i}\nabla_i^2\psi + V\psi &= W\psi.
\end{aligned}
\right\}
\qquad (12\text{-}7)
$$

The se?ond of these is often written in the form

$$
\sum_{i=1}^{N}\frac{1}{m_i}\nabla_i^2\psi + \frac{8\pi^2}{h^2}(W - V)\psi = 0. \qquad (12\text{-}8)
$$

This is Schrödinger's amplitude equation for a conservative system of point particles.

The auxiliary conditions which must be satisfied by a solution of the amplitude equation in order that it be an acceptable wave function are given in Section 9c. These conditions must hold throughout configuration space, that is, for all values between $-\infty$ and $+\infty$ for each of the $3N$ Cartesian coordinates of the system. Just as for the one-dimensional case, it is found that acceptable solutions exist only for certain values of the energy parameter W. These values may form a discrete set, a continuous set, or both.

It is usually found convenient to represent the various successive values of the energy parameter and the corresponding amplitude functions by the use of $3N$ integers, which represent $3N$ quantum numbers $n_1 \cdots n_{3N}$, associated with the $3N$ coordinates. The way in which this association occurs will be made clear in the detailed discussion of examples in the following sections of this chapter and in later chapters. For the present let us represent all of the quantum numbers $n_1 \cdots n_{3N}$ by the one letter n, and write instead of $W_{n_1} \cdots {}_{n_{3N}}$ and $\psi_{n_1} \cdots {}_{n_{3N}}$ the simpler symbols W_n and ψ_n.

The equation for $\varphi(t)$ gives on integration

$$
\varphi(t) = e^{-2\pi i\frac{W_n}{h}t}, \qquad (12\text{-}9)
$$

exactly as for the one-dimensional system. The various particular solutions of the wave equation are hence

$$\Psi_n(x_1 \cdots z_N, t) = \psi_n(x_1 \cdots z_N)e^{-2\pi i\frac{W_n}{h}t}. \quad (12\text{-}10)$$

These represent the various stationary states of the system. The general solution of the wave equation is

$$\Psi(x_1 \cdots z_N, t) = \sum_n a_n\Psi_n(x_1 \cdots z_N, t) =$$

$$\sum_n a_n\psi_n(x_1 \cdots z_N)e^{-2\pi i\frac{W_n}{h}t}, \quad (12\text{-}11)$$

in which the quantities a_n are constants. The symbol $\sum_n$ repre-

sents summation for all discrete values of W_n and integration over all continuous ranges of values.

12c. The Complex Conjugate Wave Function $\Psi^*(x_1 \cdots z_N, t)$.—The complex conjugate wave function $\Psi^*(x_1 \cdots z_N, t)$ is a solution of the conjugate wave equation

$$-\frac{h^2}{8\pi^2}\sum_{i=1}^{N}\frac{1}{m_i}\nabla_i^2\Psi^*(x_1 \cdots z_N, t) +$$

$$V^*(x_1 \cdots z_N, t)\Psi^*(x_1 \cdots z_N, t) =$$

$$\frac{h}{2\pi i}\frac{\partial}{\partial t}\Psi^*(x_1 \cdots z_N, t). \quad (12\text{-}12)$$

The general solution of this equation for a conservative system is

$$\Psi^*(x_1 \cdots z_N, t) = \sum_n a_n^*\Psi_n^*(x_1 \cdots z_N, t) =$$

$$\sum_n a_n^*\psi_n^*(x_1 \cdots z_N)e^{2\pi i\frac{W_n}{h}t}. \quad (12\text{-}13)$$

12d. The Physical Interpretation of the Wave Functions.— The physical interpretation of the wave functions for this general system is closely analogous to that for the one-dimensional system discussed in Section 10. We first make the following postulate, generalizing that of Section 10a:

The quantity $\Psi^*(x_1 \cdots z_N, t)\Psi(x_1 \cdots z_N, t)dx_1 \cdots dz_N$ *is the probability that the system in the physical situation represented by the wave function* $\Psi(x_1 \cdots z_N, t)$ *have at the time t the configuration represented by a point in the volume element* $dx_1 \cdots dz_N$ *of configuration space.* $\Psi^*\Psi$ thus serves as a probability distribution function for the configuration of the system.

The function $\Psi(x_1 \cdots z_N, t)$ must then be normalized to unity, satisfying the equation

$$\int \Psi^*(x_1 \cdots z_N, t)\Psi(x_1 \cdots z_N, t)d\tau = 1, \qquad (12\text{--}14)$$

in which the symbol $d\tau$ is used to represent the volume element $dx_1 \cdots dz_N$ in configuration space, and the integral is to be taken over the whole of configuration space. (In the remaining sections of this book the simple integral sign followed by $d\tau$ is to be considered as indicating an integral over the whole of configuration space.) It is also convenient to normalize the amplitude functions $\psi_n(x_1 \cdots z_N)$, according to the equation

$$\int \psi_n^*(x_1 \cdots z_N)\psi_n(x_1 \cdots z_N)d\tau = 1. \qquad (12\text{--}15)$$

It is found, as shown in Appendix III, that the independent solutions of any amplitude equation (just as for the one-dimensional case) can be chosen in such a way that any two of them are orthogonal, satisfying the orthogonality equation

$$\int \psi_m^*(x_1 \cdots z_N)\psi_n(x_1 \cdots z_N)d\tau = 0, \qquad m \neq n. \quad (12\text{--}16)$$

A wave function $\Psi(x_1 \cdots z_N, t) = \sum_n a_n \Psi_n(x_1 \cdots z_N, t)$ is then normalized if the coefficients a_n satisfy the equation

$$\sum_n a_n^* a_n = 1. \qquad (12\text{--}17)$$

An argument analogous to that of Section 10b shows that the wave functions $\Psi_n(x_1 \cdots z_N, t) = \psi_n(x_1 \cdots z_N)e^{-2\pi i \frac{W_n}{h} t}$ give probability distribution functions which are independent of the time and hence correspond to stationary states.

A more general physical interpretation can be given the wave functions, along the lines indicated in Section 10c, by making the postulate that the average value of the dynamical function $G(p_{x_1} \cdots p_{z_N}, x_1 \cdots z_N, t)$ predicted for a system in the physical situation represented by the wave function $\Psi(x_1 \cdots z_N, t)$ is given by the integral

$$\bar{G} = \int \Psi^*(x_1 \cdots z_N, t)G\left(\frac{h}{2\pi i}\frac{\partial}{\partial x_1} \cdots \frac{h}{2\pi i}\frac{\partial}{\partial z_N}, x_1 \cdots z_N, t\right)$$
$$\Psi(x_1 \cdots z_N, t)d\tau, \quad (12\text{--}18)$$

in which the operator G, obtained from $G(p_{x_1} \cdots p_{z_N}, x_1 \cdots z_{N_1}$

t) by replacing $p_{x_1} \cdots p_{z_N}$ by $\dfrac{h}{2\pi i}\dfrac{\partial}{\partial x_1} \cdots \dfrac{h}{2\pi i}\dfrac{\partial}{\partial z_N}$, respectively,
operates on the function $\Psi(x_1 \cdots z_N, t)$ and the integration is extended throughout the configuration space of the system. Further discussion of the physical interpretation of the wave functions will be found in Chapter XV.

13. THE FREE PARTICLE

A particle of mass m moving in a field-free space provides the simplest application of the Schrödinger equation in three dimensions. Since V is constant (we choose the value zero for convenience), the amplitude equation 12–8 assumes the following form:

$$\nabla^2\psi + \frac{8\pi^2 m}{h^2}W\psi = 0, \tag{13-1}$$

or, in Cartesian coordinates,

$$\frac{\partial^2\psi}{\partial x^2} + \frac{\partial^2\psi}{\partial y^2} + \frac{\partial^2\psi}{\partial z^2} + \frac{8\pi^2 m}{h^2}W\psi = 0. \tag{13-2}$$

This is a partial differential equation in three independent variables x, y, and z. In order to solve such an equation it is usually necessary to obtain three total differential equations, one in each of the three variables, using the method of separation of variables which we have already employed to solve the Schrödinger time equation (Sec. 9b). We first investigate the possibility that a solution may be written in the form

$$\psi(x, y, z) = X(x) \cdot Y(y) \cdot Z(z), \tag{13-3}$$

where $X(x)$ is a function of x alone, $Y(y)$ a function of y alone, and $Z(z)$ a function of z alone. If we substitute this expression in Equation 13–2, we obtain, after dividing through by ψ, the equation

$$\frac{1}{X}\frac{d^2X}{dx^2} + \frac{1}{Y}\frac{d^2Y}{dy^2} + \frac{1}{Z}\frac{d^2Z}{dz^2} + \frac{8\pi^2 m}{h^2}W = 0. \tag{13-4}$$

Since X is a function only of x, the first term does not change its value when y and z change. Likewise the second term is independent of x and z and the third term of x and y. Nevertheless, the sum of these three terms must be equal to the constant $-\dfrac{8\pi^2 m}{h^2}W$ for any choice of x, y, z. By holding y and z

fixed and varying x, only the first term can vary, since the others do not depend upon x. However, since the sum of all the terms is equal to a constant, we are led to the conclusion that $\frac{1}{X}\frac{d^2X}{dx^2}$ is independent of x as well as of y and z, and is therefore itself equal to a constant. Applying an identical argument to the other terms, we obtain the three ordinary differential equations

$$\frac{1}{X}\frac{d^2X}{dx^2} = k_x, \qquad \frac{1}{Y}\frac{d^2Y}{dy^2} = k_y, \qquad \text{and} \qquad \frac{1}{Z}\frac{d^2Z}{dz^2} = k_z, \quad (13\text{-}5)$$

with the condition

$$k_x + k_y + k_z = -\frac{8\pi^2 m}{h^2}W. \quad (13\text{-}6)$$

It is convenient to put $k_x = -\frac{8\pi^2 m}{h^2}W_x$, which gives the equation in x the form

$$\frac{d^2X}{dx^2} + \frac{8\pi^2 m}{h^2}W_xX = 0. \quad (13\text{-}7)$$

This is now a total differential equation, which can be solved by familiar methods. As may be verified by insertion in the equation, a solution is

$$X(x) = N_x \sin\left\{\frac{2\pi}{h}\sqrt{2mW_x}(x - x_0)\right\}. \quad (13\text{-}8)$$

Since it contains two independent arbitrary constants N_x and x_0, it is the general solution. It is seen that the constant x_0 defines the location of the zeros of the sine function. The equations for Y and Z are exactly analogous to Equation 13-7, and have the solutions

$$\left.\begin{array}{l}Y(y) = N_y \sin\left\{\frac{2\pi}{h}\sqrt{2mW_y}(y - y_0)\right\}, \\[2mm] Z(z) = N_z \sin\left\{\frac{2\pi}{h}\sqrt{2mW_z}(z - z_0)\right\}.\end{array}\right\} \quad (13\text{-}9)$$

The fact that we have been able to obtain the functions X, Y, and Z justifies the assumption inherent in Equation 13-3. It can also be proved[1] that no other solutions satisfying the

[1] The necessary theorems are given in R. Courant and D. Hilbert, "Methoden der mathematischen Physik," 2d ed., Julius Springer, Berlin, 1931.

boundary conditions can be found which are linearly independent of these, i.e., which cannot be expressed as a linear combination of these solutions.

The function ψ must now be examined to see for what values of $W = W_x + W_y + W_z$ it satisfies the conditions for an acceptable wave function given in Section 9c. Since the sine function is continuous, single-valued, and finite for all real values of its argument, the only restriction that is placed on W is that W_x, W_y, W_z and therefore W be positive. We have thus reached the conclusion that the free particle has a continuous spectrum of allowed energy values, as might have been anticipated from the argument of Section 9c.

The complete expression for the wave function corresponding to the energy value

$$W = W_x + W_y + W_z \tag{13-10}$$

is

$$\psi(x, y, z) = N \sin\left\{\frac{2\pi}{h}\sqrt{2mW_x}(x - x_0)\right\}$$
$$\cdot \sin\left\{\frac{2\pi}{h}\sqrt{2mW_y}(y - y_0)\right\} \cdot \sin\left\{\frac{2\pi}{h}\sqrt{2mW_z}(z - z_0)\right\}, \tag{13-11}$$

in which N is a normalization constant. The problem of the normalization of wave functions of this type, the value of which remains appreciable over an infinite volume of configuration space (corresponding to a continuous spectrum of energy values), is a complicated one. Inasmuch as we shall concentrate our attention on problems of atomic and molecular structure, with little mention of collision problems and other problems involving free particles, we shall not discuss the question further, contenting ourselves with reference to treatments in other books.[1]

In discussing the physical interpretation of the wave functions for this system, let us first consider that the physical situation is represented by a wave function as given in Equation 13–11 with W_y and W_z equal to zero and W_x equal to W. The func-

[1] A. SOMMERFELD, "Wave Mechanics," English translation by H. L. Brose, pp. 293–295, E. P. Dutton & Co., Inc., New York, 1929; RUARK and UREY, "Atoms, Molecules, and Quanta," p. 541, McGraw-Hill Book Company, Inc., New York, 1930.

tion[1] $\Psi(x, y, z, t) = N \sin \left\{ \dfrac{2\pi}{h} \sqrt{2mW} (x - x_0) \right\} e^{-2\pi i \frac{W}{h} t}$ is then a

set of standing waves with wave fronts normal to the x axis.
The wave length is seen to be given by the equation

$$\lambda = \frac{h}{\sqrt{2mW}}. \tag{13-12}$$

In classical mechanics the speed v of a free particle of mass m
moving with total energy W is given by the equation $\frac{1}{2}mv^2 = W$.
A further discussion of this system shows that a similar interpretation of W holds in the quantum mechanics. Introducing
v in place of W in Equation 13–12, we obtain

$$\lambda = \frac{h}{mv}. \tag{13-13}$$

This is the de Broglie expression[2] for the wave length associated
with a particle of mass m moving with speed v.

It is the sinusoidal nature of the wave functions for the free
particle and the similar nature of the wave functions for other
systems which has caused the name *wave mechanics* to be applied
to the theory of mechanics which forms the subject of this book.
This sinusoidal character of wave functions gives rise to experimental phenomena which are closely similar to those associated
in macroscopic fields with wave motions. Because of such
experiments, many writers have considered the wavelike character of the electron to be more fundamental than its corpuscular
character, but we prefer to regard the electron as a particle and
to consider the wavelike properties as manifestations of the
sinusoidal nature of the associated wave functions. Neither
view is without logical difficulties, inasmuch as waves and
particles are macroscopic concepts which are difficult to apply to
microscopic phenomena. We shall, however, in discussing the
results of wave-mechanical calculations, adhere to the particle
concept throughout, since we believe it is the simplest upon
which to base an intuitive feeling for the mathematical results
of wave mechanics.

---

[1] It can be shown that the factors involving y and x in Equation 13–11
approach a constant value in this limiting case.

[2] L. DE BROGLIE, Thesis, 1924; *Ann. de phys.* **3**, 22 (1925).

The wave function which we have been discussing corresponds to a particle moving along the x axis, inasmuch as a calculation of the kinetic energy $T_x = \dfrac{1}{2m}p_x^2$ associated with this motion shows that the total energy of the system is kinetic energy of motion in the x direction. This calculation is made by the general method of Section 12d. The average value of T_x is

$$\begin{aligned}
\overline{T}_x &= \frac{1}{2m}\int \Psi^*\left(\frac{h}{2\pi i}\right)^2 \frac{\partial^2}{\partial x^2}\Psi d\tau \\
&= \frac{1}{2m}\left(\frac{h}{2\pi i}\right)^2\left(\frac{2\pi i}{h}\sqrt{2mW_x}\right)^2 \int \Psi^*\Psi d\tau \\
&= W_x,
\end{aligned}$$

or, since in this case we have assumed W_x to equal W,

$$\overline{T}_x = W.$$

Similarly we find $\overline{T}_x^r = W^r = (\overline{T}_x)^r$, which shows, in accordance with the discussion of Section 10c, that the kinetic energy of motion along the x axis has the definite value W, its probability distribution function vanishing except for this value.

On the other hand, the average value of p_x itself is found on calculation to be zero. The wave function

$$N\sin\left\{\frac{2\pi}{h}\sqrt{2mW}(x-x_0)\right\}e^{-2\pi i\frac{W}{h}t}$$

hence cannot be interpreted as representing a particle in motion in either the positive or negative direction along the x axis but rather a particle in motion along the x axis in either direction, the two directions of motion having equal probability.

The wave function $N\cos\left\{\dfrac{2\pi}{h}\sqrt{2mW}(x-x_0)\right\}e^{-2\pi i\frac{W}{h}t}$ differs from the sine function only in the phase, the energy being the same. The sum and difference of this function and the sine function with coefficient i are the complex functions

$$N'e^{\frac{2\pi i}{h}\sqrt{2mW}(x-x_0)}e^{-\frac{2\pi iW}{h}t} \quad \text{and} \quad N'e^{-\frac{2\pi i}{h}\sqrt{2mW}(x-x_0)}e^{-\frac{2\pi iW}{h}t},$$

which are also solutions of the wave equation equivalent to the sine and cosine functions. These complex wave functions represent physical situations of the system in which the particle is moving along the x axis in the positive direction with the definite momentum $p_x = \sqrt{2mW}$ or $p_x = -\sqrt{2mW}$, the motion in the positive direction corresponding to the first of the complex wave functions and in the negative direction to the second. This is easily verified by calculation of $\overline{p_x}$ and $\overline{p_x^r}$ for these wave functions.

The more general wave function of Equation 13–11 also represents a set of standing plane waves with wave length $\lambda = h/\sqrt{2mW}$, the line normal to the wave fronts having the direction cosines $\sqrt{W_x/W}$, $\sqrt{W_y/W}$, and $\sqrt{W_z/W}$ relative to the x, y, and z axes.

Problem 13–1. Verify the statements of the next to the last paragraph regarding the value of p_x.

14. THE PARTICLE IN A BOX[1]

Let us now consider a particle constrained to stay inside of a rectangular box, with edges a, b, and c in length. We can represent this system by saying that the potential function $V(x, y, z)$ has the constant value zero within the region $0 < x < a$, $0 < y < b$, and $0 < z < c$, and that it increases suddenly in value at the boundaries of this region, remaining infinitely large everywhere outside of the boundaries. It will be found that for this system the stationary states no longer correspond to a continuous range of allowed energy values, but instead to a discrete set, the values depending on the size and shape of the box.

Let us represent a potential function of the type described as

$$V(x, y, z) = V_x(x) + V_y(y) + V_z(z), \qquad (14\text{–}1)$$

the function $V_x(x)$ being equal to zero for $0 < x < a$ and to infinity for $x < 0$ or $x > a$, and the functions $V_y(y)$ and $V_z(z)$ showing a similar behavior. The wave equation

$$\frac{\partial^2\psi}{\partial x^2} + \frac{\partial^2\psi}{\partial y^2} + \frac{\partial^2\psi}{\partial z^2} + \frac{8\pi^2 m}{h^2}$$
$$\{W - V_x(x) - V_y(y) - V_z(z)\}\psi = 0 \quad (14\text{–}2)$$

is separated by the same substitution

$$\psi(x, y, z) = X(x) \cdot Y(y) \cdot Z(z) \qquad (14\text{–}3)$$

[1] Treated in Section 6d by the methods of the old quantum theory.

as for the free particle, giving three total differential equations, that in x being

$$\frac{d^2X}{dx^2} + \frac{8\pi^2 m}{h^2}\{W_x - V_x(x)\}X = 0. \tag{14-4}$$

In the region $0 < x < a$ the general solution of the wave equation is a sine function of arbitrary amplitude, frequency, and phase, as for the free particle. Several such functions are represented in Figure 14-1. All of these are not acceptable wave

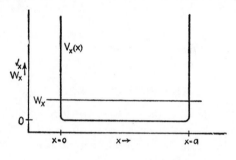

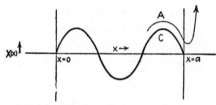

Fig. 14–1.—The potential-energy function $V_x(x)$ and the behavior of $X(x)$ near the point $x = a$.

functions, however; instead only those sine functions whose value falls to zero at the two points $x = 0$ and $x = a$ behave properly at the boundaries. To show this, let us consider the behavior of Curve A as x approaches and passes the value a, using the type of argument of Section 9c. Curve A has a finite positive value as x approaches a, and a finite slope. Its curvature is given by the equation

$$\frac{d^2X}{dx^2} = -\frac{8\pi^2 m}{h^2}\{W_x - V_x(x)\}X. \tag{14-5}$$

At the point $x = a$ the value of $V(x)$ increases very rapidly and without limit, so that, no matter how large a value the constant W_x has, $W_x - V_x$ becomes negative and of unbounded magnitude. The curvature or rate of change of the slope consequently becomes extremely great, and the curve turns sharply upward and experiences the infinity catastrophe. This can be avoided in only one way; the function $X(x)$ itself must have the value zero at the point $x = a$, in order that it may then remain bounded (and, in fact, have the value zero) for all larger values of x.

Similarly the sine function must fall to zero at $x = 0$, as shown by Curve C. An acceptable wave function $X(x)$ is hence a sine function with a zero at $x = 0$ and another zero at $x = a$,

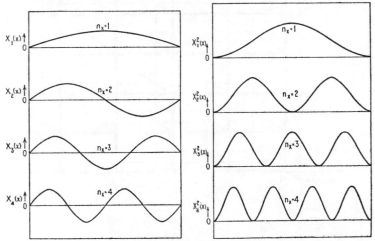

Fig. 14–2.—The wave functions $X_{n_x}(x)$ and probability distribution functions $[X_{n_x}(x)]^2$ for the particle in a box.

thus having an integral number of loops in this region. The phase and frequency (or wave length) are consequently fixed, and the amplitude is determined by normalizing the wave function to unity. Introducing the quantum number n_x as the number of loops in the region between 0 and a, the wave length becomes $2a/n_x$, and the normalized $X(x)$ function is given by the expression

$$X_{n_x}(x) = \sqrt{\frac{2}{a}} \sin \frac{n_x \pi x}{a}, \qquad n_x = 1, 2, 3, \cdots ,$$

$$0 < x < a, \quad (14\text{–}6)$$

with

$$W_{n_x} = \frac{n_x^2 h^2}{8ma^2}. \qquad (14\text{–}7)$$

The first four wave functions $X_1(x)$, $\cdots$, $X_5(x)$ are represented in Figure 14–2, together with the corresponding probability distribution functions $\{X_{n_x}(x)\}^2$.

A similar treatment of the y and z equations leads to similar expressions for $Y_{n_y}(y)$ and $Z_{n_z}(z)$ and for W_y and W_z. The com-

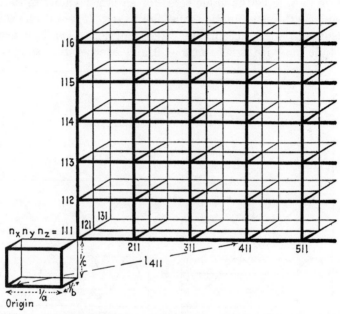

Fig. 14–3.—A geometrical representation of the energy levels for a particle in a rectangular box.

plete wave function $\psi_{n_x n_y n_z}(x, y, z)$ has the form, for values of x, y, and z inside the box,

$$\psi_{n_x n_y n_z}(x, y, z) = \sqrt{\frac{8}{abc}} \sin \frac{n_x \pi x}{a} \sin \frac{n_y \pi y}{b} \sin \frac{n_z \pi z}{c}, \quad (14\text{–}8)$$

with $n_x = 1, 2, 3, \cdots$; $n_y = 1, 2, 3, \cdots$; $n_z = 1, 2, 3, \cdots$; and

$$W_{n_x n_y n_z} = W_{n_x} + W_{n_y} + W_{n_z} = \frac{h^2}{8m}\left(\frac{n_x^2}{a^2} + \frac{n_y^2}{b^2} + \frac{n_z^2}{c^2}\right). \quad (14\text{–}9)$$

The wave function $\Psi_{n_x n_y n_z}$ can be described as consisting of standing waves along the x, y, and z directions, with $n_x + 1$ equally spaced nodal planes perpendicular to the x axis (beginning with $x = 0$ and ending with $x = a$), $n_y + 1$ nodal planes perpendicular to the y axis, and $n_z + 1$ nodal planes perpendicular to the z axis.

The various stationary states with their energy values may be conveniently represented by means of a geometrical analogy. Using a system of Cartesian coordinates, let us consider the

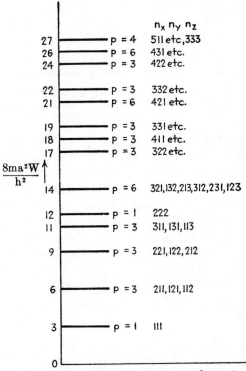

Fɪɢ. 14–4.—Energy levels, degrees of degeneracy, and quantum numbers for a particle in a cubic box.

lattice whose points have the coordinates n_x/a, n_y/b, and n_z/c, with $n_x = 1, 2, \cdots$; $n_y = 1, 2, \cdots$; and $n_z = 1, 2 \cdots$. This is the lattice defined in one octant about the origin by the translations $1/a$, $1/b$, and $1/c$, respectively; it divides the octant into unit cells of volume $1/abc$ (Fig. 14–3). Each point of the lattice represents a wave function. The corresponding energy value is

$$W_{n_x n_y n_z} = \frac{h^2}{8m} l^2_{n_x n_y n_z},$$ (14–10)

in which $l_{n_x n_y n_z}$ is the distance from the origin to the lattice point $n_x n_y n_z$, given by the equation

$$l_{n_x n_y n_z} = \sqrt{\frac{n_x^2}{a^2} + \frac{n_y^2}{b^2} + \frac{n_z^2}{c^2}}. \qquad (14\text{--}11)$$

In case that no two of the edges of the box a, b, and c are in the ratio of integers, the energy levels corresponding to various sets of values of the three quantum numbers are all different, with one and only one wave function associated with each. Energy levels of this type are said to be *non-degenerate*. If, however, there exists an integral relation among a, b, and c, there will occur certain values of the energy corresponding to two or more distinct sets of values of the three quantum numbers and to two or more independent wave functions. Such an energy level is said to be *degenerate*, and the corresponding state of the system is called a *degenerate state*. For example, if the box is a cube, with $a = b = c$, most of the energy levels will be degenerate. The lowest level, with quantum numbers 111 (for n_x, n_y, n_z, respectively) is non-degenerate, with energy $3h^2/8ma^2$. The next level, with quantum numbers 211, 121, and 112 and energy $6h^2/8ma^2$, is triply degenerate. Successive levels, with sets of quantum numbers and degrees of degeneracy (represented by p), are shown in Figure 14–4. The degree of degeneracy (the number of independent wave functions associated with a given energy level) is often called the *quantum weight* of the level.

15. THE THREE-DIMENSIONAL HARMONIC OSCILLATOR IN CARTESIAN COORDINATES

Another three-dimensional problem which is soluble in Cartesian coordinates is the three-dimensional harmonic oscillator, a special case of which, the isotropic oscillator, we have treated in Section 1a by the use of classical mechanics. The more general system consists of a particle bound to the origin by a force whose components along the x, y, and z axes are equal to $-k_x x$, $-k_y y$, and $-k_z z$, respectively, where k_x, k_y, k_z are the force constants in the three directions and x, y, z are the components of the displacement along the three axes. The potential energy is thus

$$V = \tfrac{1}{2}k_x x^2 + \tfrac{1}{2}k_y y^2 + \tfrac{1}{2}k_z z^2, \qquad (15\text{--}1)$$

which, on introducing instead of the constants k_x, k_y, k_z their expressions in terms of the classical frequencies ν_x, ν_y, ν_z, becomes

$$V = 2\pi^2 m(\nu_x^2 x^2 + \nu_y^2 y^2 + \nu_z^2 z^2), \tag{15-2}$$

since

$$\left.\begin{aligned} k_x &= 4\pi^2 m \nu_x^2, \\ k_y &= 4\pi^2 m \nu_y^2, \\ k_z &= 4\pi^2 m \nu_z^2. \end{aligned}\right\} \tag{15-3}$$

The general wave equation 12–8 thus assumes for this problem the form

$$\frac{\partial^2 \psi}{\partial x^2} + \frac{\partial^2 \psi}{\partial y^2} + \frac{\partial^2 \psi}{\partial z^2} + \frac{8\pi^2 m}{h^2} \{ W - 2\pi^2 m(\nu_x^2 x^2 + \nu_y^2 y^2 + \nu_z^2 z^2) \}\psi = 0, \tag{15-4}$$

which, on introducing the abbreviations

$$\lambda = \frac{8\pi^2 m}{h^2} W, \tag{15-5a}$$

$$\alpha_x = \frac{4\pi^2 m}{h} \nu_x, \;=\; \frac{m\,\omega_x}{h} \tag{15-5b}$$

$$\alpha_y = \frac{4\pi^2 m}{h} \nu_y, \tag{15-5c}$$

and

$$\alpha_z = \frac{4\pi^2 m}{h} \nu_z, \tag{15-5d}$$

simplifies to the equation

$$\frac{\partial^2 \psi}{\partial x^2} + \frac{\partial^2 \psi}{\partial y^2} + \frac{\partial^2 \psi}{\partial z^2} + (\lambda - \alpha_x^2 x^2 - \alpha_y^2 y^2 - \alpha_z^2 z^2)\psi = 0. \tag{15-6}$$

To solve this equation we proceed in exactly the same manner as in the case of the free particle (Sec. 13); namely, we attempt to separate variables by making the substitution

$$\psi(x, y, z) = X(x) \cdot Y(y) \cdot Z(z). \tag{15-7}$$

This gives, on substitution in Equation 15–6 and division of the result by ψ, the equation

$$\left(\frac{1}{X}\frac{d^2 X}{dx^2} - \alpha_x^2 x^2\right) + \left(\frac{1}{Y}\frac{d^2 Y}{dy^2} - \alpha_y^2 y^2\right) + \left(\frac{1}{Z}\frac{d^2 Z}{dz^2} - \alpha_z^2 z^2\right) + \lambda = 0. \tag{15-8}$$

It is evident that this equation has been separated into terms each of which depends upon one variable only; each term is therefore equal to a constant, by the argument used in Section 13. We obtain in this way three total differential equations similar to the following one:

$$\frac{d^2X(x)}{dx^2} + (\lambda_x - \alpha_x^2 x^2)X(x) = 0, \qquad (15\text{--}9)$$

in which λ_x is a separation constant, such that

$$\lambda_x + \lambda_y + \lambda_z = \lambda. \qquad (15\text{--}10)$$

Equation 15–9 is the same as the wave equation 11–2 for the one-dimensional harmonic oscillator which was solved in Section 11. Referring to that section, we find that $X(x)$ is given by the expression

$$X(x) = N_{n_x} e^{-\frac{\alpha_x x^2}{2}} H_n\left(\sqrt{\alpha_x} x\right) \qquad (15\text{--}11)$$

and that λ_x is restricted by the relation

$$\lambda_x = (2n_x + 1)\alpha_x, \qquad (15\text{--}12)$$

in which the quantum number n_x can assume the values 0, 1, 2, $\cdots$. Exactly similar expressions hold for $Y(y)$ and $Z(z)$ and for λ_y and λ_z. The total energy is thus given by the equation

$$W_{n_x n_y n_z} = h\{(n_x + \tfrac{1}{2})\nu_x + (n_y + \tfrac{1}{2})\nu_y + (n_z + \tfrac{1}{2})\nu_z\}, \quad (15\text{--}13)$$

and the complete wave function by the expression

$$\psi_{n_x n_y n_z}(x, y, z) =$$
$$N_{n_x n_y n_z} e^{-\frac{1}{2}(\alpha_x x^2 + \alpha_y y^2 + \alpha_z z^2)} H_{n_x}\left(\sqrt{\alpha_x} x\right) H_{n_y}\left(\sqrt{\alpha_y} y\right) H_{n_z}\left(\sqrt{\alpha_z} z\right). \quad (15\text{--}14)$$

The normalizing factor has the value

$$N_{n_x n_y n_z} = \left\{\frac{(\alpha_x \alpha_y \alpha_z)^{\frac{1}{2}}}{\pi^{\frac{3}{2}} 2^{n_x + n_y + n_z} n_x! n_y! n_z!}\right\}^{\frac{1}{2}}. \qquad (15\text{--}15)$$

For the special case of the isotropic oscillator, in which $\nu_x = \nu_y = \nu_z = \nu_0$ and $\alpha_x = \alpha_y = \alpha_z$, Equation 15–13 for the energy reduces to the form

$$W = (n_x + n_y + n_z + \tfrac{3}{2})h\nu_0 = (n + \tfrac{3}{2})h\nu_0. \quad (15\text{--}16)$$

$n = n_x + n_y + n_z$ may be called the total quantum number. Since the energy for this system depends only on the sum of the

quantum numbers, all the energy levels for the isotropic oscillator, except the lowest one, are degenerate, with the quantum
weight $\dfrac{(n + 1)(n + 2)}{2}$. Figure 15-1 shows the first few energy

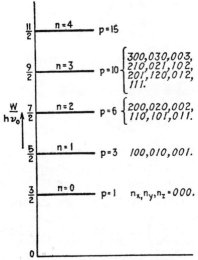

FIG. 15-1.—Energy levels, degrees of degeneracy, and quantum numbers for the
three-dimensional isotropic harmonic oscillator.

levels, together with their quantum weights and quantum
numbers.

16. CURVILINEAR COORDINATES

In Chapter I we found that curvilinear coordinates, such as
spherical polar coordinates, are more suitable than Cartesian
coordinates for the solution of many problems of classical
mechanics. In the applications of wave mechanics, also, it is
very frequently necessary to use different kinds of coordinates.
In Sections 13 and 15 we have discussed two different systems,
the free particle and the three-dimensional harmonic oscillator,
whose wave equations are separable in Cartesian coordinates.
Most problems cannot be treated in this manner, however, since
it is usually found to be impossible to separate the equation into
three parts, each of which is a function of one Cartesian coordinate only. In such cases there may exist other coordinate
systems in terms of which the wave equation is separable, so
that by first transforming the differential equation into the proper

coordinates the same technique of solution may often be applied.

In order to make such a transformation, which may be represented by the transformation equations

$$x = f(u, v, w), \qquad (16\text{-}1a)$$
$$y = g(u, v, w), \qquad (16\text{-}1b)$$
$$z = h(u, v, w), \qquad (16\text{-}1c)$$

it is necessary to know what form the Laplace operator ∇^2 assumes in the new system, since this operator has been defined only in Cartesian coordinates by the expression

$$\nabla^2 \equiv \frac{\partial^2}{\partial x^2} + \frac{\partial^2}{\partial y^2} + \frac{\partial^2}{\partial z^2}. \qquad (16\text{-}2)$$

The process of transforming these second partial derivatives is a straightforward application of the principles of the theory of partial derivatives and leads to the result that the operator ∇^2 in the orthogonal coordinate system uvw has the form

$$\nabla^2 = \frac{1}{q_u q_v q_w} \left\{ \frac{\partial}{\partial u}\left(\frac{q_v q_w}{q_u} \frac{\partial}{\partial u} \right) + \frac{\partial}{\partial v}\left(\frac{q_u q_w}{q_v} \frac{\partial}{\partial v} \right) + \frac{\partial}{\partial w}\left(\frac{q_u q_v}{q_w} \frac{\partial}{\partial w} \right) \right\},$$

$$(16\text{-}3)$$

in which

$$
\left.
\begin{aligned}
q_u^2 &= \left(\frac{\partial x}{\partial u}\right)^2 + \left(\frac{\partial y}{\partial u}\right)^2 + \left(\frac{\partial z}{\partial u}\right)^2, \\
q_v^2 &= \left(\frac{\partial x}{\partial v}\right)^2 + \left(\frac{\partial y}{\partial v}\right)^2 + \left(\frac{\partial z}{\partial v}\right)^2, \\
q_w^2 &= \left(\frac{\partial x}{\partial w}\right)^2 + \left(\frac{\partial y}{\partial w}\right)^2 + \left(\frac{\partial z}{\partial w}\right)^2.
\end{aligned}
\right\} \qquad (16\text{-}4)
$$

Equation 16-3 is restricted to coordinates u, v, w which are orthogonal, that is, for which the coordinate surfaces represented by the equations $u = $ constant, $v = $ constant, and $w = $ constant intersect at right angles. All the common systems are of this type.

The volume element $d\tau$ for a coordinate system of this type is also determined when q_u, q_v, and q_w are known. It is given by the expression

$$d\tau = dx\,dy\,dz = q_u q_v q_w\, du\,dv\,dw. \qquad (16\text{-}5)$$

In Appendix IV, q_u, q_v, q_w, and ∇^2 itself are given for a number of important coordinate systems.

Mathematicians[1] have studied the conditions under which the wave equation is separable, obtaining the result that the three-dimensional wave equation can be separated only in a limited number of coordinate systems (listed in Appendix IV) and then only if the potential energy is of the form

$$V = q_u\Phi_u(u) + q_v\Phi_v(v) + q_w\Phi_w(w),$$

in which $\Phi_u(u)$ is a function of u alone, $\Phi_v(v)$ of v alone, and $\Phi_w(w)$ of w alone.

17. THE THREE-DIMENSIONAL HARMONIC OSCILLATOR IN CYLINDRICAL COORDINATES

The isotropic harmonic oscillator in space is soluble by separation of variables in several coordinate systems, including Car-

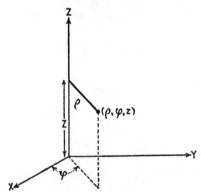

FIG. 17–1.—Diagram showing cylindrical coordinates.

tesian, cylindrical polar, and spherical polar coordinates. We shall use the cylindrical system in this section, comparing the results with those obtained in Section 15 with Cartesian coordinates.

Cylindrical polar coordinates ρ, φ, z, which are shown in Figure 17–1, are related to Cartesian coordinates by the equations of transformation

$$\left.\begin{array}{l} x = \rho\cos\varphi, \\ y = \rho\sin\varphi, \\ z = z. \end{array}\right\} \qquad (17\text{–}1)$$

[1] H. P. ROBERTSON, *Mathematische Annalen* **98**, 749 (1928); L. P. EISEN-HART, *Ann. Mathematics* **35**, 284 (1934).

Reference to Appendix IV shows that ∇^2 in terms of ρ, φ, z has the form

$$\nabla^2 \equiv \frac{1}{\rho} \frac{\partial}{\partial\rho}\left(\rho\frac{\partial}{\partial\rho}\right) + \frac{1}{\rho^2} \frac{\partial^2}{\partial\varphi^2} + \frac{\partial^2}{\partial z^2}. \qquad (17\text{–}2)$$

Consequently, the wave equation 15–4 for the three-dimensional harmonic oscillator becomes

$$\frac{1}{\rho} \frac{\partial}{\partial\rho}\left(\rho\frac{\partial\psi}{\partial\rho}\right) + \frac{1}{\rho^2} \frac{\partial^2\psi}{\partial\varphi^2} + \frac{\partial^2\psi}{\partial z^2} + \frac{8\pi^2 m}{h^2}$$
$$\{W - 2\pi^2 m(\nu_0^2\rho^2 + \nu_z^2 z^2)\}\psi = 0, \qquad (17\text{–}3)$$

when we make $\nu_x = \nu_y = \nu_0$ (only in this case is the wave equation separable in these coordinates). Making the substitutions

$$\lambda = \frac{8\pi^2 m}{h^2}W, \qquad (17\text{–}4a)$$

$$\alpha = \frac{4\pi^2 m}{h}\nu_0, \qquad (17\text{–}4b)$$

and

$$\alpha_z = \frac{4\pi^2 m}{h}\nu_z, \qquad (17\text{–}4c)$$

we obtain the equation

$$\frac{1}{\rho} \frac{\partial}{\partial\rho}\left(\rho\frac{\partial\psi}{\partial\rho}\right) + \frac{1}{\rho^2} \frac{\partial^2\psi}{\partial\varphi^2} + \frac{\partial^2\psi}{\partial z^2} + (\lambda - \alpha^2\rho^2 - \alpha_z^2 z^2)\psi = 0. \qquad (17\text{–}5)$$

Pursuing the method used in Section 15, we try the substitution

$$\psi = \mathrm{P}(\rho) \cdot \Phi(\varphi) \cdot Z(z), \qquad (17\text{–}6)$$

in which $\mathrm{P}(\rho)$ depends only on ρ, $\Phi(\varphi)$ only on φ, and $Z(z)$ only on z. Introduction of this into Equation 17–5 and division by ψ leads to the expression

$$\frac{1}{\mathrm{P}\rho} \frac{d}{d\rho}\left(\rho\frac{d\mathrm{P}}{d\rho}\right) + \frac{1}{\rho^2\Phi} \frac{d^2\Phi}{d\varphi^2} + \frac{1}{Z} \frac{d^2Z}{dz^2} + \lambda - \alpha^2\rho^2 - \alpha_z^2 z^2 = 0. \qquad (17\text{–}7)$$

The terms of this equation may be divided into two classes: those which depend only on z and those which depend only on ρ and φ. As before, since the two parts of the equation are functions of different sets of variables and since their sum is constant, each of the two parts must be constant. Therefore, we obtain two equations

$$\frac{d^2Z}{dz^2} + (\lambda_z - \alpha_z^2 z^2)Z = 0 \tag{17-8}$$

and

$$\frac{1}{P\rho}\frac{d}{d\rho}\left(\rho\frac{dP}{d\rho}\right) + \frac{1}{\rho^2\Phi}\frac{d^2\Phi}{d\varphi^2} - \alpha^2\rho^2 + \lambda' = 0, \tag{17-9}$$

with

$$\lambda' + \lambda_z = \lambda$$

The first of these is the familiar one-dimensional harmonic oscillator equation whose solutions

$$Z_{n_z}(z) = N_{n_z}e^{-\frac{\alpha_z}{2}z^2}H_{n_z}(\sqrt{\alpha_z}z) \tag{17-10}$$

are the Hermite orthogonal functions discussed in Section 11c. As in the one-dimensional problem, the requirement that the wave function satisfy the conditions of Section 9c restricts the parameter λ_z to the values

$$\lambda_z = (2n_z + 1)\alpha_z, \qquad n_z = 0, 1, 2, \cdots. \tag{17-11}$$

Equation 17–9, the second part of the wave equation, is a function of ρ and φ and so must be further separated. This may be accomplished by multiplying through by ρ^2. The second term of the resulting equation is independent of ρ; it is therefore equal to a constant, which we shall call $-m^2$. The two equations we obtain are the following:

$$\frac{d^2\Phi}{d\varphi^2} + m^2\Phi = 0 \tag{17-12}$$

and

$$\frac{1}{\rho}\frac{d}{d\rho}\left(\rho\frac{dP}{d\rho}\right) + \left(\lambda' - \alpha^2\rho^2 - \frac{m^2}{\rho^2}\right)P = 0. \tag{17-13}$$

The first of these is a familiar equation whose normalized solution is[1]

$$\Phi(\varphi) = \frac{1}{\sqrt{2\pi}}e^{im\varphi}. \tag{17-14}$$

Inasmuch as $e^{im\varphi}$ is equal to $\cos m\varphi + i \sin m\varphi$, we see that for arbitrary values of the separation constant m this function is

---

[1] Instead of the exponential, the forms $\Phi(\varphi) = N \cos m\varphi$ and $N \sin m\varphi$ may be used. See Section 18b, Chapter V.

not single-valued; that is, Φ does not have the same value for $\varphi = 0$ and for $\varphi = 2\pi$, which correspond to the same point in space. Only when m is a positive or negative integer or zero is Φ single-valued, as is required in order that it be an acceptable wave function (Sec. 9c); m must therefore be restricted to such values. φ is called a *cyclic coordinate* (or *ignorable coordinate*), these names being applied to a variable which does not occur anywhere in the wave equation (although derivatives with respect to it do appear). Such a coordinate always enters the wave function as an exponential factor of the type given in Equation 17–14.[1]

The equation for $P(\rho)$ may be treated by the same general method as was employed for the equation of the linear harmonic oscillator in Section 11a. The first step is to obtain an asymptotic solution for large values of ρ, in which region Equation 17–13 becomes approximately

$$\frac{d^2P}{d\rho^2} - \alpha^2\rho^2P = 0. \qquad (17\text{–}15)$$

The asymptotic solution of this is $e^{\pm\frac{\alpha}{2}\rho^2}$, since this function satisfies the equation

$$\frac{d^2e^{\pm\frac{\alpha}{2}\rho^2}}{d\rho^2} - (\alpha^2\rho^2 \pm \alpha)e^{\pm\frac{\alpha}{2}\rho^2} = 0,$$

which reduces to 17–15 for large values of ρ. Following the reasoning of Section 11a, we make the substitution

$$P(\rho) = e^{-\frac{\alpha}{2}\rho^2}f(\rho) \qquad (17\text{–}16)$$

in Equation 17–13. From this we find that f must satisfy the equation

$$f'' - 2\alpha\rho f' + \frac{1}{\rho}f' + (\lambda' - 2\alpha)f - \frac{m^2}{\rho^2}f = 0. \qquad (17\text{–}17)$$

As before, it is convenient to replace ρ by the variable

$$\xi = \sqrt{\alpha}\rho \qquad (17\text{–}18)$$

and $f(\rho)$ by $F(\xi)$, a process which gives the equation

$$\frac{d^2F}{d\xi^2} - 2\xi\frac{dF}{d\xi} + \frac{1}{\xi}\frac{dF}{d\xi} + \left(\frac{\lambda'}{\alpha} - 2 - \frac{m^2}{\xi^2}\right)F = 0. \qquad (17\text{–}19)$$

[1] Condon and Morse, "Quantum Mechanics," p. 72.

We could expand F directly as a power series in ξ, as in Section 11a. This is not very convenient, however, because the first few coefficients would turn out to be zero. Instead, we make the substitution

$$F(\xi) = \xi^s \sum_{\nu=0}^{\infty} a_\nu \xi^\nu = a_0 \xi^s + a_1 \xi^{s+1} + \cdots, \qquad (17\text{-}20)$$

in which s is an undetermined parameter and a_0 is not equal to zero.

This substitution is, indeed, called for by the character of the differential equation.[1] Equation 17–19 is written in the standard form

$$\frac{d^2F}{d\xi^2} + p(\xi)\frac{dF}{d\xi} + q(\xi)F = 0,$$

the coefficient of $\dfrac{d^2F}{d\xi^2}$ being unity. The coefficients p and q in Equation 17–19 possess singularities[2] at $\xi = 0$. The singular point $\xi = 0$ is a *regular point*, however, inasmuch as $p(\xi)$ is of order $1/\xi$ and $q(\xi)$ of order $1/\xi^2$. To solve a differential equation possessing a regular point at the origin, the substitution 17–20 is made in general. It is found that it leads to an *indicial equation* from which the index s can be determined.

Since we are interested only in acceptable wave functions, we shall ignore negative values of s. For this reason we could assume $F(\xi)$ to contain only positive powers of ξ. Occasionally, however, the indicial equation leads to non-integral values of s, in which case the treatment is greatly simplified by the substitution 17–20.

If we introduce the series 17–20 into Equation 17–19 and group together coefficients of equal powers of ξ, we obtain the equation

$$(s^2 - m^2)a_0\xi^{s-2} + \{(s+1)^2 - m^2\}a_1\xi^{s-1}$$
$$+ \left[\{(s+2)^2 - m^2\}a_2 + \left\{\frac{\lambda'}{\alpha} - 2(s+1)\right\}a_0\right]\xi^s + \cdots$$
$$+ \left[\{(s+\nu)^2 - m^2\}a_\nu + \left\{\frac{\lambda'}{\alpha} - 2(s+\nu-1)\right\}a_{\nu-2}\right]\xi^{s+\nu-2}$$
$$+ \cdots = 0. \qquad (17\text{-}21)$$

Since this is an identity in ξ, that is, an equation which is true for all values of ξ, we can show that the coefficient of each power

---

[1] See the standard treatments of the theory of linear differential equations; for example, Whittaker and Watson, "Modern Analysis," Chap. X.

[2] A singularity for a function $p(\xi)$ is a point at which $p(\xi)$ becomes infinite.

of ξ must be itself equal to zero.   This argument gives the set of equations

$$(s^2 - m^2)a_0 = 0, \qquad (17\text{--}22a)$$

$$\{(s + 1)^2 - m^2\}a_1 = 0, \qquad (17\text{--}22b)$$

. . . .

$$\{(s + \nu)^2 - m^2\}a_\nu + \left\{\frac{\lambda'}{\alpha} - 2(s + \nu - 1)\right\}a_{\nu-2} = 0, \quad (17\text{--}22c)$$

etc.

The first of these, 17–22a, is the indicial equation.   From it we see that s is equal to $+m$ or $-m$, inasmuch as a_0 is not equal to zero.   In order to obtain a solution of the form of Equation 17–20 which is finite at the origin, we must have s positive, so that we choose $s = +|m|$.   This value of s inserted in Equation 17–22b leads to the conclusion that a_1 must be zero.   Since the general recursion relation 17–22c connects coefficients whose subscripts differ by two, and since a_1 is zero, all odd coefficients are zero.   The even coefficients may be obtained in terms of a_0 by the use of 17–22c.

However, just as in the case of the linear harmonic oscillator, the infinite series so obtained is not a satisfactory wave function for general values of λ', because its value increases so rapidly with increasing ξ as to cause the total wave function to become infinite as ξ increases without limit.   In order to secure an acceptable wave function it is necessary to cause the series to break off after a finite number of terms.   The condition that the series break off at the term $a_{n'}\xi^{n'+|m|}$, where n' is an even integer, is obtained from 17–22c by putting $n' + 2$ in place of ν and equating the coefficient of $a_{n'}$ to zero.   This yields the result

$$\lambda' = 2(|m| + n' + 1)\alpha. \qquad (17\text{--}23)$$

Combining the expressions for λ_z and λ' given by Equations 17–11 and 17–23, we obtain the result

$$\lambda = \lambda' + \lambda_z = 2(|m| + n' + 1)\alpha + 2(n_z + \tfrac{1}{2})\alpha_z, \quad (17\text{--}24)$$

or, on insertion of the expressions for λ, α, and α_z,

$$W_{mn'n_z} = (|m| + n' + 1)h\nu_0 + (n_z + \tfrac{1}{2})h\nu_z. \qquad (17\text{--}25)$$

In the case of the isotropic harmonic oscillator, with $\nu_z = \nu_0$, this becomes

$$W_n = (n + \tfrac{3}{2})h\nu_0, \qquad n = |m| + n' + n_z. \quad (17\text{-}26)$$

The quantum numbers are restricted as follows:

$$m = 0, \pm 1, \pm 2, \cdots,$$
$$n' = 0, 2, 4, 6, \cdots,$$
$$n_z = 0, 1, 2, \cdots.$$

These lead to the same quantum weights for the energy levels as found in Section 15.

The wave functions have the form

$$\psi_{n'mn_z}(\rho, \varphi, z) = N e^{im\varphi} e^{-\frac{\alpha}{2}\rho^2} F_{|m|,n'}(\sqrt{\alpha}\rho) e^{-\frac{\alpha_z}{2}z^2} H_{n_z}(\sqrt{\alpha_z}z),$$

$$(17\text{-}27)$$

in which N is the normalization constant and $F_{|m|,n'}(\sqrt{\alpha}\rho)$ is a polynomial in ρ obtained from Equation 17–20 by the use of the recursion relations 17–22 for the coefficients a_ν. It contains only odd powers of ρ if $|m|$ is odd, and only even powers if $|m|$ is even.

Problem 17–1. The equation for the free particle is separable in many coordinate systems. Using cylindrical polar coordinates, set up and separate the wave equation, obtain the solutions in φ and z, and obtain the recursion formula for the coefficients in the series solution of the ρ equation. *Hint:* In applying the polynomial method, omit the step of finding the asymptotic solution.

Problem 17–2. Calculate $\overline{p_z^2}$ for a harmonic oscillator in a state represented by $\psi_{n'mn_z}$ of Equation 17–27. Show that $\overline{p_x}$ is zero in the same state. *Hint:* Transform $\dfrac{\partial}{\partial x}$ into cylindrical polar coordinates.

Problem 17–3. The equation for the isotropic harmonic oscillator is separable also in spherical polar coordinates. Set up the equation in these coordinates and carry out the separation of variables, obtaining the three total differential equations.

CHAPTER V

THE HYDROGEN ATOM

The problem of the structure of the hydrogen atom is the most important problem in the field of atomic and molecular structure, not only because the theoretical treatment of this atom is simpler than that of other atoms and of molecules, but also because it forms the basis for the discussion of more complex atomic systems. The wave-mechanical treatment of polyelectronic atoms and of molecules is usually closely related in procedure to that of the hydrogen atom, often being based on the use of hydrogen-like or closely related wave functions. Moreover, almost without exception the applications of qualitative and semiquantitative wave-mechanical arguments to chemistry involve the functions which occur in the treatment of the hydrogen atom.

The hydrogen atom has held a prominent place in the development of physical theory. The first spectral series expressed by a simple formula was the Balmer series of hydrogen. Bohr's treatment of the hydrogen atom marked the beginning of the old quantum theory of atomic structure, and wave mechanics had its inception in Schrödinger's first paper, in which he gave the solution of the wave equation for the hydrogen atom. Only in Heisenberg's quantum mechanics was there extensive development of the theory (by Heisenberg, Born, and Jordan) before the treatment of the hydrogen atom, characterized by its difficulty, was finally given by Pauli. In later developments, beyond the scope of this book, the hydrogen atom retains its important position; Dirac's relativistic quantum theory of the electron is applicable only to one-electron systems, its extension to more complicated systems not yet having been made.

The discussion of the hydrogen atom given in this chapter is due to Sommerfeld, differing in certain minor details from that of Schrödinger. It is divided into four sections. In the first, Section 18, the wave equation is separated and solved by the polynomial method, and the energy levels are discussed. Sections 19 and 20 include the definition of certain functions, the Legendre and Laguerre functions, which occur in the hydrogen-atom wave functions, and the discussion of their properties. A detailed description of the wave functions themselves is given in Section 21.

18. THE SOLUTION OF THE WAVE EQUATION BY THE POLYNOMIAL METHOD AND THE DETERMINATION OF THE ENERGY LEVELS

18a. The Separation of the Wave Equation. The Translational Motion.—We consider the hydrogen atom as a system of two interacting point particles, the interaction being that due to the Coulomb attraction of their electrical charges. Let us for generality ascribe to the nucleus the charge $+Ze$, the charge of the electron being $-e$. The potential energy of the system, in the absence of external fields, is $-\dfrac{Ze^2}{r}$, in which r is the distance between the electron and the nucleus.

If we write for the Cartesian coordinates of the nucleus and the electron x_1, y_1, z_1 and x_2, y_2, z_2, and for their masses m_1 and m_2, respectively, the wave equation has the form

$$\frac{1}{m_1}\left(\frac{\partial^2\psi_T}{\partial x_1^2} + \frac{\partial^2\psi_T}{\partial y_1^2} + \frac{\partial^2\psi_T}{\partial z_1^2}\right) + \frac{1}{m_2}\left(\frac{\partial^2\psi_T}{\partial x_2^2} + \frac{\partial^2\psi_T}{\partial y_2^2} + \frac{\partial^2\psi_T}{\partial z_2^2}\right)$$
$$+ \frac{8\pi^2}{h^2}(W_T - V)\psi_T = 0, \quad (18\text{--}1)$$

in which

$$V = -\frac{Ze^2}{\sqrt{(x_2 - x_1)^2 + (y_2 - y_1)^2 + (z_2 - z_1)^2}}$$

Here the subscript T (signifying total) is written for W and ψ to indicate that these quantities refer to the complete system, with six coordinates.

This equation can be immediately separated into two, one of which represents the translational motion of the molecule as a whole and the other the relative motion of the two particles. In fact, this separation can be accomplished in a somewhat more general case, namely, when the potential energy V is a general function of the relative positions of the two particles, that is, $V = V(x_2 - x_1, y_2 - y_1, z_2 - z_1)$. This includes, for example, the hydrogen atom in a constant electric field, the potential energy due to the field then being $eEz_2 - eEz_1 = eE(z_2 - z_1)$, in which E is the strength of the field, considered as being in the direction of the z axis.

To effect the separation, we introduce the new variables x, y, and z, which are the Cartesian coordinates of the center of mass of the system, and r, ϑ, and φ, the polar coordinates of the second particle relative to the first. These coordinates are related to the Cartesian coordinates of the two particles by the equations

$$x = \frac{m_1 x_1 + m_2 x_2}{m_1 + m_2}, \qquad (18\text{-}2a)$$

$$y = \frac{m_1 y_1 + m_2 y_2}{m_1 + m_2}, \qquad (18\text{-}2b)$$

$$z = \frac{m_1 z_1 + m_2 z_2}{m_1 + m_2}, \qquad (18\text{-}2c)$$

$$r \sin \vartheta \cos \varphi = x_2 - x_1, \qquad (18\text{-}2d)$$

$$r \sin \vartheta \sin \varphi = y_2 - y_1, \qquad (18\text{-}2e)$$

$$r \cos \vartheta = z_2 - z_1. \qquad (18\text{-}2f)$$

The introduction of these new independent variables in Equation 18–1 is easily made in the usual way. The resultant wave equation is

$$\frac{1}{m_1 + m_2}\left(\frac{\partial^2 \psi_T}{\partial x^2} + \frac{\partial^2 \psi_T}{\partial y^2} + \frac{\partial^2 \psi_T}{\partial z^2}\right) + \frac{1}{\mu}\left\{\frac{1}{r^2}\frac{\partial}{\partial r}\left(r^2 \frac{\partial \psi_T}{\partial r}\right)\right.$$
$$+ \frac{1}{r^2 \sin^2 \vartheta}\frac{\partial^2 \psi_T}{\partial \varphi^2} + \frac{1}{r^2 \sin \vartheta}\frac{\partial}{\partial \vartheta}\left(\sin \vartheta \frac{\partial \psi_T}{\partial \vartheta}\right)\right\}$$
$$+ \frac{8\pi^2}{h^2}\{W_T - V(r, \vartheta, \varphi)\}\psi_T = 0. \quad (18\text{-}3)$$

In this equation the symbol μ has been introduced to represent the quantity

$$\mu = \frac{m_1 m_2}{m_1 + m_2} \quad \left(\text{or } \frac{1}{\mu} = \frac{1}{m_1} + \frac{1}{m_2}\right); \qquad (18\text{-}4)$$

μ is the reduced mass of the system, already discussed in Section 2d in the classical treatment of this problem.

It will be noticed that the quantity in the first set of parentheses is the Laplacian of ψ_T in the Cartesian coordinates x, y, and z, and the quantity in the first set of braces is the Laplacian in the polar coordinates r, ϑ, and φ (Appendix IV).

We now attempt to separate this equation by expressing ψ_T as the product of a function of x, y, z and a function of r, ϑ, φ, writing

$$\psi_T(x, y, z, r, \vartheta, \varphi) = F(x, y, z)\psi(r, \vartheta, \varphi). \qquad (18\text{--}5)$$

On introducing this in Equation 18–3 and dividing through by $\psi_T = F\psi$, it is found that the equation is the sum of two parts, one of which is dependent only on x, y, and z and the other only on r, ϑ, and φ. Each part must hence be equal to a constant. The resulting equations are

$$\frac{\partial^2 F}{\partial x^2} + \frac{\partial^2 F}{\partial y^2} + \frac{\partial^2 F}{\partial z^2} + \frac{8\pi^2(m_1 + m_2)}{h^2} W_{tr}F = 0, \qquad (18\text{--}6)$$

and

$$\frac{1}{r^2}\frac{\partial}{\partial r}\left(r^2\frac{\partial\psi}{\partial r}\right) + \frac{1}{r^2\sin^2\vartheta}\frac{\partial^2\psi}{\partial\varphi^2} + \frac{1}{r^2\sin\vartheta}\frac{\partial}{\partial\vartheta}\left(\sin\vartheta\frac{\partial\psi}{\partial\vartheta}\right)$$
$$+ \frac{8\pi^2\mu}{h^2}\{W - V(r, \vartheta, \varphi)\}\psi = 0, \qquad (18\text{--}7)$$

with

$$W_{tr} + W = W_T. \qquad (18\text{--}8)$$

Equation 18–6 is identical with Equation 13–2 of Section 13, representing the motion of a free particle; hence the translational motion of the system is the same as that of a particle with mass $m_1 + m_2$ equal to the sum of the masses of the two particles. In most problems the state of translational motion is not important, and a knowledge of the translational energy W_{tr} is not required. In our further discussion we shall refer to W, the energy of the system aside from the translational energy, simply as the energy of the system.

Equation 18–7 is identical with the wave equation of a single particle of mass μ under the influence of a potential function $V(r, \vartheta, \varphi)$. This identity corresponds to the classical identity of Section 2d (Eqs. 2–25).

If we now restrict ourselves to the case in which the potential function V is a function of r alone,

$$V = V(r),$$

Equation 18–7 can be further separated. We write

$$\psi(r, \vartheta, \varphi) = R(r) \cdot \Theta(\vartheta) \cdot \Phi(\varphi); \qquad (18\text{--}9)$$

on introducing this in Equation 18–7 and dividing by $R\Theta\Phi$, it becomes

$$\frac{1}{r^2R}\frac{d}{dr}\left(r^2\frac{dR}{dr}\right) + \frac{1}{r^2\sin^2\vartheta\,\Phi}\frac{d^2\Phi}{d\varphi^2} + \frac{1}{r^2\sin\vartheta\,\Theta}\frac{d}{d\vartheta}\left(\sin\vartheta\frac{d\Theta}{d\vartheta}\right)$$
$$+ \frac{8\pi^2\mu}{h^2}\{W - V(r)\} = 0. \quad (18\text{–}10)$$

On multiplying through by $r^2\sin^2\vartheta$, the remaining part of the second term, $\frac{1}{\Phi}\frac{d^2\Phi}{d\varphi^2}$, which could only be a function of the independent variable φ, is seen to be equal to terms independent of φ. Hence this term must be equal to a constant, which we call $-m^2$:

$$\frac{d^2\Phi}{d\varphi^2} = -m^2\Phi. \quad (18\text{–}11)$$

The equation in ϑ and r then can be written as

$$\frac{1}{R}\frac{d}{dr}\left(r^2\frac{dR}{dr}\right) - \frac{m^2}{\sin^2\vartheta} + \frac{1}{\sin\vartheta\,\Theta}\frac{d}{d\vartheta}\left(\sin\vartheta\frac{d\Theta}{d\vartheta}\right) + \frac{8\pi^2\mu r^2}{h^2}\{W - V(r)\} = 0.$$

The part of this equation containing the second and third terms is independent of r and the remaining part is independent of ϑ, so that we can equate each to a constant. If we set the ϑ terms equal to the constant $-\beta$, and the r terms equal to $+\beta$, we obtain the following equations, after multiplication by Θ and by R/r^2, respectively:

$$\frac{1}{\sin\vartheta}\frac{d}{d\vartheta}\left(\sin\vartheta\frac{d\Theta}{d\vartheta}\right) - \frac{m^2}{\sin^2\vartheta}\Theta + \beta\Theta = 0 \quad (18\text{–}12)$$

and

$$\frac{1}{r^2}\frac{d}{dr}\left(r^2\frac{dR}{dr}\right) - \frac{\beta}{r^2}R + \frac{8\pi^2\mu}{h^2}\{W - V(r)\}R = 0. \quad (18\text{–}13)$$

Equations 18–11, 18–12, and 18–13 are now to be solved in order to determine the allowed values of the energy. The sequence of solution is the following: We first find that Equation 18–11 possesses acceptable solutions only for certain values of the parameter m. Introducing these in Equation 18–12, we find that it then possesses acceptable solutions only for certain values of β. Finally, we introduce these values of β

in Equation 18–13 and find that this equation then possesses acceptable solutions only for certain values of W. These are the values of the energy for the stationary states of the system.

It may be mentioned that the wave equation for the hydrogen atom can also be separated in coordinate systems other than the polar coordinates r, ϑ, and φ which we have chosen, and for some purposes another coordinate system may be especially appropriate, as, for example, in the treatment of the Stark effect, for which (as shown by Schrödinger in his third paper) it is convenient to use parabolic coordinates.

18b. The Solution of the φ Equation.—As was discussed in Section 17, the solutions of Equation 18–11, involving the cyclic coordinate φ, are

$$\Phi_m(\varphi) = \frac{1}{\sqrt{2\pi}}e^{im\varphi}. \qquad (18\text{--}14)$$

In order for the function to be single-valued at the point $\varphi = 0$ (which is identical with $\varphi = 2\pi$), the parameter m must be equal to an integer. The independent acceptable solutions of the φ equation are hence given by Equation 18–14, with $m = 0$, $+1$, $+2$, $\cdots$, -1, -2, $\cdots$; these values are usually written as 0, ± 1, ± 2, $\cdots$, it being understood that positive and negative values correspond to distinct solutions.

The constant m is called the *magnetic quantum number*. It is the analogue of the same quantum number in the old-quantum-theory treatment (Sec. 7b).

The factor $1/\sqrt{2\pi}$ is introduced in order to normalize the functions $\Phi_m(\varphi)$, which then satisfy the equation

$$\int_0^{2\pi}\Phi^*(\varphi)\Phi_m(\varphi)d\varphi = 1. \qquad (18\text{--}15)$$

It may be pointed out that for a given value of $|m|$ (the absolute value of m), the two functions $\Phi_{|m|}(\varphi)$ and $\Phi_{-|m|}(\varphi)$ satisfy the same differential equation, with the same value of the parameter, and that any linear combination of them also satisfies the equation. The sum and the difference of these two functions are the cosine and sine functions. It is sometimes convenient to use these in place of the complex exponential functions as the independent solutions of the wave equation, the normalized solutions then being

$$\Phi_0(\varphi) = \frac{1}{\sqrt{2\pi}},$$

$$\Phi_{|m|}(\varphi) = \begin{cases} \dfrac{1}{\sqrt{\pi}} \cos |m|\,\varphi, \\ \dfrac{1}{\sqrt{\pi}} \sin |m|\,\varphi, \quad |m| = 1, 2, 3, \cdots \end{cases} \qquad (18\text{–}16)$$

There is only one solution for $|m| = 0$. These functions are normalized and are mutually orthogonal.

It is sometimes convenient to use the symbol m to represent the absolute value of the magnetic quantum number as well as the quantum number itself. To avoid confusion, however, we shall not adopt this practice but shall write $|m|$ for the absolute value of m.

18c. The Solution of the ϑ Equation.—In order to solve the ϑ equation 18–12, it is convenient for us to introduce the new independent variable

$$z = \cos \vartheta, \qquad (18\text{–}17)$$

which varies between the limits -1 and $+1$, and at the same time to replace $\Theta(\vartheta)$ by the function $P(z)$ to which it is equal:

$$P(z) = \Theta(\vartheta). \qquad (18\text{–}18)$$

Noting that $\sin^2 \vartheta = 1 - z^2$ and that

$$\frac{d\Theta}{d\vartheta} = \frac{dP}{dz}\frac{dz}{d\vartheta} = -\frac{dP}{dz}\sin \vartheta,$$

we see that our equation becomes

$$\frac{d}{dz}\left\{(1 - z^2)\frac{dP(z)}{dz}\right\} + \left\{\beta - \frac{m^2}{1 - z^2}\right\}P(z) = 0. \qquad (18\text{–}19)$$

On attempting to solve this equation by the polynomial method, it is found that the recursion formula involves more than two terms. If, however, a suitable substitution is made, the equation can be reduced to one to which the polynomial method can be applied.

The equation has singular points at $z = \pm 1$, both of which are regular points (see Sec. 17), so that it is necessary to discuss the indicial equation at each of these points. In order to study the behavior near $z = +1$, it is convenient to make the substitution $x = 1 - z$, $R(x) = P(z)$, bringing this point to the origin of x. The resulting equation is

$$\frac{d}{dx}\left\{x(2-x)\frac{dR}{dx}\right\} + \left\{\beta - \frac{m^2}{x(2-x)}\right\} R = 0.$$

f we substitute $R = x^s \sum_{\nu=0}^{\infty} a_\nu x^\nu$ in this equation, we find that the indicial equation (see Sec. 17) leads to the value $|m|/2$ for s. Likewise, if we investigate the point $z = -1$ by making the substitution $y = 1 + z$ and similarly study the indicial equation at the origin of y, we find the same value for the index there.

The result of these considerations is that the substitution

$$P(z) = x^{\frac{|m|}{2}} y^{\frac{|m|}{2}} G(z) = (1 - z^2)^{\frac{|m|}{2}} G(z) \qquad (18\text{--}20)$$

is required. On introducing this into Equation 18–19, the differential equation satisfied by $G(z)$—which should now be directly soluble by a power series—is found to be

$$(1 - z^2)G'' - 2(|m| + 1)zG' +$$
$$\{\beta - |m|(|m| + 1)\}G = 0, \quad (18\text{--}21)$$

in which G' represents $\frac{dG}{dz}$ and G'' represents $\frac{d^2G}{dz^2}$.

This equation we now treat by the polynomial method, the successive steps being similar to those taken in Section 11 in the discussion of the harmonic oscillator. Let

$$G = a_0 + a_1 z + a_2 z^2 + a_3 z^3 + \cdots, \qquad (18\text{--}22)$$

with G' and G'' similar series obtained from this by differentiation. On the introduction of these in Equation 18–21, it becomes

$$1 \cdot 2a_2 + 2 \cdot 3a_3 z + 3 \cdot 4a_4 z^2 + 4 \cdot 5a_5 z^3 + \cdots$$
$$- 1 \cdot 2a_2 z^2 - 2 \cdot 3a_3 z^3 - \cdots$$
$$-2(|m| + 1)a_1 z - 2 \cdot 2(|m| + 1)a_2 z^2 - 2 \cdot 3(|m| + 1)a_3 z^3 - \cdots$$
$$+\{\beta - |m|(|m| + 1)\}a_0 + \{\}a_1 z + \{\}a_2 z^2 + \{\}a_3 z^3 + \cdots = 0,$$

in which the braces $\{\}$ represent $\{\beta - |m|(|m| + 1)\}$. This equation is an identity in z, and hence the coefficients of individual powers of z must vanish; that is,

$$1 \cdot 2a_2 + \{\}a_0 = 0,$$
$$2 \cdot 3a_3 + (\{\} - 2(|m| + 1))a_1 = 0,$$
$$3 \cdot 4a_4 + (\{\} - 2 \cdot 2(|m| + 1) - 1 \cdot 2)a_2 = 0,$$
$$4 \cdot 5a_5 + (\{\} - 2 \cdot 3(|m| + 1) - 2 \cdot 3)a_3 = 0,$$

or, in general, for the coefficient of z^ν,

$$(\nu + 1)(\nu + 2)a_{\nu+2} + [\{\beta - |m|(|m| + 1)\} \\ - 2\nu(|m| + 1) - \nu(\nu - 1)]a_\nu = 0.$$

This leads to the two-term recursion formula

$$a_{\nu+2} = \frac{(\nu + |m|)(\nu + |m| + 1) - \beta}{(\nu + 1)(\nu + 2)}a_\nu \qquad (18\text{-}23)$$

between the coefficients $a_{\nu+2}$ and a_ν in the series for G.

It is found on discussion by the usual methods[1] that an infinite series with this relation between alternate coefficients converges (for any values of $|m|$ and β) for $-1 < z < 1$, but diverges for $z = +1$ or -1, and in consequence does not correspond to an acceptable wave function. In order to be satisfactory, then, our series for G must contain only a finite number of terms. Either the even or the odd series can be broken off at the term in $z^{\nu'}$ by placing

$$\beta = (\nu' + |m|)(\nu' + |m| + 1), \qquad \nu' = 0, 1, 2, \cdots,$$

and the other series can be made to vanish by equating a_1 or a_0 to zero. The characteristic values of the parameter β are thus found to be given by the above expression, the corresponding functions $G(z)$ containing only even or odd powers of z as ν' is even or odd.

It is convenient to introduce the new quantum number

$$l = \nu' + |m| \qquad (18\text{-}24)$$

in place of ν', the allowed values for l being (from its definition) $|m|, |m| + 1, |m| + 2, \cdots$. The characteristic values of β are then

$$\beta = l(l + 1), \qquad l = |m|, |m| + 1, \cdots. \qquad (18\text{-}25)$$

l is called the *azimuthal quantum number;* it is analogous to the quantum number k of the old quantum theory. Spectral states which are now represented by a given value of l were formerly represented by a value of k one unit greater, $k = 1$ corresponding to $l = 0$, and so on.

[1] R. COURANT and D. HILBERT, "Methoden der mathematischen Physik," 2d ed.,Vol. I, p. 281, Julius Springer, Berlin, 1931.

We have now shown that the allowed solutions of the ϑ equation are $\Theta(\vartheta) = (1 - z^2)^{|m|/2}G(z)$, in which $G(z)$ is defined by the recursion formula 18–23, with $\beta = l(l + 1)$. It will be shown in Section 19 that the functions $\Theta(\vartheta)$ are the associated Legendre functions. A description of the functions will be given in Section 21.

18d. The Solution of the r Equation.—Having evaluated β as $l(l + 1)$, the equation in r becomes

$$\frac{1}{r^2}\frac{d}{dr}\left(r^2\frac{dR}{dr}\right) + \left[-\frac{l(l + 1)}{r^2} + \frac{8\pi^2\mu}{h^2}\{W - V(r)\} \right]R = 0, \quad (18\text{–}26)$$

in which $V(r) = -Ze^2/r$, Z being the atomic number of the atom. It is only now, by the introduction of this expression for the potential energy, that we specialize the problem to that of the one-electron or hydrogenlike atom. The discussion up to this point is applicable to any system of two particles which interact with one another in a way expressible by a potential function $V(r)$, as, for example, the two nuclei in a diatomic molecule after the electronic interactions have been considered by the Born-Oppenheimer method (Sec. 35a).

Let us first consider the case of W negative, corresponding to a total energy insufficient to ionize the atom. Introducing the symbols

and
$$\left.\begin{array}{l} \alpha^2 = -\dfrac{8\pi^2\mu W}{h^2} \\[2ex] \lambda = \dfrac{4\pi^2\mu Ze^2}{h^2\alpha}, \end{array}\right\} \quad (18\text{–}27)$$

and the new independent variable

$$\rho = 2\alpha r, \quad (18\text{–}28)$$

the wave equation becomes

$$\frac{1}{\rho^2}\frac{d}{d\rho}\left(\rho^2\frac{dS}{d\rho}\right) + \left\{-\frac{1}{4} - \frac{l(l + 1)}{\rho^2} + \frac{\lambda}{\rho}\right\}S = 0,$$
$$0 \leqslant \rho \leqslant \infty, \quad (18\text{–}29)$$

in which $S(\rho) = R(r)$. As in the treatment of the harmonic oscillator, we first discuss the asymptotic equation. For ρ large, the equation approaches the form

$$\frac{d^2S}{d\rho^2} = \frac{1}{4}S,$$

the solutions of which are

$$S = e^{+\frac{\rho}{2}} \text{ and } S = e^{-\frac{\rho}{2}}.$$

Only the second of these is satisfactory as a wave function.

We now assume that the solution of the complete equation 18–29 has the form

$$S(\rho) = e^{-\frac{\rho}{2}}F(\rho). \tag{18–30}$$

The equation satisfied by $F(\rho)$ is found to be

$$F'' + \left(\frac{2}{\rho} - 1\right)F' + \left\{\frac{\lambda}{\rho} - \frac{l(l+1)}{\rho^2} - \frac{1}{\rho}\right\}F = 0,$$
$$0 \leqslant \rho \leqslant \infty. \tag{18–31}$$

The coefficients of F' and F possess singularities at the origin, which is a regular point (cf. Sec. 17), so that we again make the substitution

$$F(\rho) = \rho^s L(\rho), \tag{18–32}$$

in which $L(\rho)$ is a power series in ρ beginning with a non-vanishing constant term:

$$L(\rho) = \sum_\nu a_\nu \rho^\nu, \qquad a_0 \neq 0. \tag{18–33}$$

Since

$$F'(\rho) = s\rho^{s-1}L + \rho^s L'$$

and

$$F''(\rho) = s(s - 1)\rho^{s-2}L + 2s\rho^{s-1}L' + \rho^s L'',$$

Equation 18–31 becomes

$$\begin{aligned}
\rho^{s+2}L'' &+ 2s\rho^{s+1}L' + s(s - 1)\rho^s L \\
&+ 2\rho^{s+1}L' + 2s\rho^s L \\
&- \rho^{s+2}L' - s\rho^{s+1}L \\
&+ (\lambda - 1)\rho^{s+1}L - l(l + 1)\rho^s L = 0.
\end{aligned} \tag{18–34}$$

Since L begins with the term a_0, the coefficient of ρ^s is seen to be $\{s(s-1) + 2s - l(l+1)\}a_0$, and, since a_0 does not vanish, the expression in braces must vanish in order for Equation 18–34 to be satisfied as an identity in ρ. This gives as the indicial equation for s:

$$s(s+1) - l(l+1) = 0, \quad \text{or} \quad s = +l \quad \text{or} \quad -(l+1). \quad (18\text{–}35)$$

Of the two solutions of the indicial equation, the solution $s = -(l+1)$ does not lead to an acceptable wave function. We accordingly write

$$F(\rho) = \rho^l L(\rho), \quad (18\text{–}36)$$

and obtain from 18–34 the equation

$$\rho L'' + \{2(l+1) - \rho\}L' + (\lambda - l - 1)L = 0, \quad (18\text{–}37)$$

after substituting l for s and dividing by ρ^{l+1}. We now introduce the series 18–33 for L in this equation and obtain an equation involving powers of ρ, the coefficients of which must vanish individually. These conditions are successively

$$(\lambda - l - 1)a_0 + 2(l+1)a_1 = 0,$$
$$(\lambda - l - 1 - 1)a_1 + \{2 \cdot 2(l+1) + 1 \cdot 2\}a_2 = 0,$$
$$(\lambda - l - 1 - 2)a_2 + \{3 \cdot 2(l+1) + 2 \cdot 3\}a_3 = 0$$

or, for the coefficient of ρ^ν,

$$(\lambda - l - 1 - \nu)a_\nu + \{2(\nu+1)(l+1) + \nu(\nu+1)\}a_{\nu+1} = 0. \quad (18\text{–}38)$$

It can be shown by an argument similar to that used in Section 11a for the harmonic oscillator that for any values of λ and l the series whose coefficients are determined by this formula leads to a function $S(\rho)$ unacceptable as a wave function unless it breaks off. For very large values of ν the successive terms of an infinite series given by 18–38 approach the terms of the expansion of e^ρ, which accordingly represents the asymptotic behavior of the series. This corresponds to an asymptotic behavior of $S(\rho) = e^{-\frac{\rho}{2}}\rho^l L(\rho)$ similar to $e^{+\frac{\rho}{2}}$, leading to the infinity catastrophe with increasing ρ.

Consequently the series must break off after a finite number of terms. The condition that it break off after the term in $\rho^{n'}$ is seen from Equation 18–38 to be

$$\lambda - l - 1 - n' = 0$$

or

$$\lambda = n, \quad \text{where} \quad n = n' + l + 1. \quad (18\text{--}39)$$

n' is called the *radial quantum number* and n the *total quantum number*. From its nature it is seen that n' can assume the values 0, 1, 2, 3, $\cdots$. The values of n will be discussed in the next section.

In this section we have found the allowed solutions of the r equation to have the form $R(r) = e^{-\frac{\rho}{2}}\rho^l L(\rho)$, in which $L(\rho)$ is defined by the recursion formula 18–38, with $\lambda = n$. It will be shown in Section 20 that these functions are certain associated Laguerre functions, and a description of them will be given in Section 21.

18e. The Energy Levels.—Introducing for λ its value as given in Equation 18–27, and solving for W, it is found that Equation 18–39 leads to the energy expression

$$W_n = -\frac{2\pi^2\mu Z^2 e^4}{h^2 n^2} = -\frac{RhcZ^2}{n^2} = -\frac{Z^2}{n^2}W_H, \quad (18\text{--}40)$$

in which

$$R = \frac{2\pi^2\mu e^4}{h^3 c} \quad \text{and} \quad W_H = Rhc.$$

This expression is identical with that of the old quantum theory (Eq. 7–24), even to the inclusion of the reduced mass μ. It is seen that the energy of a hydrogenlike atom in the state represented by the quantum numbers n', l, and m does not depend on their individual values but only on the value of the total quantum number $n = n' + l + 1$. Inasmuch as both n' and l by their nature can assume the values 0, 1, 2, $\cdots$, we see that the allowed values of n are 1, 2, 3, 4, $\cdots$, as assumed in the old quantum theory and verified by experiment (discussed in Sec. 7b).

Except for $n = 1$, each energy level is degenerate, being represented by more than one independent solution of the wave equation. If we introduce the quantum numbers n, l, and m as subscripts (using n in preference to n'), the wave functions we have found as acceptable solutions of the wave equation may be written as

$$\psi_{nlm}(r, \vartheta, \varphi) = R_{nl}(r)\Theta_{lm}(\vartheta)\Phi_m(\varphi), \quad (18\text{--}41)$$

the functions themselves being those determined in Sections 18*b*, 18*c*, and 18*d*. The wave functions corresponding to distinct sets of values for *n*, *l*, and *m* are independent. The allowed values of these quantum numbers we have determined to be

$$m = 0, \pm 1, \pm 2, \cdots,$$
$$l = |m|, |m| + 1, |m| + 2, \cdots,$$
$$n = l + 1, l + 2, l + 3, \cdots.$$

This we may rewrite as

total quantum number $n = 1, 2, 3, \cdots,$
azimuthal quantum number $l = 0, 1, 2, \cdots, n - 1,$
magnetic quantum number $m = -l, -l + 1, \cdots, -1, 0,$
$$+1, \cdots, +l - 1, +l.$$

There are consequently $2l + 1$ independent wave functions with given values of *n* and *l*, and n^2 independent wave functions with a given value of *n*, that is, with the same energy value. The $2l + 1$ wave functions with the same *n* and *l* are said to form a *completed subgroup*, and the n^2 wave functions with the same *n* a *completed group*. The wave functions will be described in the following sections of this chapter.

A similar treatment applied to the wave equation with *W* positive leads to the result that there exist acceptable solutions for all positive values of the energy, as indicated by the general discussion of Section 9*c*. It is a particularly pleasing feature of the quantum mechanics that a unified treatment can be given the continuous as well as the discrete spectrum of energy values. Because of the rather complicated nature of the discussion of the wave functions for the continuous spectrum (in particular their orthogonality and normalization properties) and of their minor importance for most chemical problems, we shall not treat them further.[1]

19. LEGENDRE FUNCTIONS AND SURFACE HARMONICS

The functions of ϑ which we have obtained by solution of the ϑ equation are well known to mathematicians under the name of *associated Legendre functions*.[2] The functions of ϑ and φ are

---

[1] See SOMMERFELD, "Wave Mechanics," p. 290.

[2] The functions of ϑ for $m = 0$ are called Legendre functions. The associated Legendre functions include the Legendre functions and additional

called *surface harmonics* (or, in case cosine and sine functions of φ are used instead of exponential functions, *tesseral harmonics*). We could, of course, proceed to develop the properties of these functions from the recursion formulas for the coefficients in the polynomials obtained in the foregoing treatment. This would be awkward and laborious, however; it is simpler for us to define the functions anew by means of differential expressions or generating functions, and to discuss their properties on this basis, ultimately proving the identity of these functions with those obtained earlier by application of the polynomial method.

19a. The Legendre Functions or Legendre Polynomials.—The Legendre functions or Legendre polynomials $P_l(\cos \vartheta) = P_l(z)$ may be defined by means of a generating function $T(t, z)$ such that

$$T(t, z) \equiv \sum_{l=0}^{\infty} P_l(z) t^l \equiv \frac{1}{\sqrt{1 - 2tz + t^2}}. \qquad (19\text{--}1)$$

As in the case of the Hermite polynomials (Sec. 11c), we obtain relations among the polynomials and their derivatives by differentiating the generating function with respect to t and to z. Thus on differentiation with respect to t, we write

$$\frac{\partial T}{\partial t} \equiv \sum_{l=0}^{\infty} l P_l t^{l-1} \equiv -\frac{\frac{1}{2}(-2z + 2t)}{(1 - 2zt + t^2)^{3/2}}$$

or

$$(1 - 2zt + t^2) \sum_l l P_l t^{l-1} \equiv (z - t) \sum_l P_l t^l$$

(the right side having been transformed with the use of Equation 19–1), and consequently, by equating coefficients of given powers of t on the two sides, we obtain the recursion formula for the Legendre polynomials

$$(l + 1)P_{l+1}(z) - (2l + 1)zP_l(z) + lP_{l-1}(z) = 0. \quad (19\text{--}2)$$

Similarly, by differentiation with respect to z, there is obtained

$$\frac{\partial T}{\partial z} \equiv \sum_l P_l' t^l \equiv \frac{t}{(1 - 2zt + t^2)^{3/2}}$$

---

functions (corresponding to $|m| > 0$) conveniently defined in terms of the Legendre functions.

or

$$(1 - 2zt + t^2)\sum_l P'_l t^l \equiv t\sum_l P_l t^l,$$

which gives the relation

$$P'_{l+1}(z) - 2zP'_l(z) + P'_{l-1}(z) - P_l(z) = 0 \qquad (19\text{-}3)$$

involving the derivatives of the polynomials. Somewhat simpler relations may be obtained by combining these. From 19–2 and 19–3, after differentiating the former, we find

$$zP'_l(z) - P'_{l-1}(z) - lP_l(z) = 0 \qquad (19\text{-}4)$$

and

$$P'_{l+1}(z) - zP'_l(z) - (l+1)P_l(z) = 0. \qquad (19\text{-}5)$$

We can now easily find the differential equation which $P_l(z)$ satisfies. Reducing the subscript l to $l - 1$ in 19–5, and subtracting 19–4 after multiplication by z, we obtain

$$(1 - z^2)P'_l + lzP_l - lP_{l-1} = 0,$$

which on differentiation becomes

$$\frac{d}{dz}\left\{(1 - z^2)\frac{dP_l(z)}{dz}\right\} + lP_l(z) + lzP'_l(z) - lP'_{l-1}(z) = 0.$$

The terms in P'_l and P'_{l-1} may be replaced by l^2P_l, from 19–4, and there then results the differential equation for the Legendre polynomials

$$\frac{d}{dz}\left\{(1 - z^2)\frac{dP_l(z)}{dz}\right\} + l(l+1)P_l(z) = 0. \qquad (19\text{-}6)$$

19b. The Associated Legendre Functions.—We define the associated Legendre functions of degree l and order $|m|$ (with values $l = 0, 1, 2, \cdots$ and $|m| = 0, 1, 2, \cdots, l$) in terms of the Legendre polynomials by means of the equation

$$P_l^{|m|}(z) = (1 - z^2)^{|m|/2}\frac{d^{|m|}}{dz^{|m|}}P_l(z). \qquad (19\text{-}7)$$

[It is to be noted that the order $|m|$ is restricted to positive values (and zero); we are using the rather clumsy symbol $|m|$ to represent the order of the associated Legendre function so that we may later identify m with the magnetic quantum number previously introduced.] The differential equation satisfied by these functions may be found in the following way. On differentiating Equation 19–6 $|m|$ times, there results

$$(1 - z^2)\frac{d^{|m|+2}P_l(z)}{dz^{|m|+2}} - 2(|m| + 1)z\frac{d^{|m|+1}P_l(z)}{dz^{|m|+1}}$$
$$+ \{l(l + 1) - |m|(|m| + 1)\}\frac{d^{|m|}P_l(z)}{dz^{|m|}} = 0 \quad (19\text{–}8)$$

as the differential equation satisfied by $\dfrac{d^{|m|}P_l(z)}{dz^{|m|}}$. With the use of Equation 19–7 this equation is easily transformed into

$$(1 - z^2)\frac{d^2 P_l^{|m|}(z)}{dz^2} - 2z\frac{dP_l^{|m|}(z)}{dz} + \left\{l(l + 1) - \frac{m^2}{1 - z^2}\right\}P_l^{|m|}(z)$$
$$= 0, \quad (19\text{–}9)$$

which is the differential equation satisfied by the associated Legendre function $P_l^{|m|}(z)$.

This result enables us to identify[1] the ϑ functions of Section 18c (except for constant factors) with the associated Legendre functions, inasmuch as Equation 19–9 is identical with Equation 18–19, except that $P(z)$ is replaced by $P_l^{|m|}(z)$ and β is replaced by $l(l + 1)$, which was found in Section 18c to represent the characteristic values of β. Hence the wave functions in ϑ corresponding to given values of the azimuthal quantum number l and the magnetic quantum number m are the associated Legendre functions $P_l^{|m|}(z)$.

The associated Legendre functions are most easily tabulated by the use of the recursion formula 19–2 and the definition 19–7, together with the value $P_0^0(z) = 1$ as the starting point. A detailed discussion of the functions is given in Section 21.

For some purposes the generating function for the associated Legendre functions is useful. It is found from that for the Legendre polynomials to be

$$T_{|m|}(z, t) \equiv \sum_{l=|m|}^{\infty} P_l^{|m|}(z)t^l \equiv \frac{(2|m|)!(1 - z^2)^{|m|/2}t^{|m|}}{2^{|m|}(|m|)!(1 - 2zt + t^2)^{|m|+\frac{1}{2}}}.$$
$$(19\text{–}10)$$

[1] The identification is completed by the fact that both functions are formed from polynomials of the same degree.

In Appendix VI it is shown that

$$\int_{-1}^{+1} P_l^{|m|}(z)P_{l'}^{|m|}(z)dz = \begin{cases} 0 \text{ for } l' \neq l, \\ \dfrac{2}{(2l+1)} \dfrac{(l+|m|)!}{(l-|m|)!} \text{ for } l' = l. \end{cases} \quad (19\text{--}11)$$

Using this result, we obtain the constant necessary to normalize the part of the wave function which depends on ϑ. The final form for $\Theta(\vartheta)$ is

$$\Theta(\vartheta) = \sqrt{\frac{(2l+1)}{2} \frac{(l-|m|)!}{(l+|m|)!}} P_l^{|m|}(\cos \vartheta). \quad (19\text{--}12)$$

Problem 19-1. Prove that the definition of the Legendre polynomials

$$\left. \begin{array}{c} P_0(z) = 1, \\ P_l(z) = \dfrac{1}{2^l l!} \dfrac{d^l(z^2-1)^l}{dz^l}, \quad l = 1, 2, \cdots, \end{array} \right\} \quad (19\text{--}13)$$

is equivalent to that of Equation 19-1.

Problem 19-2. Derive the following relations involving the associated Legendre functions:

$$(1-z^2)^{\frac{1}{2}}P_l^{|m|-1}(z) = \frac{1}{(2l+1)}P_{l+1}^{|m|}(z) - \frac{1}{(2l+1)}P_{l-1}^{|m|}(z), \quad (19\text{--}14)$$

$$(1-z^2)^{\frac{1}{2}}P_l^{|m|+1}(z) = \frac{(l+|m|)(l+|m|+1)}{(2l+1)}P_{l-1}^{|m|}(z) -$$
$$\frac{(l-|m|)(l-|m|+1)}{(2l+1)}P_{l+1}^{|m|}(z), \quad (19\text{--}15)$$

and

$$zP_l^{|m|}(z) = \frac{(l+|m|)}{(2l+1)}P_{l-1}^{|m|}(z) + \frac{(l-|m|+1)}{(2l+1)}P_{l+1}^{|m|}(z). \quad (19\text{--}16)$$

20. THE LAGUERRE POLYNOMIALS AND ASSOCIATED LAGUERRE FUNCTIONS

20a. The Laguerre Polynomials.—The Laguerre polynomials of a variable ρ, within the limits $0 \leqslant \rho \leqslant \infty$, may be defined by means of the generating function

$$U(\rho, u) \equiv \sum_{r=0}^{\infty} \frac{L_r(\rho)}{r!}u^r \equiv \frac{e^{-\frac{\rho u}{1-u}}}{1-u}. \quad (20\text{--}1)$$

To find the differential equation satisfied by these polynomials $L_r(\rho)$, we follow the now familiar procedure of differentiating the generating function with respect to u and to ρ. From $\dfrac{\partial U}{\partial u}$ we obtain

$$\sum_r \frac{L_r(\rho)}{(r-1)!} u^{r-1} = \frac{e^{-\frac{\rho u}{1-u}}}{1-u}\left(-\frac{\rho}{1-u} - \frac{\rho u}{(1-u)^2} + \frac{1}{1-u}\right)$$

or

$$(1 - 2u + u^2)\sum_r \frac{L_r(\rho)}{(r-1)!} u^{r-1} = (1 - u - \rho)\sum_r \frac{L_r(\rho)}{r!} u^r,$$

from which there results the recursion formula

$$L_{r+1}(\rho) + (\rho - 1 - 2r)L_r(\rho) + r^2 L_{r-1}(\rho) = 0. \quad (20\text{-}2)$$

Similarly from $\dfrac{\partial U}{\partial \rho}$ we have

$$\sum_r \frac{L_r'(\rho)}{r!} u^r = -\frac{u}{1-u}\sum_r \frac{L_r(\rho)}{r!} u^r,$$

or

$$L_r'(\rho) - rL_{r-1}'(\rho) + rL_{r-1}(\rho) = 0, \quad (20\text{-}3)$$

in which the prime denotes the derivative with respect to ρ. Equation 20-3 may be rewritten and differentiated, giving

$$L_{r+1}'(\rho) = (r + 1)\{L_r'(\rho) - L_r(\rho)\}$$

and

$$L_{r+1}''(\rho) = (r + 1)\{L_r''(\rho) - L_r'(\rho)\},$$

with similar equations for $L_{r+2}'(\rho)$ and $L_{r+2}''(\rho)$. Replacing r by $r + 1$ in Equation 20-2 and differentiating twice, we obtain the equation

$$L_{r+2}''(\rho) + (\rho - 3 - 2r)L_{r+1}''(\rho) + (r + 1)^2 L_r''(\rho) + 2L_{r+1}'(\rho) = 0.$$

With the aid of the foregoing expressions this is then transformed into an equation in $L_r(\rho)$ alone,

$$\rho L_r''(\rho) + (1 - \rho)L_r'(\rho) + rL_r(\rho) = 0, \quad (20\text{-}4)$$

which is the differential equation for the rth Laguerre polynomial.

Problem 20-1. Show that $L_r(\rho) = e^\rho \dfrac{d^r}{d\rho^r}(\rho^r e^{-\rho})$.

20b. The Associated Laguerre Polynomials and Functions.—
The sth derivative of the rth Laguerre polynomial is called the
associated Laguerre polynomial of degree $r - s$ and order s:

$$L_r^s(\rho) = \frac{d^s}{d\rho^s} L_r(\rho). \qquad (20\text{--}5)$$

The differential equation satisfied by $L_r^s(\rho)$ is found by differentiating Equation 20–4 to be

$$\rho L_r^s{}''(\rho) + (s + 1 - \rho)L_r^s{}'(\rho) + (r - s)L_r^s(\rho) = 0. \qquad (20\text{--}6)$$

If we now replace r by $n + l$ and s by $2l + 1$, Equation 20–6
becomes

$$\rho L_{n+l}^{2l+1}{}''(\rho) + \{2(l + 1) - \rho\}L_{n+l}^{2l+1}{}'(\rho)$$
$$+ (n - l - 1)L_{n+l}^{2l+1}(\rho) = 0. \qquad (20\text{--}7)$$

On comparing this with Equation 18–37 obtained in the treatment of the r equation for the hydrogen atom by the polynomial
method, we see that the two equations are identical when
$L_{n+l}^{2l+1}(\rho)$ is identified with $L(\rho)$ and the parameter λ is replaced
by its characteristic value n. The polynomials obtained in the
solution of the r equation for the hydrogen atom are hence the
associated Laguerre polynomials of degree $n - l - 1$ and of
order $2l + 1$. Moreover, the wave functions in r are, except for
normalizing factors, the functions

$$e^{-\frac{\rho}{2}} \rho^l L_{n+l}^{2l+1}(\rho).$$

These functions are called the *associated Laguerre functions.*
We shall discuss them in detail in succeeding sections.

It is easily shown from Equation 20–1 that the generating
function for the associated Laguerre polynomials of order s is[1]

$$U_s(\rho, u) \equiv \sum_{r=s}^{\infty} \frac{L_r^s(\rho)}{r!} u^r \equiv (-1)^s \frac{e^{-\frac{\rho u}{1-u}}}{(1 - u)^{s+1}} u^s. \qquad (20\text{--}8)$$

The polynomials can also be expressed explicitly:

$$L_{n+l}^{2l+1}(\rho) = \sum_{k=0}^{n-l-1} (-1)^{k+1} \frac{\{(n + l)!\}^2}{(n - l - 1 - k)!(2l + 1 + k)!k!} \rho^k.$$
$$(20\text{--}9)$$

[1] This was given by Schrödinger in his third paper, *Ann. d. Phys* **80,** **485**
(1926).

In Appendix VII, it is shown that the normalization integral
for the associated Laguerre function has the value

$$\int_0^\infty e^{-\rho}\rho^{2l}\{L_{n+l}^{2l+1}(\rho)\}^2\rho^2 d\rho = \frac{2n\{(n+l)!\}^3}{(n-l-1)!}, \quad (20\text{--}10)$$

the factor ρ^2 arising from the volume element in polar coordinates.
From this it follows that the normalized radial factor of the wave
function for the hydrogen atom is

$$R_{nl}(r) = -\sqrt{\left(\frac{2Z}{na_0}\right)^3 \frac{(n-l-1)!}{2n\{(n+l)!\}^3}}e^{-\frac{\rho}{2}}\rho^l L_{n+l}^{2l+1}(\rho), \quad (20\text{--}11)$$

with

$$\rho = 2\alpha r = \frac{8\pi^2\mu Z e^2}{nh^2}r = \frac{2Z}{na_0}r. \quad (20\text{--}12)$$

Problem 20–2. Derive relations for the associated Laguerre polynomials
and functions corresponding to those of Equations 20–2 and 20–3.

21. THE WAVE FUNCTIONS FOR THE HYDROGEN ATOM

21a. Hydrogenlike Wave Functions.—We have now found
the wave functions for the discrete stationary states of a one-
electron or hydrogenlike atom. They are

$$\psi_{nlm}(r, \vartheta, \varphi) = R_{nl}(r)\Theta_{lm}(\vartheta)\Phi_m(\varphi), \quad (21\text{--}1)$$

with

$$\Phi_m(\varphi) = \frac{1}{\sqrt{2\pi}}e^{im\varphi}, \quad (21\text{--}2)$$

$$\Theta_{lm}(\vartheta) = \left\{\frac{(2l+1)(l-|m|)!}{2(l+|m|)!}\right\}^{\frac{1}{2}}P_l^{|m|}(\cos\vartheta), \quad (21\text{--}3)$$

and

$$R_{nl}(r) = -\left[\left(\frac{2Z}{na_0}\right)^3 \frac{(n-l-1)!}{2n\{(n+l)!\}^3}\right]^{\frac{1}{2}}e^{-\frac{\rho}{2}}\rho^l L_{n+l}^{2l+1}(\rho), \quad (21\text{--}4)$$

in which

$$\rho = \frac{2Z}{na_0}r \quad (21\text{--}5)$$

and

$$a_0 = \frac{h^2}{4\pi^2\mu e^2},$$

a_0 being the quantity interpreted in the old quantum theory as
the radius of the smallest orbit in the hydrogen atom. The

functions $P_l^{|m|}(\cos\vartheta)$ are the associated Legendre functions discussed in Section 19, and the functions $L_{n+l}^{2l+1}(\rho)$ are the associated Laguerre polynomials of Section 20. The minus sign in Equation 21–4 is introduced for convenience to make the function positive for small values of r.

The wave functions as written here are normalized, so that

$$\int_0^\infty \int_0^\pi \int_0^{2\pi} \psi_{nlm}^*(r,\vartheta,\varphi)\psi_{nlm}(r,\vartheta,\varphi)r^2\sin\vartheta d\varphi d\vartheta dr = 1. \quad (21\text{–}6)$$

Moreover, the functions in r, ϑ, and φ are separately normalized to unity:

$$\left.\begin{aligned}
\int_0^{2\pi} \Phi_m^*(\varphi)\Phi_m(\varphi)d\varphi &= 1,\\
\int_0^\pi \{\Theta_{lm}(\vartheta)\}^2\sin\vartheta d\vartheta &= 1,\\
\int_0^\infty \{R_{nl}(r)\}^2 r^2 dr &= 1.
\end{aligned}\right\} \quad (21\text{–}7)$$

They are also mutually orthogonal, the integral

$$\int_0^\infty \int_0^\pi \int_0^{2\pi} \psi_{nlm}^*(r,\vartheta,\varphi)\psi_{n'l'm'}(r,\vartheta,\varphi)r^2\sin\vartheta d\varphi d\vartheta dr$$

vanishing except for $n = n'$, $l = l'$, and $m = m'$; inasmuch as if $m \neq m'$, the integral in φ vanishes; if $m = m'$, but $l \neq l'$, the integral in ϑ vanishes; and if $m = m'$ and $l = l'$, but $n \neq n'$, the integral in r vanishes.

Expressions for the normalized wave functions for all sets of quantum numbers out to $n = 6$, $l = 5$ are given in Tables 21–1, 21–2, and 21–3.

The functions $\Phi_m(\varphi)$ are given in both the complex and the real form, either set being satisfactory. (For some purposes one is more convenient, for others the other.)

TABLE 21–1.—THE FUNCTIONS $\Phi_m(\varphi)$

$$\Phi_0(\varphi) = \frac{1}{\sqrt{2\pi}} \qquad \text{or} \qquad \Phi_0(\varphi) = \frac{1}{\sqrt{2\pi}}$$

$$\Phi_1(\varphi) = \frac{1}{\sqrt{2\pi}}e^{i\varphi} \qquad \text{or} \qquad \Phi_{1\cos}(\varphi) = \frac{1}{\sqrt{\pi}}\cos\varphi$$

$$\Phi_{-1}(\varphi) = \frac{1}{\sqrt{2\pi}}e^{-i\varphi} \qquad \text{or} \qquad \Phi_{1\sin}(\varphi) = \frac{1}{\sqrt{\pi}}\sin\varphi$$

TABLE 21-1.—THE FUNCTIONS $\Phi_m(\varphi)$.—(Continued)

$$\Phi_2(\varphi) = \frac{1}{\sqrt{2\pi}}e^{i2\varphi} \quad \text{or} \quad \Phi_{2\cos}(\varphi) = \frac{1}{\sqrt{\pi}}\cos 2\varphi$$

$$\Phi_{-2}(\varphi) = \frac{1}{\sqrt{2\pi}}e^{-i2\varphi} \quad \text{or} \quad \Phi_{2\sin}(\varphi) = \frac{1}{\sqrt{\pi}}\sin 2\varphi$$

Etc.

TABLE 21-2.—THE WAVE FUNCTIONS $\Theta_{lm}(\vartheta)$
(The associated Legendre functions normalized to unity)

$l = 0$, s orbitals:

$$\Theta_{00}(\vartheta) = \frac{\sqrt{2}}{2}$$

$l = 1$, p orbitals:

$$\Theta_{10}(\vartheta) = \frac{\sqrt{6}}{2}\cos\vartheta$$

$$\Theta_{1\pm1}(\vartheta) = \frac{\sqrt{3}}{2}\sin\vartheta$$

$l = 2$, d orbitals:

$$\Theta_{20}(\vartheta) = \frac{\sqrt{10}}{4}(3\cos^2\vartheta - 1)$$

$$\Theta_{2\pm1}(\vartheta) = \frac{\sqrt{15}}{2}\sin\vartheta\cos\vartheta$$

$$\Theta_{2\pm2}(\vartheta) = \frac{\sqrt{15}}{4}\sin^2\vartheta$$

$l = 3$, f orbitals:

$$\Theta_{30}(\vartheta) = \frac{3\sqrt{14}}{4}\left(\frac{5}{3}\cos^3\vartheta - \cos\vartheta\right)$$

$$\Theta_{3\pm1}(\vartheta) = \frac{\sqrt{42}}{8}\sin\vartheta(5\cos^2\vartheta - 1)$$

$$\Theta_{3\pm2}(\vartheta) = \frac{\sqrt{105}}{4}\sin^2\vartheta\cos\vartheta$$

$$\Theta_{3\pm3}(\vartheta) = \frac{\sqrt{70}}{8}\sin^3\vartheta$$

$l = 4$, g orbitals:

$$\Theta_{40}(\vartheta) = \frac{9\sqrt{2}}{16}\left(\frac{35}{3}\cos^4\vartheta - 10\cos^2\vartheta + 1\right)$$

$$\Theta_{4\pm1}(\vartheta) = \frac{9\sqrt{10}}{8}\sin\vartheta\left(\frac{7}{3}\cos^3\vartheta - \cos\vartheta\right)$$

TABLE 21-2.—THE WAVE FUNCTIONS $\Theta_{lm}(\vartheta)$.—(*Continued*)

$$\Theta_{4\pm2}(\vartheta) = \frac{3\sqrt{5}}{8} \sin^2 \vartheta (7 \cos^2 \vartheta - 1)$$

$$\Theta_{4\pm3}(\vartheta) = \frac{3\sqrt{70}}{8} \sin^3 \vartheta \cos \vartheta$$

$$\Theta_{4\pm4}(\vartheta) = \frac{3\sqrt{35}}{16} \sin^4 \vartheta$$

$l = 5$, h orbitals:

$$\Theta_{50}(\vartheta) = \frac{15\sqrt{22}}{16}\left(\frac{21}{5} \cos^5 \vartheta - \frac{14}{3} \cos^3 \vartheta + \cos \vartheta\right)$$

$$\Theta_{5\pm1}(\vartheta) = \frac{\sqrt{165}}{16} \sin \vartheta (21 \cos^4 \vartheta - 14 \cos^2 \vartheta + 1)$$

$$\Theta_{5\pm2}(\vartheta) = \frac{\sqrt{1155}}{8} \sin^2 \vartheta (3 \cos^3 \vartheta - \cos \vartheta)$$

$$\Theta_{5\pm3}(\vartheta) = \frac{\sqrt{770}}{32} \sin^3 \vartheta (9 \cos^2 \vartheta - 1)$$

$$\Theta_{5\pm4}(\vartheta) = \frac{3\sqrt{385}}{16} \sin^4 \vartheta \cos \vartheta$$

$$\Theta_{5\pm5}(\vartheta) = \frac{3\sqrt{154}}{32} \sin^5 \vartheta$$

TABLE 21-3.—THE HYDROGENLIKE RADIAL WAVE FUNCTIONS

$n = 1$, K shell:

$l = 0$, $1s$ $R_{10}(r) = (Z/a_0)^{3/2} \cdot 2e^{-\frac{\rho}{2}}$

$n = 2$, L shell:

$l = 0$, $2s$ $R_{20}(r) = \dfrac{(Z/a_0)^{3/2}}{2\sqrt{2}}(2 - \rho)e^{-\frac{\rho}{2}}$

$l = 1$, $2p$ $R_{21}(r) = \dfrac{(Z/a_0)^{3/2}}{2\sqrt{6}}\rho e^{-\frac{\rho}{2}}$

$n = 3$, M shell:

$l = 0$, $3s$ $R_{30}(r) = \dfrac{(Z/a_0)^{3/2}}{9\sqrt{3}}(6 - 6\rho + \rho^2)e^{-\frac{\rho}{2}}$

$l = 1$, $3p$ $R_{31}(r) = \dfrac{(Z/a_0)^{3/2}}{9\sqrt{6}}(4 - \rho)\rho e^{-\frac{\rho}{2}}$

$l = 2$, $3d$ $R_{32}(r) = \dfrac{(Z/a_0)^{3/2}}{9\sqrt{30}}\rho^2 e^{-\frac{\rho}{2}}$

TABLE 21-3.—THE HYDROGENLIKE RADIAL WAVE FUNCTIONS.—(*Continued*)

$n = 4$, N shell:

$$l = 0, \ 4s \quad R_{40}(r) = \frac{(Z/a_0)^{3/2}}{96}(24 - 36\rho + 12\rho^2 - \rho^3)e^{-\frac{\rho}{2}}$$

$$l = 1, \ 4p \quad R_{41}(r) = \frac{(Z/a_0)^{3/2}}{32\sqrt{15}}(20 - 10\rho + \rho^2)\rho e^{-\frac{\rho}{2}}$$

$$l = 2, \ 4d \quad R_{42}(r) = \frac{(Z/a_0)^{3/2}}{96\sqrt{5}}(6 - \rho)\rho^2 e^{-\frac{\rho}{2}}$$

$$l = 3, \ 4f \quad R_{43}(r) = \frac{(Z/a_0)^{3/2}}{96\sqrt{35}}\rho^3 e^{-\frac{\rho}{2}}$$

$n = 5$, O shell:

$$l = 0, \ 5s \quad R_{50}(r) = \frac{(Z/a_0)^{3/2}}{300\sqrt{5}}(120 - 240\rho + 120\rho^2 - 20\rho^3 + \rho^4)e^{-\frac{\rho}{2}}$$

$$l = 1, \ 5p \quad R_{51}(r) = \frac{(Z/a_0)^{3/2}}{150\sqrt{30}}(120 - 90\rho + 18\rho^2 - \rho^3)\rho e^{-\frac{\rho}{2}}$$

$$l = 2, \ 5d \quad R_{52}(r) = \frac{(Z/a_0)^{3/2}}{150\sqrt{70}}(42 - 14\rho + \rho^2)\rho^2 e^{-\frac{\rho}{2}}$$

$$l = 3, \ 5f \quad R_{53}(r) = \frac{(Z/a_0)^{3/2}}{300\sqrt{70}}(8 - \rho)\rho^3 e^{-\frac{\rho}{2}}$$

$$l = 4, \ 5g \quad R_{54}(r) = \frac{(Z/a_0)^{3/2}}{900\sqrt{70}}\rho^4 e^{-\frac{\rho}{2}}$$

$n = 6$, P shell:

$$l = 0, \ 6s \quad R_{60}(r) = \frac{(Z/a_0)^{3/2}}{2160\sqrt{6}}(720 - 1800\rho + 1200\rho^2 - 300\rho^3 + 30\rho^4 - \rho^5)e^{-\frac{\rho}{2}}$$

$$l = 1, \ 6p \quad R_{61}(r) = \frac{(Z/a_0)^{3/2}}{432\sqrt{210}}(840 - 840\rho + 252\rho^2 - 28\rho^3 + \rho^4)\rho e^{-\frac{\rho}{2}}$$

$$l = 2, \ 6d \quad R_{62}(r) = \frac{(Z/a_0)^{3/2}}{864\sqrt{105}}(336 - 168\rho + 24\rho^2 - \rho^3)\rho^2 e^{-\frac{\rho}{2}}$$

$$l = 3, \ 6f \quad R_{63}(r) = \frac{(Z/a_0)^{3/2}}{2592\sqrt{35}}(72 - 18\rho + \rho^2)\rho^3 e^{-\frac{\rho}{2}}$$

$$l = 4, \ 6g \quad R_{64}(r) = \frac{(Z/a_0)^{3/2}}{12960\sqrt{7}}(10 - \rho)\rho^4 e^{-\frac{\rho}{2}}$$

$$l = 5, \ 6h \quad R_{65}(r) = \frac{(Z/a_0)^{3/2}}{12960\sqrt{77}}\rho^5 e^{-\frac{\rho}{2}}$$

The wave functions $\Theta_{lm}(\vartheta)$ given in Table 21–2 are the associated Legendre functions $P_l^{|m|}$ (cos ϑ) normalized to unity. The functions $P_l^{|m|}$ (cos ϑ) as usually written and as defined by Equations 19–1 and 19–7 consist of the term $\sin^{|m|}\vartheta$ and the polynomial in cos ϑ multiplied by the factor

$$\frac{(l + |m|)!}{2^l\left(\dfrac{l + |m|}{2}\right)!\left(\dfrac{l - |m|}{2}\right)!} \quad \text{or} \quad \frac{(l + |m| + 1)!}{2^l\left(\dfrac{l + |m| + 1}{2}\right)!\left(\dfrac{(l - |m| - 1)}{2}\right)!},$$

as $m + l$ is even or odd. Expressions for additional associated Legendre functions are given in many books, as, for example, by Byerly.[1] Numerical tables for the Legendre polynomials are given by Byerly and by Jahnke and Emde.[2]

Following Mulliken, we shall occasionally refer to one-electron orbital wave functions such as the hydrogenlike wave functions of this chapter as *orbitals*. In accordance with spectroscopic practice, we shall also use the symbols s, p, d, f, g, $\cdots$ to refer to states characterized by the values 0, 1, 2, 3, 4, $\cdots$, respectively, of the azimuthal quantum number l, speaking, for example, of an s orbital to mean an orbital with $l = 0$.

In the table of hydrogenlike radial wave functions the polynomial contained in parentheses represents for each function the associated Laguerre polynomial $L_{n+l}^{2l+1}(\rho)$, as defined by Equations 20–1 and 20–5, except for the factor

$$-(n + l)!/(n - l - 1)!,$$

which has been combined with the normalizing factor and reduced to the simplest form. It is to be borne in mind that the variable ρ is related to r in different ways for different values of n.

The complete wave functions $\psi_{nlm}(r, \vartheta, \varphi)$ for the first three shells are given in Table 21–4. Here for convenience the variable $\rho = 2Zr/na_0$ has been replaced by the new variable σ, such that

$$\sigma = \frac{n}{2}\rho = \frac{Z}{a_0}r.$$

[1] W. E. Byerly, "Fourier's Series and Spherical Harmonics," pp. 151, 159, 198, Ginn and Company, Boston, 1893.

[2] W. E. Byerly, *ibid.*, pp. 278–281; Jahnke and Emde, "Funktionentafeln," B. G. Teubner, Leipzig, 1933.

The relation between σ and r is the same for all values of the quantum numbers. The real form of the φ functions is used. The symbols p_x, p_y, p_z, d_{x+y}, d_{y+z}, d_{x+z}, d_{xy}, and d_z are introduced for convenience. It is easily shown that the functions ψ_{np_x}, ψ_{np_y}, and ψ_{np_z} are identical except for orientation in space, the three being equivalently related to the x, y, and z axes, respectively. Similarly the four functions $\psi_{nd_{x+y}}$, $\psi_{nd_{y+z}}$, $\psi_{nd_{x+z}}$, and $\psi_{nd_{xy}}$ are identical except for orientation. The fifth d function ψ_{nd_z} is different.

TABLE 21-4.—HYDROGENLIKE WAVE FUNCTIONS

K Shell

$n = 1, l = 0, m = 0$:

$$\psi_{1s} = \frac{1}{\sqrt{\pi}} \left(\frac{Z}{a_0}\right)^{3/2} e^{-\sigma}$$

L Shell

$n = 2, l = 0, m = 0$:

$$\psi_{2s} = \frac{1}{4\sqrt{2\pi}} \left(\frac{Z}{a_0}\right)^{3/2} (2 - \sigma)e^{-\frac{\sigma}{2}}$$

$n = 2, l = 1, m = 0$:

$$\psi_{2p_z} = \frac{1}{4\sqrt{2\pi}} \left(\frac{Z}{a_0}\right)^{3/2} \sigma e^{-\frac{\sigma}{2}} \cos \vartheta$$

$n = 2, l = 1, m = \pm 1$:

$$\psi_{2p_x} = \frac{1}{4\sqrt{2\pi}} \left(\frac{Z}{a_0}\right)^{3/2} \sigma e^{-\frac{\sigma}{2}} \sin \vartheta \cos \varphi$$

$$\psi_{2p_y} = \frac{1}{4\sqrt{2\pi}} \left(\frac{Z}{a_0}\right)^{3/2} \sigma e^{-\frac{\sigma}{2}} \sin \vartheta \sin \varphi$$

M Shell

$n = 3, l = 0, m = 0$:

$$\psi_{3s} = \frac{1}{81\sqrt{3\pi}} \left(\frac{Z}{a_0}\right)^{3/2} (27 - 18\sigma + 2\sigma^2)e^{-\frac{\sigma}{3}}$$

$n = 3, l = 1, m = 0$:

$$\psi_{3p_z} = \frac{\sqrt{2}}{81\sqrt{\pi}} \left(\frac{Z}{a_0}\right)^{3/2} (6 - \sigma)\sigma e^{-\frac{\sigma}{3}} \cos \vartheta$$

$n = 3, l = 1, m = \pm 1$:

$$\psi_{3p_x} = \frac{\sqrt{2}}{81\sqrt{\pi}} \left(\frac{Z}{a_0}\right)^{3/2} (6 - \sigma)\sigma e^{-\frac{\sigma}{3}} \sin \vartheta \cos \varphi$$

TABLE 21-4.—HYDROGENLIKE WAVE FUNCTIONS.—(*Continued*)

$$\psi_{3p_y} = \frac{\sqrt{2}}{81\sqrt{\pi}}\left(\frac{Z}{a_0}\right)^{3/2}(6 - \sigma)\sigma e^{-\frac{\sigma}{3}}\sin\vartheta\sin\varphi$$

$n = 3, l = 2, m = 0$:

$$\psi_{3d_z} = \frac{1}{81\sqrt{6\pi}}\left(\frac{Z}{a_0}\right)^{3/2}\sigma^2 e^{-\frac{\sigma}{3}}(3\cos^2\vartheta - 1)$$

$n = 3, l = 2, m = \pm1$:

$$\psi_{3d_{z+s}} = \frac{\sqrt{2}}{81\sqrt{\pi}}\left(\frac{Z}{a_0}\right)^{3/2}\sigma^2 e^{-\frac{\sigma}{3}}\sin\vartheta\cos\vartheta\cos\varphi$$

$$\psi_{3d_{y+s}} = \frac{\sqrt{2}}{81\sqrt{\pi}}\left(\frac{Z}{a_0}\right)^{3/2}\sigma^2 e^{-\frac{\sigma}{3}}\sin\vartheta\cos\vartheta\sin\varphi$$

$n = 3, l = 2, m = \pm2$:

$$\psi_{3d_{xy}} = \frac{1}{81\sqrt{2\pi}}\left(\frac{Z}{a_0}\right)^{3/2}\sigma^2 e^{-\frac{\sigma}{3}}\sin^2\vartheta\cos 2\varphi$$

$$\psi_{3d_{x+y}} = \frac{1}{81\sqrt{2\pi}}\left(\frac{Z}{a_0}\right)^{3/2}\sigma^2 e^{-\frac{\sigma}{3}}\sin^2\vartheta\sin 2\varphi$$

with $\sigma = \dfrac{Z}{a_0}r$.

21b. The Normal State of the Hydrogen Atom.

—The properties of the hydrogen atom in its normal state ($1s$, with $n = 1$, $l = 0$, $m = 0$) are determined by the wave function

$$\psi_{100} = \frac{1}{\sqrt{\pi a_0^3}}e^{-\frac{r}{a_0}}.$$

The physical interpretation postulated for the wave function requires that $\psi^*\psi = \dfrac{1}{\pi a_0^3}e^{-\frac{2r}{a_0}}$ be a probability distribution function for the electron relative to the nucleus. Since this expression is independent of ϑ and φ, the normal hydrogen atom is spherically symmetrical. The chance that the electron be in the volume element $r^2 dr \sin\vartheta d\vartheta d\varphi$ is $\dfrac{1}{\pi a_0^3}e^{-\frac{2r}{a_0}}r^2 dr \sin\vartheta d\vartheta d\varphi$, which is seen to be independent of ϑ and φ for a given size of the volume element. This spherical symmetry is a property not possessed by the normal Bohr atom, for the Bohr orbit was restricted to a single plane.

By integrating over ϑ and φ (over the surface of a sphere), we obtain the expression

$$D(r)dr = \frac{4}{a_0^3}r^2e^{-\frac{2r}{a_0}}dr$$

as the probability that the electron lie between the distances r and $r + dr$ from the nucleus. The *radial distribution function* $D_{100}(r) = \frac{4}{a_0^3}r^2e^{-\frac{2r}{a_0}}$ is shown in Figure 21–1 (together with ψ_{100} and ψ_{100}^2) as a function of r, the distance from the nucleus. It

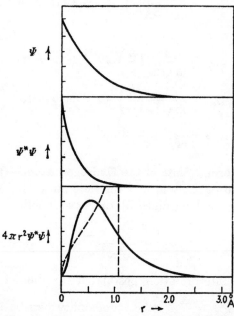

Fig. 21–1.—The functions ψ, $\psi^*\psi$, and $4\pi r^2\psi^*\psi$ for the normal hydrogen atom. The dashed curve represents the probability distribution function for a Bohr orbit.

is seen that the probability that the electron remain within about 1 Å of the nucleus is large; that is, the "size" of the hydrogen atom is about the same as given by the Bohr theory. Indeed, there is a close relation; the most probable distance of the electron from the nucleus, which is the value of r at which $D(r)$ has its maximum value, is seen from Figure 21–1 to be $a_0 = 0.529$Å, which is just the radius of the normal Bohr orbit for hydrogen.

The distribution function itself is not at all similar to that for a circular Bohr orbit of radius a_0, which would be zero everywhere except at the point $r = a_0$. The function ψ_{100}^2 has its maximum value at $r = 0$, showing that the most probable position for the electron is in the immediate neighborhood of the nucleus; that is, the chance that the electron lie in a small volume element very near the nucleus is larger than the chance that it lie in a volume element of the same size at a greater distance from the nucleus.[1] It may be pointed out that a Bohr orbit in the form of a degenerate line ellipse, obtained by giving the azimuthal quantum number k of the old quantum theory the value 0 instead of the value 1, leads to a distribution function resembling the wave-mechanical one a little more closely. This is shown in Figure 21-1 by the dashed curve. The average distance of the electron from the nucleus, given by the equation

$$\bar{r}_{nlm} = \int\int\int \psi_{nlm}^* r \psi_{nlm} r^2 dr \sin \vartheta d\vartheta d\varphi, \qquad (21\text{-}8)$$

is found in this case to be equal to $\frac{3}{2}a_0$. This is also the value calculated for the Bohr orbit with $k = 0$; in fact, it will be shown in the next section that for any stationary state of the hydrogen atom the average value of r as given by the quantum mechanics is the same as for the Bohr orbit with the same value of n and with k^2 equal to $l(l + 1)$. It will also be shown in Chapter XV that the normal hydrogen atom has no orbital angular momentum. This corresponds to a Bohr orbit with $k = 0$ but not with $k = 1$. The root-mean-square linear momentum of the electron is shown in the next section to have the value $2\pi\mu e^2/h$, which is the same as for the Bohr orbit. We may accordingly form a rough picture of the normal hydrogen atom as consisting of an electron moving about a nucleus in somewhat the way corresponding to the Bohr orbit with $n = 1$, $k = 0$, the motion being essentially radial (with no angular momentum), the amplitude of the motion being sufficiently variable to give rise to a radial distribution function $D(r)$ extending to infinity, though falling off rapidly with increasing r outside of a radius of 1 or 2Å, the speed of the electron being about the same as in the lowest Bohr orbit, and the orientation of the orbit being

---

[1] The difference between the statement of the preceding paragraph and this statement is the result of the increase in size of the volume element $4\pi r^2 dr$ for the former case with increasing r.

sufficiently variable to make the atom spherically symmetrical. Great significance should not be attached to such a description. We shall, however, make continued use of the comparison of wave-mechanical calculations for the hydrogen atom with the corresponding calculations for Bohr orbits for the sake of convenience.

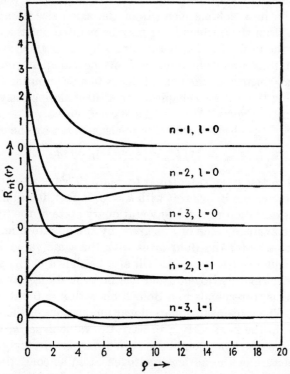

Fig. 21–2.—Hydrogen-atom radial wave functions $R_{nl}(r)$ for $n = 1, 2,$ and 3 and $l = 0$ and 1.

21c. Discussion of the Hydrogenlike Radial Wave Functions. The radial wave functions $R_{nl}(r)$ for $n = 1, 2,$ and 3 and $l = 0$ and 1 are shown plotted in Figure 21–2. The abscissas represent values of ρ; hence the horizontal scale should be increased by the factor n in order to show $R(r)$ as functions of the electron-nucleus distance r. It will be noticed that only for s states (with $l = 0$) is the wave function different from zero at $r = 0$. The wave function crosses the ρ axis $n - l - 1$ times in the region between $\rho = 0$ and $\rho = \infty$.

The radial distribution function

$$D_{nl}(r) = r^2\{R_{nl}(r)\}^2 \qquad (21\text{-}9)$$

is represented as a function of ρ for the same states in Figure 21-3. It is seen from Figures 21-2 and 21-3 that the probability distribution function $\psi^*\psi$, which is spherically symmetrical for s states, falls off for these states from a maximum value at $r = 0$. We might say that over a period of time the electron

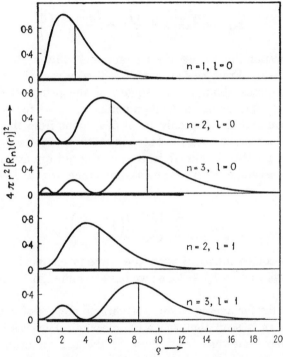

Fig. 21-3.—Electron distribution functions $4\pi r^2[R_{nl}(r)]^2$ for the hydrogen atom

may be considered in a hydrogen atom in the normal state to form a ball about the nucleus, in the $2s$ state to form a ball and an outer shell, in the $3s$ state to form a ball and two concentric shells, etc. The region within which the radial distribution function differs largely from zero is included between the values of r at perihelion and aphelion for the Bohr orbit with the same value of n and with $k^2 = l(l + 1)$, as is shown by the heavy horizontal line for each curve in Figure 21-3, drawn between the minimum and maximum values of the electron-nucleus distance

for this Bohr orbit in each case. For these s orbits (with $k = 0$)
the heavy line extends to $r = 0$, corresponding to a line ellipse
with vanishingly small minor axis, in agreement with the large
value of $\psi^*\psi$ at $r = 0$. For states with $l > 0$, on the other hand,
$\psi^*\psi$ vanishes at $r = 0$, and similarly the minimum value of r
for the Bohr orbits with $k = \sqrt{l(l + 1)}$ is greater than zero.

The average distance of the electron from the nucleus, as given
by Equation 21–8, is found on evaluating the integral to be

$$\bar{r}_{nlm} = \frac{n^2 a_0}{Z}\left[1 + \frac{1}{2}\left\{1 - \frac{l(l + 1)}{n^2}\right\}\right] \qquad (21\text{–}10)$$

The corresponding values of ρ are represented by vertical lines in
Figure 21–3.   From this expression it is seen that the size of the
atom increases about as the square of the principal quantum
number n, $\bar{r}_{nlm}$ being in fact proportional to n^2 for the states
with $l = 0$ and showing only small deviations from this proportionality for other states.   This variation of size of orbit with
quantum number is similar to that of the old quantum theory,
the time-average electron-nucleus distance for a Bohr orbit
being

$$\bar{r}_{nk} = \frac{n^2 a_0}{Z}\left\{1 + \frac{1}{2}\left(1 - \frac{k^2}{n^2}\right)\right\}, \qquad (21\text{–}11)$$

which becomes identical with the wave-mechanical expression
if k^2 is replaced by $l(l + 1)$, as we have assumed in the foregoing
discussion.

Formulas for average values of various powers of r are given
below.[1] It is seen that the wave-mechanical expressions as a
rule differ somewhat from those of the old quantum theory,
even when k^2 is replaced by $l(l + 1)$.

AVERAGE VALUES* OF r^s

Wave Mechanics

$$\overline{r^2} = \frac{a_0^2 n^4}{Z^2}\left[1 + \frac{3}{2}\left\{1 - \frac{l(l + 1) - \frac{1}{3}}{n^2}\right\}\right]$$

* Expressions for $\bar{r}$ are given in Equations 21–10 and 21–11.

---

[1] I. WALLER, *Z. f. Phys.* **38**, 635 (1926); expressions for $\overline{\left(\dfrac{1}{r^5}\right)}$ and $\overline{\left(\dfrac{1}{r^6}\right)}$
are given by J. H. Van Vleck, *Proc. Roy. Soc.* **A 143**, 679 (1934).

AVERAGE VALUES OF r^s.—(*Continued*)

$$\overline{\left(\frac{1}{r}\right)} = \frac{Z}{a_0 n^2}$$

$$\overline{\left(\frac{1}{r^2}\right)} = \frac{Z^2}{a_0^2 n^3 (l + \frac{1}{2})}$$

$$\overline{\left(\frac{1}{r^3}\right)} = \frac{Z^3}{a_0^3 n^3 l (l + \frac{1}{2})(l + 1)}$$

$$\overline{\left(\frac{1}{r^4}\right)} = \frac{\frac{3}{2} Z^4 \left\{ 1 - \frac{l(l + 1)}{3n^2} \right\}}{a_0^4 n^3 (l + \frac{3}{2})(l + 1)(l + \frac{1}{2}) l (l - \frac{1}{2})}$$

Old Quantum Theory

$$\overline{r^2} = \frac{a_0^2 n^4}{Z^2} \left\{ 1 + \frac{3}{2} \left(1 - \frac{k^2}{n^2} \right) \right\}$$

$$\overline{\left(\frac{1}{r}\right)} = \frac{Z}{a_0 n^2}$$

$$\overline{\left(\frac{1}{r^2}\right)} = \frac{Z^2}{a_0^2 n^3 k}$$

$$\overline{\left(\frac{1}{r^3}\right)} = \frac{Z^3}{a_0^3 n^3 k^3}$$

$$\overline{\left(\frac{1}{r^4}\right)} = \frac{\frac{3}{2} Z^4 \left(1 - \frac{k^2}{3n^2} \right)}{a_0^4 n^3 k^5}$$

To illustrate the use of these formulas, let us calculate the average potential energy of the electron in the field of the nucleus. It is

$$\bar{V}_{nlm} = - \int \int \int \psi_{nlm}^* \frac{Ze^2}{r} \psi_{nlm} r^2 dr \sin \vartheta d\vartheta d\varphi$$

$$= -Ze^2 \overline{\left(\frac{1}{r}\right)}_{nlm}$$

$$= -\frac{Z^2 e^2}{a_0 n^2}. \qquad\qquad (21\text{--}12)$$

Now the total energy W, which is the sum of the average kinetic energy $\bar{T}$ and the average potential energy $\bar{V}$, is equal to $-Z^2 e^2 / 2a_0 n^2$. Hence we have shown that the total energy is just one-half of the average potential energy, and that the average kinetic energy is equal to the total energy with the sign changed, i.e.,

$$\bar{T}_{nlm} = \frac{Z^2 e^2}{2a_0 n^2}.$$ (21-13)

This relation connecting the average potential energy, the average kinetic energy, and the total energy for a system of particles with Coulomb interaction holds also in classical mechanics, being there known as the *virial theorem* (Sec. 7a).

Now we may represent the kinetic energy as

$$T = \frac{1}{2\mu}(p_x^2 + p_y^2 + p_z^2),$$

in which p_x, p_y, and p_z represent components of linear momentum of the electron and nucleus relative to the center of mass (that is, the components of linear momentum of the electron alone if the small motion of the nucleus be neglected). Hence the average value of the square of the total linear momentum $p^2 = p_x^2 + p_y^2 + p_z^2$ is equal to 2μ times the average value of the kinetic energy, which is itself given by Equation 21–13 for both wave mechanics and old quantum theory. We thus obtain

$$\overline{p_{nlm}^2} = \frac{2\mu Z^2 e^2}{2a_0 n^2} = \left(\frac{2\pi Z \mu e^2}{nh}\right)^2$$ (21-14)

as the equation representing the average squared linear momentum for a hydrogenlike atom in the wave mechanics as well as in the old quantum theory. This corresponds to a root-mean-square speed of the electron of

$$\sqrt{\overline{v_{nlm}^2}} = \frac{2\pi Z e^2}{nh},$$ (21-15)

which for the normal hydrogen atom has the value 2.185×10^8 cm/sec.

Problem 21–1. Using recursion formulas similar to Equation 20–2 (or in some other way) derive the expression for $\bar{r}_{nlm}$.

21d. Discussion of the Dependence of the Wave Functions on the Angles ϑ and φ.—In discussing the angular dependence of hydrogenlike wave functions, we shall first choose the complex form of the functions $\Phi(\varphi)$ rather than the real form. It will be shown in Chapter XV that there is a close analogy between the stationary states represented by these wave functions and the Bohr orbits of the old quantum theory in regard to the orbital angular momentum of the electron about the nucleus. The square of the total angular momentum for a given value of l is $l(l + 1)\dfrac{h^2}{4\pi^2}$, and the component of angular momentum along the z axis is $mh/2\pi$, whereas the corresponding values for a Bohr orbit with quantum numbers nkm are $k^2h^2/4\pi^2$ and $mh/2\pi$, respectively. We interpret the wave functions with a given value of l and different values of m as representing states in which the total angular momentum is the same, but with different orientations in space.

It can be shown by a simple extension of the wave equation to include electromagnetic phenomena (a subject which will not be discussed in this book) that the magnetic moment associated with the orbital motion of an electron is obtained from the orbital angular momentum by multiplication by the factor $e/2m_0c$, just as in the classical and old quantum theory (Sec. 7d). The component of orbital magnetic moment along the z axis is hence $m\dfrac{he}{4\pi m_0 c}$, and the energy of magnetic interaction of this moment with a magnetic field of strength H parallel to the z axis is $m\dfrac{he}{4\pi m_0 c}H$.

In the old quantum theory this spatial quantization was supposed to determine the plane of the orbit relative to the fixed direction of the z axis, the plane being normal to the z axis for $m = \pm k$ and inclined at various angles for other values of m. We may interpret the probability distribution function $\psi^*\psi$ in a similar manner. For example, in the states with $m = \pm l$ the component of angular momentum along the z axis, $mh/2\pi$, is nearly equal to the total angular momentum, $\sqrt{l(l + 1)}\,h/2\pi$, so that, by analogy with the Bohr orbit whose plane would be nearly normal to the z axis, we expect the probability distribution function to be large at $\vartheta = 90°$ and small at $\vartheta = 0°$ and $180°$. This is found to be the case, as is shown in Figure 21–4, in which there is represented the function $\{\Theta_{lm}(\vartheta)\}^2$ for $m = \pm l$ and for $l = 0, 1, 2, 3, 4,$ and 5. It is seen that as l increases the probability distribution function becomes more and more concentrated about the xy plane.

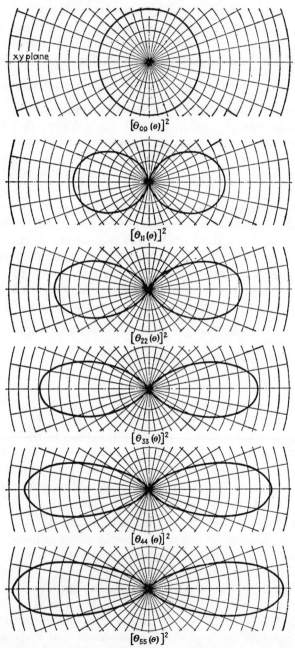

FIG. 21-4.—Polar graphs of the function $[\Theta_{lm}(\vartheta)]^2$ for $m = \pm l$ and $l = 0$, 1, 2, 3, 4, and 5, showing the concentration of the function about the xy plane with increasing l.

The behavior of the distribution function for other values of m is similarly shown in Figure 21–5, representing the same function for $l = 3$ and $m = 0$, ± 1, ± 2, ± 3. It is seen that the function tends to be concentrated in directions corresponding to the plane of the oriented Bohr orbit (this plane being determined only to the extent that its angle with the z axis is fixed). With the complex form of the φ functions, these figures represent completely the angular dependence of the probability

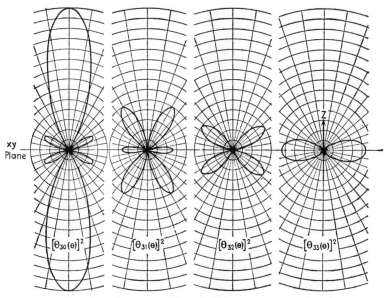

Fig. 21–5.—Polar graphs of the function $[\Theta_{lm}(\vartheta)]^2$ for $l = 3$ and $m = 0$, ± 1, ± 2, and ± 3.

distribution function, which is independent of φ. The alternative sine and cosine functions of φ correspond to probability distribution functions dependent on φ in the way corresponding to the functions $\sin^2 m\varphi$ and $\cos^2 m\varphi$. The angular dependence of the probability distribution function for s and p orbitals in the real form (as given in Table 21–4) is illustrated in Figure 21–6. It is seen that, as mentioned before, the function s is spherically symmetric, and the functions p_x, p_y, and p_z are equivalent except for orientation. The conditions determining the choice of wave functions representing degenerate states of a system will be discussed in the following chapter.

A useful theorem, due to Unsöld,[1] states that the sum of the probability distribution functions for a given value of l and all values of m is a constant; that is,

$$\sum_{m=-l}^{+l} \Theta_{lm}(\vartheta)\Phi_m{}^*(\varphi)\Theta_{lm}(\vartheta)\Phi_m(\varphi) = \text{constant.} \quad (21\text{--}16)$$

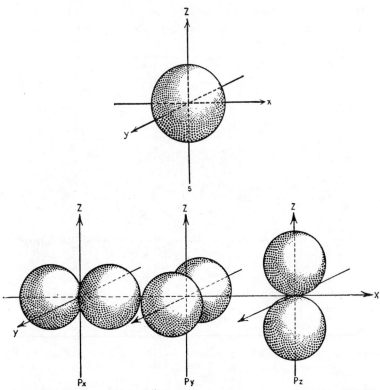

Fig. 21–6.—Polar representation of the absolute values of the angular wave functions for s and p orbitals. The squares of these are the probability distribution functions.

The significance of this will be discussed in the chapter dealing with many-electron atoms (Chap. IX).

Problem 21–2. Prove Unsöld's theorem (Eq. 21–16).

[1] A. Unsöld, *Ann. d. Phys.* **82**, 355 (1927).

CHAPTER VI

PERTURBATION THEORY[1]

In case that the wave equation for a system of interest can be treated by the methods described in the preceding chapters, or can be rigorously treated by any amplification of these methods, a complete wave-mechanical discussion of the system can be given. Very often, however, such a procedure cannot be carried out, the wave equation being of such a nature as to resist accurate solution. Thus even the simplest many-electron systems, the helium atom and the hydrogen molecule, lead to wave equations which have not been rigorously solved. In order to permit the discussion of these systems, which more often than not are those involved in a physical or especially a chemical problem, various methods of approximate solution of the wave equation have been devised, leading to the more or less accurate approximate evaluation of energy values and wave functions. Of these methods the first and in many respects the most interesting is the beautiful and simple wave-mechanical perturbation theory, developed by Schrödinger in his third paper in the spring of 1926. It is especially fortunate that this theory is very much easier to handle than the perturbation theory which is necessary for the treatment of general problems in classical dynamics.

Before we can discuss this method, however, we need certain mathematical results concerning the possibility of expanding arbitrary functions in infinite series of normalized orthogonal functions. These results, which are of great generality and widespread utility, we shall discuss in the next section without attempting any complete proof.

22. EXPANSIONS IN SERIES OF ORTHOGONAL FUNCTIONS

The use of power series to represent certain types of functions is discussed in elementary courses in mathematics, and the theorems which state under what conditions the infinite series

---

[1] A generalized perturbation theory will be discussed in Section 27a.

obtained by formal methods converge to the functions they are meant to represent are also well known. An almost equally useful type of infinite series, which we shall use very frequently, is a series the terms of which are members of a set of normalized orthogonal functions each multiplied by a constant coefficient. If $f_0(x)$, $f_1(x)$, $f_2(x)$, $\cdots$ are members of such a set of normalized orthogonal functions, we might write as the series

$$\varphi(x) = a_0 f_0(x) + a_1 f_1(x) + a_2 f_2(x) + \cdots$$
$$\backsim \sum_{n=0}^{\infty} a_n f_n(x). \tag{22-1}$$

If the series converges and has a definite sum $\varphi(x)$, we may express Equation 22-1 by saying that the infinite series on the right of the equation represents the function $\varphi(x)$ in a certain region of values of x. We may ask if it is possible to find the coefficients a_n for the series which represents any given function $\varphi(x)$. A very simple formal answer may be given to this question. If we multiply both sides of Equation 22-1 by $f_k^*(x)$ and then integrate, assuming that the series is properly convergent so that the term-by-term integration of the series is justified, then we obtain the result

$$\int_a^b \varphi(x) f_k^*(x) dx = a_k, \tag{22-2}$$

since

$$\int_a^b f_k^*(x) f_n(x) dx = 0 \text{ if } n \neq k, \left.\begin{matrix} \\ \end{matrix}\right\}$$
$$= 1 \text{ if } n = k. \left.\begin{matrix} \\ \end{matrix}\right\} \tag{22-3}$$

$a \leqslant x \leqslant b$ defines the orthogonality interval for the functions $f_n(x)$.

In many cases the assumptions involved in carrying out this formal process are not justified, since the series obtained may either not converge at all or converge to a function other than $\varphi(x)$. Mathematicians have studied in great detail the conditions under which such series converge and have proved theorems which enable one to make a decision in all ordinary cases. For our purposes, however, we need only know that such theorems exist and may be used to justify all the expansions which occur in this and later chapters.

The familiar Fourier series is only one special form of an expansion in terms of orthogonal functions. Figure 22–1, which gives a plot of the function

$$\varphi(x) = 1 \text{ for } 0 < x < \pi, \\ \varphi(x) = -1 \text{ for } \pi < x < 2\pi, \bigg\} \qquad (22\text{–}4)$$

together with the first, third, and fifth approximations of its Fourier-series expansion

$$\varphi(x) = a_0 + a_1 \sin x + b_1 \cos x + a_2 \sin 2x +$$
$$b_2 \cos 2x + \cdots , \qquad (22\text{–}5)$$

illustrates that a series of orthogonal functions may represent even a discontinuous function except at the point of discontinuity.

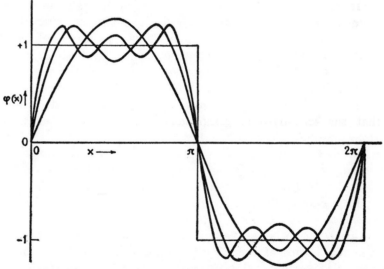

Fig. 22–1.—The function $\varphi(x) = +1$ for $0 < x < \pi$, -1 for $\pi < x < 2\pi$, and the first, third, and fifth Fourier-series approximations to it, involving terms to $\sin x$, $\sin 3x$, and $\sin 5x$, respectively.

If we had evaluated more and more terms of Equation 22–5, the series would have approached more and more closely to the function $\varphi(x)$, except in the neighborhood of the discontinuity.

The most useful sets of orthogonal functions for our purposes are the wave functions belonging to a given wave equation. In preceding chapters we have shown that the solutions of certain wave equations form sets of normalized orthogonal functions, such as for example the Hermite orthogonal functions which are the solutions of the harmonic oscillator problem (Sec. 11). In Appendix III it is shown that the solutions of any wave equation form such a set of orthogonal functions.

In making expansions in terms of orthogonal functions, it is necessary to be sure that the set of functions is *complete*. Thus in the example of Equations 22–4 and 22–5, if we had used the set $\cos x$, $\cos 2x$, $\cdots$, without the sine terms, the series obtained would have converged, but not to the function $\varphi(x)$, because the set of functions $\cos x$, $\cos 2x$, $\cdots$ is not complete. This requirement of completeness necessitates that all the solutions of the wave equation be included when using these solutions for an expansion of an arbitrary function. Since many wave equations lead to a continuous spectrum of energy levels as well as a discrete spectrum, it is necessary to include the wave functions belonging to the continuous levels when making an expansion. The quantum numbers for the continuous spectra do not have discrete values but may vary continuously, so that the part of the expansion involving these wave functions becomes an integral instead of a sum as in Equation 22–1.

However, in many special cases it is easy to see that certain of the coefficients a_k will be zero so that in those cases an expansion is possible in a set of functions which is not complete. Thus if the function $\varphi(x)$ which we are attempting to represent is an even function[1] of x, and if the orthogonal set we are using for the expansion contains both even and odd functions, the coefficients of all the odd functions $f_k(x)$ will vanish, as may be seen from the consideration of Equation 22–2.

All the ideas which have been discussed in this section can be generalized without difficulty to systems of several variables. Normalized orthogonal functions in several variables x_1, y_1, $\cdots$, z_N satisfy the condition

$$\int \cdots \int f_n^*(x_1, y_1, \cdots, z_N) f_m(x_1, y_1, \cdots, z_N) d\tau$$
$$\begin{aligned} &= 0 \text{ if } n \neq m, \\ &= 1 \text{ if } n = m, \end{aligned} \right\} \quad (22\text{–}6)$$

in which the integration is carried out over the whole of the configuration space for the system, and $d\tau$ is the volume element

---

[1] The function $f(x)$ is called an even function of x if $f(-x)$ is equal to $f(x)$ for all values of x, and an odd function of x if $f(-x)$ is equal to $-f(x)$ for all values of x.

for the particular coordinate system in which the integral is expressed. Orthogonal functions in several variables usually are distinguished by several indices, which may however be symbolized by a single letter. An example of a three-dimensional set of normalized orthogonal functions is the set of solutions of the wave equation for the hydrogen atom. We have obtained in Chapter V the solutions belonging to the discrete levels; the quantum numbers nlm provide the indices for these functions. The solutions for the continuous spectrum of the atom, i.e., the system resulting when the electron has been completely removed from the nucleus, must be included if a complete set is desired.[1]

The coefficients in the expansion of an arbitrary function of several variables are obtained from an equation entirely analogous to Equation 22–2,

$$a_k = \int \cdots \int \varphi(x_1, y_1, \cdots, z_N) f_k^*(x_1, y_1, \cdots, z_N) d\tau, \quad (22\text{–}7)$$

in which the limits of integration and the meaning of $d\tau$ are the same as in Equation 22–6.

A function φ which is expressed in terms of the normalized functions of a complete orthogonal set is itself normalized if the coefficients in the expansion satisfy the relation $\sum_n a_n^* a_n = 1$.

It may be mentioned that in some cases it is convenient to make use of complete sets of functions which are not mutually orthogonal. An arbitrary function can be expanded in terms of the functions of such a set; the determination of the values of the coefficients is, however, not so simple as for orthogonal functions. An example of an expansion of this type occurs in Section 24.

In certain applications of expansions in terms of orthogonal functions, we shall obtain expressions of the form

$$\sum_n a_n f_n(x) = 0.$$

By multiplying by $f_k^*(x)$ and integrating, we see that the coefficient of each term must be zero; i.e., $a_n = 0$ for all values of n.

[1] For a discussion of the wave functions for the continuous spectrum of hydrogen, see Sommerfeld, "Wave Mechanics," p. 290.

Problem 22–1. Obtain the first four coefficients in the expansion $\varphi(x) =$ $(2 + x)^{-3/2} = \sum_k a_k \sqrt{\dfrac{2k+1}{2}} P_k(x)$, where $P_k(x)$ is the kth Legendre polynomial given in Section 19. This expansion is valid only for $|x| \leqslant 1$. Plot $\varphi(x)$ and the approximations to it given by including the first, second, third, and fourth terms of the expansion. If possible, obtain a general expression for a_k, using the generating function for $P_k(x)$.

23. FIRST-ORDER PERTURBATION THEORY FOR A NON-DEGENERATE LEVEL

In discussing many problems which cannot be directly solved, a solution can be obtained of a wave equation which differs from the true one only in the omission of certain terms whose effect on the system is small. Perturbation theory provides a method of treating such problems, whereby the approximate equation is first solved and then the small additional terms are introduced as corrections.

Let us write the true wave equation in the form

$$H\psi - W\psi = 0, \tag{23-1}$$

in which H represents the operator

$$H = -\frac{h^2}{8\pi^2} \sum_i \frac{1}{m_i} \nabla_i^2 + V. \tag{23-2}$$

We assume that it is possible to expand H in terms of some parameter λ, yielding the expression

$$H = H^0 + \lambda H' + \lambda^2 H'' + \cdots, \tag{23-3}$$

in which λ has been chosen in such a way that the equation to which 23–1 reduces when $\lambda \to 0$,

$$H^0\psi^0 - W^0\psi^0 = 0, \tag{23-4}$$

can be directly solved. This equation is said to be *the wave equation for the unperturbed system*, while the terms

$$\lambda H' + \lambda^2 H'' + \cdots$$

are called the *perturbation*. As an illustration, we might mention the problem of the Stark effect in atomic hydrogen, in which an electric field is applied to the atom. In this problem the field strength E provides a convenient parameter in terms of which the Hamiltonian may be expanded. When E is zero, the problem reduces to that of the ordinary hydrogen atom, which we have already solved.

The unperturbed equation 23–4 has solutions

$$\psi_0^0, \psi_1^0, \psi_2^0, \cdots, \psi_k^0, \cdots,$$

called the *unperturbed wave functions*, and corresponding energy values

$$W_0^0, W_1^0, W_2^0, \cdots, W_k^0, \cdots.$$

The functions ψ_k^0 form a complete orthogonal set as discussed in Section 22, and, if we assume that they have also been normalized, they satisfy the equation (Appendix III)

$$\left. \begin{aligned} \int \psi_i^0 \psi_j^0 d\tau &= 0 \text{ if } i \neq j, \\ &= 1 \text{ if } i = j. \end{aligned} \right\} \qquad (23\text{–}5)$$

Now let us consider the effect of the perturbation. By hypothesis it will be small, and from the continuity properties of wave functions[1] we know that the energy values and wave functions for the perturbed system will lie near those for the unperturbed system. In other words, the application of a small perturbation is not going to cause large changes. With these facts in mind we can expand the energy W and the wave function ψ for the perturbed problem in terms of λ and have reasonable assurance that the expansions will converge, writing

$$\psi_k = \psi_k^0 + \lambda \psi_k' + \lambda^2 \psi_k'' + \cdots \qquad (23\text{–}6)$$

and

$$W_k = W_k^0 + \lambda W_k' + \lambda^2 W_k'' + \cdots. \qquad (23\text{–}7)$$

If the perturbation is really a small one, the terms of these series will become rapidly smaller as we consider the coefficients of larger powers of λ; i.e., the series will converge.

We now substitute these expansions for H, ψ_k, and W_k into the wave equation 23–1, obtaining the result, after collecting coefficients of like powers of λ,

$$\begin{aligned} (H^0\psi_k^0 - W_k^0\psi_k^0) &+ (H^0\psi_k' + H'\psi_k^0 - W_k^0\psi_k' - W_k'\psi_k^0)\lambda \\ &+ (H^0\psi_k'' + H'\psi_k' + H''\psi_k^0 - W_k^0\psi_k'' - W_k'\psi_k' - W_k''\psi_k^0)\lambda^2 \\ &+ \cdots = 0. \quad (23\text{–}8) \end{aligned}$$

---

[1] Discussed, for example, in Courant and Hilbert, "Methoden der mathematischen Physik."

If this series is properly convergent, we know that in order for it to equal zero for all values of λ the coefficients of the powers of λ must vanish separately.[1] The coefficient of λ^0 when equated to zero gives Equation 23–4, so that we were justified in beginning the expansions 23–6 and 23–7 with the terms ψ^0 and W^0. The coefficient of λ gives the equation

$$H^0\psi_k' - W_k^0\psi_k' = (W_k' - H')\psi_k^0. \tag{23–9}$$

To solve this we make use of the expansion theorem discussed in the last section. We consider that the unknown functions ψ_k' can be expanded in terms of the known functions ψ_l^0, since the latter form a normalized orthogonal set, and write

$$\psi_k' = \sum_l a_l\psi_l^0. \tag{23–10}$$

(The coefficients a_l might be written as a_{lk}, but we shall assume throughout that we are interested only in the state k and therefore shall omit the second subscript.)   Using this, we obtain the result

$$H^0\psi_k' = \sum_l a_l H^0\psi_l^0 = \sum_l a_l W_l^0\psi_l^0, \tag{23–11}$$

since

$$H^0\psi_l^0 = W_l^0\psi_l^0.$$

Equation 23–9 therefore assumes the form

$$\sum_l a_l(W_l^0 - W_k^0)\psi_l^0 = (W_k' - H')\psi_k^0. \tag{23–12}$$

If we multiply by $\psi_k^0{}^*$ and integrate over configuration space, we observe that the expression on the left vanishes:

$$\int \psi_k^0{}^* \sum_l a_l(W_l^0 - W_k^0)\psi_l^0 d\tau = \sum_l a_l(W_l^0 - W_k^0)\int \psi_k^0{}^* \psi_l^0 d\tau = 0,$$

since $\int \psi_k^0{}^* \psi_l^0 d\tau$ vanishes except for $l = k$, and for this value

[1] Thus, if

$$\sum_n a_n\lambda^n \equiv \varphi(\lambda) \equiv 0,$$

then, assuming that the series is properly convergent, we can write

$$a_n = \frac{1}{n!}\left(\frac{d^n\varphi}{d\lambda^n}\right)_{\lambda=0} = 0.$$

of l the quantity $W_l^0 - W_k^0$ vanishes; and hence we obtain the
equation

$$\int \psi_k^0 {}^*(W_k' - H')\psi_k^0 d\tau = 0. \tag{23-13}$$

This solves the problem of the determination of W_k', the first-
order correction to the energy. Since W_k' is a constant in
Equation 23–13, the integration of the term containing it can be
carried out at once, giving the result, when multiplied by λ,

$$\lambda W_k' = \lambda \int \psi_k^0 {}^* H' \psi_k^0 d\tau. \tag{23-14}$$

Since the correction to the energy is $\lambda W_k'$, it is convenient to
include the parameter λ in the symbols for the first-order perturbation and the first-order energy correction, so that to the
first order it is usual to write the relations

$$\left.\begin{aligned} H &= H^0 + H', \\ \psi_k &= \psi_k^0 + \psi_k', \\ W_k &= W_k^0 + W_k', \end{aligned}\right\} \tag{23-15}$$

in which

$$W_k' = \int \psi_k^0 {}^* H' \psi_k^0 d\tau. \tag{23-16}$$

This expression for the perturbation energy can be very simply
described: *The first-order perturbation energy for a non-degenerate
state of a system is just the perturbation function averaged over the
corresponding unperturbed state of the system.*

We can also evaluate the correction ψ_k' for the wave function.
Multiplying each side of Equation 23–12 by $\psi_j^0{}^*$, we obtain, after
integration,

$$a_j(W_j^0 - W_k^0) = -\int \psi_j^0 {}^* H' \psi_k^0 d\tau, \qquad j \neq k, \tag{23-17}$$

where we have utilized the orthogonality and normalization
properties of the ψ^0's. The coefficients a_j in the expansion 23–10
of ψ' in terms of the set ψ_j^0 are thus given by the relation

$$a_j = -\frac{\int \psi_j^0 {}^* H' \psi_k^0 d\tau}{W_j^0 - W_k^0}, \qquad j \neq k. \tag{23-18}$$

The value of a_k is not given by this process; it is to be chosen
so as to normalize the resultant ψ, and, if only first-order terms
are considered (terms in λ^2 neglected), it is equal to zero. It is
convenient to introduce the symbol

$$H_{jk}' = \int \psi_j^0 {}^* H' \psi_k^0 d\tau, \tag{23-19}$$

so that the expression for the first-order wave function of the system, on introducing the above values of the coefficients a_j, becomes

$$\psi_k = \psi_k^0 - \lambda \sum_{j=0}^{\infty} {}' \frac{H_{jk}'}{W_j^0 - W_k^0} \psi_j^0, \qquad (23\text{-}20)$$

in which the prime on the summation indicates the omission of the term with $j = k$.

As mentioned before, it is customary to include λ in the definition of H' as indicated in Equation 23–15, so that we get finally for the first-order energy and the first-order wave function the expressions

$$W_k = W_k^0 + H_{kk}' \qquad (23\text{-}21)$$

and

$$\psi_k = \psi_k^0 - \sum_{j=0}^{\infty} {}' \frac{H_{jk}'}{W_j^0 - W_k^0} \psi_j^0. \qquad (23\text{-}22)$$

23a. A Simple Example: The Perturbed Harmonic Oscillator. As a simple illustration of first-order perturbation theory we shall obtain the approximate energy levels of the system whose wave equation is

$$\frac{d^2\psi}{dx^2} + \frac{8\pi^2 m}{h^2}\left(W - \frac{1}{2}kx^2 - ax^3 - bx^4 \right)\psi = 0. \qquad (23\text{-}23)$$

We recognize that if a and b were zero this would be the wave equation for the harmonic oscillator, whose solutions we already know (Sec. 11). If a and b are small, therefore, we may treat these terms as perturbations, writing

$$H' = ax^3 + bx^4. \qquad (23\text{-}24)$$

We need then to evaluate the integrals

$$H_{nn}' = a\int_{-\infty}^{+\infty} \psi_n^0 {}^* x^3 \psi_n^0 dx + b\int_{-\infty}^{+\infty} \psi_n^0 {}^* x^4 \psi_n^0 dx. \qquad (23\text{-}25)$$

Since x^3 is an odd function and $\psi_n^0 {}^* \psi_n^0$ an even function, the first of these integrals is zero, so that the first-order perturbation due to ax^3 is zero. To calculate the second integral we refer back to

Section 11c for the functions ψ_n^0 and their properties. Substituting for ψ_n^0 from Equation 11-20 we obtain the integral

$$I = \int_{-\infty}^{\infty} \psi_n^0 *x^4\psi_n^0 dx = \frac{N_n^2}{\alpha^{5/2}}\int_{-\infty}^{\infty} e^{-\xi^2}H_n^2(\xi)\xi^4 d\xi. \quad (23\text{-}26)$$

From Equation 11-15 we see that

$$\xi H_n(\xi) = \tfrac{1}{2}H_{n+1}(\xi) + nH_{n-1}(\xi), \quad (23\text{-}27)$$

so that, after applying Equation 23-27 to ξH_{n+1} and ξH_{n-1} and collecting terms, we obtain the equation

$$\xi^2 H_n(\xi) = \tfrac{1}{4}H_{n+2}(\xi) + (n + \tfrac{1}{2})H_n(\xi) + n(n - 1)H_{n-2}(\xi). \quad (23\text{-}28)$$

By this application of the recursion formula for $H_n(\xi)$ we have expressed $\xi^2 H_n(\xi)$ in terms of Hermite polynomials with constant coefficients. By squaring this we obtain an expression for $\xi^4 H_n^2(\xi)$, which enables us to express the integral in Equation 23-26 as a sum of integrals of the form

$$\int_{-\infty}^{\infty} e^{-\xi^2}H_n(\xi)H_m(\xi)d\xi = 0 \qquad \text{if } m \neq n, \\ = 2^n n!\sqrt{\pi} \text{ if } m = n, \quad (23\text{-}29)$$

evaluated in Section 11c. Thus we find for I the expression

$$I = \frac{N_n^2\sqrt{\pi}}{\alpha^{5/2}}\left\{\frac{1}{16}2^{n+2}(n + 2)! + \left(n + \frac{1}{2}\right)^2 2^n n! + \right.$$

$$\left. n^2(n - 1)^2 2^{n-2}(n - 2)!\right\}$$

$$= \frac{3}{4\alpha^2}(2n^2 + 2n + 1),$$

when the value of N_n given in Equation 11-21 is introduced.

The first-order perturbation energy for this system is therefore

$$W' = H'_{nn} = \frac{3b}{4\alpha^2}(2n^2 + 2n + 1),$$

so that the total energy becomes (to the first order)

$$W = W^0 + W' = \left(n + \frac{1}{2}\right)h\nu_0 + \frac{3}{64\pi^4}(2n^2 + 2n + 1)\frac{h^2b}{m^2\nu_0^2}. \quad (23\text{-}30)$$

In order to calculate the first-order wave function it would be necessary to evaluate all the quantities H'_{nk}. The x^3 term as well as the x^4 term will contribute to these integrals. The number of non-zero integrals is not, however, infinite in this case but quite small, only the terms with $k = n$, $n \pm 1$, $n \pm 2$, $n \pm 3$, and $n \pm 4$ being different from zero.

23b. An Example: The Normal Helium Atom.—As another example of the application of first-order perturbation theory let us discuss the normal state of the helium atom. Since the term which we shall use as the perturbation is not particularly small, we must not expect an answer of very great accuracy. The potential energy for a system of two electrons and a nucleus of charge $+Ze$ is

$$V = -\frac{Ze^2}{r_1} - \frac{Ze^2}{r_2} + \frac{e^2}{r_{12}}, \tag{23-31}$$

in which r_1 and r_2 are the distances of electrons 1 and 2, respectively, from the nucleus, and r_{12} is the separation of the two electrons. If we make the approximation of considering the nucleus at rest, which introduces no appreciable error, the wave equation (see Equation 12–8) for the two electrons becomes

$$H\psi = -\frac{h^2}{8\pi^2 m_0}\left(\frac{\partial^2\psi}{\partial x_1^2} + \frac{\partial^2\psi}{\partial y_1^2} + \frac{\partial^2\psi}{\partial z_1^2} + \frac{\partial^2\psi}{\partial x_2^2} + \frac{\partial^2\psi}{\partial y_2^2} + \frac{\partial^2\psi}{\partial z_2^2}\right)$$
$$+ \left(-\frac{Ze^2}{r_1} - \frac{Ze^2}{r_2} + \frac{e^2}{r_{12}}\right)\psi = W\psi. \tag{23-32}$$

This equation applies to He, Li$^+$, Be^{++}, etc., with $Z = 2, 3, 4$, etc., respectively. The variables x_1, y_1, z_1 are Cartesian coordinates of one electron, and x_2, y_2, z_2 those of the other; m_0 is the mass of the electron.

Since if the term e^2/r_{12} is omitted the wave equation which is obtained can be exactly solved, we choose this term as the perturbation function,

$$H' = \frac{e^2}{r_{12}}.$$

The wave equation which remains, the unperturbed equation, can then be separated into two equations by the substitutions

$$\psi^0(x_1, y_1, z_1, x_2, y_2, z_2) = u_1^0(x_1, y_1, z_1)u_2^0(x_2, y_2, z_2)$$

and
$$W^0 = W_1^0 + W_2^0,$$

the equation[1] for u_1^0 being

$$\frac{\partial^2 u_1^0}{\partial x_1^2} + \frac{\partial^2 u_1^0}{\partial y_1^2} + \frac{\partial^2 u_1^0}{\partial z_1^2} + \frac{8\pi^2 m_0}{h^2}\left(W_1^0 + \frac{Ze^2}{r_1}\right)u_1^0 = 0. \quad (23\text{-}33)$$

The equation for u_2^0 is identical except for the changed subscripts. Equation 23-33 is just the hydrogenlike wave equation discussed in the preceding chapter, with solutions $\psi_{nlm}(r_1, \vartheta_1, \varphi_1)$ and energy values $-Z^2 W_H/n^2$, in which

$$W_H = \frac{2\pi^2 m_0 e^4}{h^2} = 13.53 \text{ v.e.}$$

The unperturbed wave function for the lowest level of the two-electron atom is therefore

$$\psi_{100,100}^0 = \psi_{100}(r_1, \vartheta_1, \varphi_1)\psi_{100}(r_2, \vartheta_2, \varphi_2) = $$
$$u_{1s}(r_1, \vartheta_1, \varphi_1)u_{1s}(r_2, \vartheta_2, \varphi_2), \quad (23\text{-}34)$$

in which $r_1, \vartheta_1, \varphi_1$ and $r_2, \vartheta_2, \varphi_2$ are polar coordinates of the two electrons relative to axes with the nucleus at the origin. The corresponding energy value is

$$W_{100,100}^0 = W_1^0 + W_2^0 = -2Z^2 W_H. \quad (23\text{-}35)$$

The first-order perturbation energy W' is the average value of the perturbation function $H' = e^2/r_{12}$ over the unperturbed state of the system, with the value

$$W' = \int \psi^{0*} H' \psi^0 d\tau = \int \frac{e^2}{r_{12}} \psi_{100,100}^2 d\tau. \quad (23\text{-}36)$$

From Table 21-4 of Chapter V we obtain for u_{1s} the expression

$$u_{1s} = \psi_{100} = \sqrt{\frac{Z^3}{\pi a_0^3}} e^{-\frac{\rho}{2}}, \quad (23\text{-}37)$$

in which $\rho = 2Zr/a_0$ and $a_0 = h^2/4\pi^2 m_0 e^2$. Using this in Equation 23-34, we find for $\psi_{100,100}$ the expression

$$\psi_{100,100} = \frac{Z^3}{\pi a_0^3} e^{-\frac{\rho_1}{2}} e^{-\frac{\rho_2}{2}}.$$

[1] The symbol u will be used for the wave function for a single electron in a many-electron atom, with subscripts $1s$, $2s$, $2p$, etc.

The volume element is

$$d\tau = r_1^2 dr_1 \sin \vartheta_1 d\vartheta_1 d\varphi_1 \cdot r_2^2 dr_2 \sin \vartheta_2 d\vartheta_2 d\varphi_2,$$

so that the integral for W' becomes

$$W' = \frac{Ze^2}{2^5 \pi^2 a_0} \int_0^{2\pi} \int_0^\pi \int_0^\infty \int_0^{2\pi} \int_0^\pi \int_0^\infty \frac{e^{-\rho_1 - \rho_2}}{\rho_{12}} \rho_1^2 d\rho_1 \sin \vartheta_1$$
$$d\vartheta_1 d\varphi_1 \rho_2^2 d\rho_2 \sin \vartheta_2 d\vartheta_2 d\varphi_2, \quad (23\text{-}38)$$

in which $\rho_{12} = 2Zr_{12}/a_0$.

The value of this integral is easily obtained, inasmuch as it corresponds to the electrostatic interaction energy of two spherically symmetrical distributions of electricity, with density functions $e^{-\rho_1}$ and $e^{-\rho_2}$, respectively. In Appendix V it is shown[1] that

$$W' = \tfrac{5}{4} Z W_H. \quad (23\text{-}39)$$

This treatment thus gives for the total energy the value

$$W = -(2Z^2 - \tfrac{5}{4}Z)W_H. \quad (23\text{-}40)$$

This may be compared with the experimental values of the total energy, which are obtained by adding the first and the second ionization energies. Table 23-1 contains, for He, Li$^+$, Be^{++}, B^{3+}, and C^{4+}, the experimental energy $W_{exp.}$, the unperturbed energy W^0, the total energy calculated by first-order perturbation theory $W^0 + W'$, the difference $\Delta^0 = W_{exp.} - W^0$, the difference $\Delta' = W_{exp.} - W^0 - W'$, and finally the ratio $-\Delta'/\Delta^0$.

It is seen that the error Δ' remains roughly constant in absolute value as the nuclear charge increases, which means that the percentage error decreases, since the total energy is larger for larger Z. This result is to be expected, inasmuch as for large nuclear charge the contribution of the attraction of the nucleus is relatively more important than that of the repulsion of the two electrons. It is pleasing that even in this problem, in which the perturbation function e^2/r_{12} is not small, the simple first-order perturbation treatment leads to a value of the total energy of the atom which is in error by only a small amount, varying from 5 per cent for He to 0.4 per cent for C^{4+}.

---

[1] This problem was first treated by A. Unsöld, *Ann. d. Phys.* **82**, 355 (1927).

TABLE 23-1:—CALCULATED AND OBSERVED VALUES OF THE ENERGY OF
HELIUMLIKE ATOMS AND IONS

	$-W_{exp.}$,v.e.	$-W^0$, v.e.	$\dfrac{-W^0-W'}{\text{v.e.}}$	Δ^0, v.e.	Δ', v.e.	$-\Delta'/\Delta^0$
He	78.62	108.24	74.42	29.62	−4.20	0.142
Li+	197.14	243.54	192.80	46.40	−4.34	.094
Be++	369.96	432.96	365.31	63.00	−4.65	.074
B³+	596.4	676.50	591.94	80.1	−4.5	.056
C⁴+	876.2	974.16	872.69	98.0	−3.5	.036

Problem 23–1. Calculate the first-order energy correction for a one-dimensional harmonic oscillator upon which the perturbation $H'(x)$ acts, where $H'(x)$ is zero unless $|x| < \epsilon$ and $H'(x) = b$ for $|x| < \epsilon$, with ϵ a quantity which is allowed to approach zero at the same time that b approaches infinity, in such a way that the product $2\epsilon b = c$. Compare the effect on the odd and even levels of the oscillator. What would be the effect of a perturbation which had a very large value at some point outside the classically allowed range of the oscillator and a zero value elsewhere?

Problem 23–2. The wave functions and energy levels of a particle in a one-dimensional box are given in Equations 14–6 and 14–7. Calculate the first-order perturbation energy for such a system with a perturbation H' such that $H' = b$ for $(a/k) - \epsilon \leqslant x \leqslant (a/k) + \epsilon$ and $H' = 0$ elsewhere, with $\epsilon \to 0$ as $b \to \infty$ in such a way that $2\epsilon b = c$, k being a given integer. With $k = 5$, determine which energy levels are the most and which are the least perturbed and explain. With $k = 2$, give the expression for the perturbed wave function, to the first order.

Problem 23–3. Let H' be a perturbation, such that $H'(x) = -b$ for $0 \leqslant x \leqslant a/2$ and $H'(x) = +b$ for $a/2 \leqslant x \leqslant a$, which is applied to a particle in a one-dimensional box (Eqs. 14–6 and 14–7). Obtain the first-order wave function. Show qualitatively that this function is such that the probability of finding the particle in the right-hand half of the box has been increased and explain in terms of classical theory. (*Hint:* Use the symmetry about the point $x = a/2$.)

24. FIRST-ORDER PERTURBATION THEORY FOR A DEGENERATE LEVEL

The methods which we have used in Section 23 to obtain the first-order perturbation energy are not applicable when the energy level of the unperturbed system is degenerate, for the reason that in carrying out the treatment we assumed that the perturbed wave function differs only slightly from one function ψ_k^0 which is the solution of the unperturbed wave equation for a given energy value whereas now there are several such functions, all belonging to the same energy level, and we do not know which one (if any) approximates closely to the solution of the perturbed wave equation.

An energy level W_k is called *α-fold degenerate* (see Sec. 14) when for $W = W_k$ there exist α linearly independent wave functions $\psi_{k1}, \psi_{k2}, \psi_{k3}, \cdots, \psi_{k\alpha}$ satisfying the wave equation.[1] Each of these is necessarily orthogonal to all wave functions for the system corresponding to other values of the energy (see Appendix III) but is not necessarily orthogonal to the other functions corresponding to the same value of the energy. Any linear combination $\sum\limits_{j=1}^{\alpha} \kappa_j \psi_{kj}$ of the wave functions of a degenerate set such as $\psi_{k1}, \psi_{k2}, \cdots, \psi_{k\alpha}$ is itself a solution of the wave equation and is a satisfactory wave function corresponding to the energy W_k. We might therefore construct α such combinations χ_{ki} by choosing sets of values for κ_j such that the different combinations thus formed are linearly independent. The set of functions so obtained,

$$\chi_{ki} = \sum_{j=1}^{\alpha} \kappa_{ij} \psi_{kj}, \qquad i = 1, 2, 3, \cdots, \alpha, \qquad (24\text{--}1)$$

is entirely equivalent to the original set $\psi_{k1}, \psi_{k2}, \cdots, \psi_{k\alpha}$. This indicates that there is nothing unique about any particular set of solutions for a degenerate level, since we can always construct an infinite number of other sets, such as $\chi_{k1}, \cdots, \chi_{k\alpha}$, which are equally good wave functions. The transformation expressed by Equation 24–1 is called a linear transformation with constant coefficients.

It is usually convenient to deal with wave functions which are normalized to unity and which are mutually orthogonal. Since the coefficients κ_{ij} can always be chosen in such a way as to make the set χ_{ki} possess these properties, we shall ultimately assume that this has been done.

Using these ideas, we can now investigate the application of perturbation theory to degenerate levels. We write the wave equation in the form

---

[1] The functions $\psi_{k1}, \psi_{k2}, \cdots, \psi_{k\alpha}$ are said to be linearly independent if there exists no relation of the form $a_1\psi_{k1} + a_2\psi_{k2} + \cdots + a_\alpha\psi_{k\alpha} = 0$ (in which $a_1, a_2, \cdots, a_\alpha$ are constant coefficients) which is satisfied for all values of the independent variables.

$$H\psi - W\psi = 0 \qquad (24\text{--}2)$$

with

$$H = H^0 + \lambda H' + \lambda^2 H'' + \cdots$$

as before. The wave equation for the unperturbed system is

$$H^0\psi^0 - W^0\psi^0 = 0, \qquad (24\text{--}3)$$

the solutions of which are

$$\psi_{01}^0,\ \psi_{02}^0,\ \cdots\ ;\ \psi_{11}^0,\ \psi_{12}^0,\ \cdots\ ;\ \cdots\ ;\ \psi_{k1}^0,\ \psi_{k2}^0,\ \cdots\ ,$$
$$\psi_{k\alpha}^0;\ \cdots\ ;$$

corresponding to the energy levels

$$W_0^0;\ W_1^0;\ \cdots\ ;\ W_k^0;\ \cdots\ .$$

Now let us consider a particular wave function for the perturbed equation 24–2. It is known, in consequence of the properties of continuity of characteristic-value differential equations, that as the perturbation function $\lambda H' + \cdots$ becomes smaller and smaller the energy value W of Equation 24–2 will approach an energy level of the unperturbed equation 24–3, W_k^0, say. The wave function under consideration will also approach more and more closely a wave function satisfying Equation 24–3. However, this limiting wave function need not be any one of the functions $\psi_{k1}^0,\ \cdots\ ,\ \psi_{k\alpha}^0$; it may be (and generally is) some linear combination of them. The first problem which must be solved in the treatment of a degenerate system is the determination of the set of unperturbed wave functions to which the perturbed functions reduce when the perturbation vanishes; that is, the evaluation of the coefficients in the linear transformation converting the initially chosen wave functions into the *correct zeroth-order wave functions*. These correct combinations, given by

$$\chi_{kl}^0 = \sum_{l'=1}^{\alpha} \kappa_{ll'}\psi_{kl'}^0, \qquad l = 1, 2, \cdots, \alpha, \qquad (24\text{--}4)$$

provide the first term of the expansion of ψ_{kl} in powers of λ, since by definition they are the functions to which the ψ_{kl}'s reduce when $\lambda \to 0$. Therefore

$$\psi_{kl} = \chi_{kl}^0 + \lambda\psi_{kl}' + \lambda^2\psi_{kl}'' + \cdots \qquad (24\text{--}5)$$

and

$$W_{kl} = W_k^0 + \lambda W_{kl}' + \lambda^2 W_{kl}'' + \cdots, \qquad (24\text{--}6)$$

where $l(= 1, 2, \cdots, \alpha)$ designates the particular one of the α degenerate wave functions in question, are the equations which are analogous to Equations 23–6 and 23–7. (As in Equation 23–10 we sometimes omit the subscript k; e.g., we write $\kappa_{ll'}$ for $\kappa_{kll'}$; it must be borne in mind that throughout we are considering the kth degenerate level.)

Substituting the expansions for ψ, W, and H into the wave equation 24–2, we obtain an equation entirely analogous to Equation 23–8 of the non-degenerate treatment,

$$(H^0\chi_{kl}^0 - W_k^0\chi_{kl}^0) + (H^0\psi_{kl}' + H'\chi_{kl}^0 - W_k^0\psi_{kl}' - W_{kl}'\chi_{kl}^0)\lambda + \cdots = 0, \quad (24\text{–}7)$$

from which, on equating the coefficient of λ to zero as before, there results the equation (cf. Eq. 23–9)

$$H^0\psi_{kl}' - W_k^0\psi_{kl}' = W_{kl}'\chi_{kl}^0 - H'\chi_{kl}^0. \quad (24\text{–}8)$$

So far our treatment differs from the previous discussion of non-degenerate levels only in the use of χ_{kl}^0 instead of ψ_{ki}^0; i.e., in the introduction of a general expression for unperturbed functions instead of the arbitrary set ψ_{ki}^0. In the next step we likewise follow the previous treatment, in which the quantities ψ_k' and $H^0\psi_k'$ were expanded in terms of the complete set of orthogonal functions ψ_k^0. Here, however, we must in addition express χ_{kl}^0 in terms of the set $\psi_{kl'}^0$, by means of Equation 24–4, in which the coefficients $\kappa_{ll'}$ are so far arbitrary. Therefore we introduce the expansions

$$\psi_{kl}' = \sum_{k'l'} a_{klk'l'}\psi_{k'l'}^0 \quad (24\text{–}9)$$

and

$$H^0\psi_{kl}' = \sum_{k'l'} a_{klk'l'}H^0\psi_{k'l'}^0 = \sum_{k'l'} a_{klk'l'}W_{k'}^0\psi_{k'l'}^0 \quad (24\text{–}10)$$

into Equation 24–8 together with the expression for χ_{kl}^0 given by Equation 24–4. The result is

$$\sum_{k'l'} a_{klk'l'}(W_{k'}^0 - W_k^0)\psi_{k'l'}^0 = \sum_{l'=1}^{\alpha} \kappa_{ll'}(W_{kl}' - H')\psi_{kl'}^0, \quad (24\text{–}11)$$

in which the right-hand side involves only functions $\psi_{kl'}^0$ belonging to the degenerate level W_k^0 while the expansion on the left includes all the $\psi_{k'l'}^0$'s. If we now multiply both sides of this equation by ψ_{kj}^{0*} and integrate over configuration space, we obtain the result

$$\sum_{k'l'} a_{klk'l'}(W_{k'}^0 - W_k^0)\int\psi_{kj}^{0*}\psi_{k'l'}^0 d\tau = \sum_{l'=1}^{\alpha} \kappa_{ll'}(W_{kl}'\int\psi_{kj}^{0*}\psi_{kl'}^0 d\tau - \int\psi_{kj}^{0*}H'\psi_{kl'}^0 d\tau). \quad (24\text{--}12)$$

The left side of this equation is zero because ψ_{kj}^0 and $\psi_{k'l'}^0$ are orthogonal if $k \neq k'$ and $W_{k'}^0 - W_k^0$ is zero if $k = k'$. If we introduce the symbols

$$H_{jl'}' = \int\psi_{kj}^{0*}H'\psi_{kl'}^0 d\tau \quad (24\text{--}13)$$

and

$$\Delta_{jl'} = \int\psi_{kj}^{0*}\psi_{kl'}^0 d\tau, \quad (24\text{--}14)$$

we may express Equation 24-12 in the form

$$\sum_{l'=1}^{\alpha} \kappa_{ll'}(H_{jl'}' - \Delta_{jl'}W_{kl}') = 0, \qquad j = 1, 2, 3, \cdots, \alpha. \quad (24\text{--}15)$$

This is a system of α homogeneous linear simultaneous equations in the α unknown quantities $\kappa_{l1}, \kappa_{l2}, \cdots, \kappa_{l\alpha}$. Written out in full, these equations are

$$\left.\begin{array}{l} (H_{11}' - \Delta_{11}W_{kl}')\kappa_{l1} + (H_{12}' - \Delta_{12}W_{kl}')\kappa_{l2} + \cdots + \\ \qquad\qquad\qquad (H_{1\alpha}' - \Delta_{1\alpha}W_{kl}')\kappa_{l\alpha} = 0, \\ (H_{21}' - \Delta_{21}W_{kl}')\kappa_{l1} + (H_{22}' - \Delta_{22}W_{kl}')\kappa_{l2} + \cdots + \\ \qquad\qquad\qquad (H_{2\alpha}' - \Delta_{2\alpha}W_{kl}')\kappa_{l\alpha} = 0, \\ \cdots\cdots\cdots\cdots\cdots\cdots\cdots\cdots\cdots, \\ (H_{\alpha1}' - \Delta_{\alpha1}W_{kl}')\kappa_{l1} + (H_{\alpha2}' - \Delta_{\alpha2}W_{kl}')\kappa_{l2} + \cdots + \\ \qquad\qquad\qquad (H_{\alpha\alpha}' - \Delta_{\alpha\alpha}W_{kl}')\kappa_{l\alpha} = 0. \end{array}\right\} \quad (24\text{--}16)$$

Such a set of equations can be solved only for the ratios of the κ's; i.e., any one κ may be chosen and all of the others expressed in terms of it. For an arbitrary value of W_{kl}', however, the set of equations may have no solution except the trivial one $\kappa_{ll'} = 0$. It is only for certain values of W_{kl}' that the set of equations has non-trivial solutions; the condition that must be satisfied if such a set of homogeneous linear equations is to have non-zero solutions is that the determinant of the coefficients of the unknown quantities vanish; that is, that

$$\begin{vmatrix} H'_{11} - \Delta_{11}W'_{kl} & H'_{12} - \Delta_{12}W'_{kl} & \cdots & H'_{1\alpha} - \Delta_{1\alpha}W'_{kl} \\ H'_{21} - \Delta_{21}W'_{kl} & H'_{22} - \Delta_{22}W'_{kl} & \cdots & H'_{2\alpha} - \Delta_{2\alpha}W'_{kl} \\ \cdots & \cdots & \cdots & \cdots \\ H'_{\alpha 1} - \Delta_{\alpha 1}W'_{kl} & H'_{\alpha 2} - \Delta_{\alpha 2}W'_{kl} & \cdots & H'_{\alpha\alpha} - \Delta_{\alpha\alpha}W'_{kl} \end{vmatrix} = 0.$$

$$(24\text{-}17)$$

This determinantal equation can be expanded into an algebraic equation in W'_{kl} which can then be solved for W'_{kl}. For the types of perturbation functions which arise in most physical and chemical problems the determinant is either symmetrical about the principal diagonal, if the elements are real, or else has the property that corresponding elements on opposite sides of the principal diagonal are the complex conjugates of each other; that is, $H'_{ij} = H'_{ji}{}^*$. In consequence of this property it can be shown that the determinant possesses α real roots, $W'_{k1}, W'_{k2}, \cdots, W'_{k\alpha}$. These are the values of the first-order perturbation energy for the α wave functions which correspond to the α-fold degenerate unperturbed energy level W^0_k. It may happen, however, that not all of the roots W'_{k1}, etc., are distinct, in which case the perturbation has not completely removed the degeneracy.

The coefficients $\kappa_{ll'}$ which determine the correct zeroth-order wave function χ^0_{kl} corresponding to any perturbed level W'_{kl} may be determined by substituting the value found for W'_{kl} into the set of simultaneous equations 24–16 and solving for the other coefficients in terms of some one of them. This remaining arbitrary coefficient may be adjusted so as to normalize χ^0_{kl}. This process does not give unique results if two or more roots W'_{kl} coincide, corresponding to the fact that since there still remains a certain amount of degeneracy the wave functions for the degenerate level are not uniquely determined but are to a certain degree arbitrary.

If the original wave functions $\psi^0_{k1}, \cdots, \psi^0_{k\alpha}$ were normalized and mutually orthogonal (which we have not hitherto needed to assume), the function $\Delta_{jl'}$ is unity for $j = l'$ and zero otherwise, so that the determinantal equation 24–17 assumes the form

$$\begin{vmatrix} H'_{11} - W'_{kl} & H'_{12} & H'_{13} & \cdots & H'_{1\alpha} \\ H'_{21} & H'_{22} - W'_{kl} & H'_{23} & \cdots & H'_{2\alpha} \\ \cdots & \cdots & \cdots & & \cdots \\ H'_{\alpha 1} & H'_{\alpha 2} & H'_{\alpha 3} & \cdots & H'_{\alpha\alpha} - W'_{kl} \end{vmatrix} = 0.$$

$$(24\text{-}18)$$

An equation such as 24–17 or 24–18 is often called a *secular equation*, and a perturbation of the type requiring the solution of such an equation a *secular perturbation*.[1]

It is interesting to note that in case the secular equation has the form

$$\begin{vmatrix} H'_{11} - W'_{kl} & 0 & \cdots \cdots & 0 \\ 0 & H'_{22} - W'_{kl} & \cdots \cdots & 0 \\ \cdots & \cdots & \cdots \cdots & \cdots \\ 0 & 0 & \cdots \cdots & H'_{\alpha\alpha} - W'_{kl} \end{vmatrix} = 0, \quad (24\text{–}19)$$

then the initially assumed functions ψ^0_{k1}, ψ^0_{k2}, $\cdots$, $\psi^0_{k\alpha}$ are the correct zeroth-order functions for the perturbation H', as is seen on evaluation of the coefficients κ of Equations 24–16. A secular equation in which all the elements are zero except along the principal diagonal is said to be in *diagonal form*. The roots W'_{kl} are of course immediately obtainable from an equation in this form, since the algebraic equation equivalent to it is

$$(H'_{11} - W'_{kl})(H'_{22} - W'_{kl}) \cdots (H'_{\alpha\alpha} - W'_{kl}) = 0, \quad (24\text{–}20)$$

with the roots $W'_{kl} = H'_{11}, H'_{22}, \cdots , H'_{\alpha\alpha}$.

[1] In this sense secular means "accomplished in a long period of time" (Latin *saeculum* = generation, age). The term *secular perturbation* was introduced in classical mechanics to describe a perturbation which produces a slow, cumulative effect on the orbit. If a system of sun and planet, for which the unperturbed orbits are ellipses of fixed size, shape, and orientation, were perturbed in such a way as to change the law of force slightly from the inverse square, as is done, for example, by the relativistic change of mass with change of speed, the position of the major axis in space would change by a small amount with each revolution of the planet, and the orbit would carry out a slow precession in its own plane, with a period which would be very long if the magnitude of the perturbation were small. Such a perturbation of the orbit is called a secular perturbation.

On the other hand we might have a system composed of a wheel in a gravitational field rotating about a horizontal frictionless axle passing through its center of mass and perturbed by the addition of a small weight at some point on its periphery in such a way as to accelerate the motion as the weight moves down and to decelerate it as the weight moves up. Such a perturbation, which produces a small effect on the motion with the high frequency characteristic of the original unperturbed motion of the system, is not a secular perturbation.

The significance of the use of the word *secular* in quantum mechanics will be seen after the study of the perturbation theory involving the time (given in Chap. XI).

This equation 24-19 illustrates, in addition, that the integrals H'_{mn} depend on the set of zeroth-order functions ψ^0_{kl} which is used to define them. Very often it is possible to guess in advance which set of degenerate ψ^0's to use for a given perturbation in order to obtain the simplest secular equation. In particular, in case that the perturbation is a function of one variable (x, say) alone, and each function of the initial set of unperturbed wave functions can be expressed as the product of a function of x and a function of the other variables, the individual functions being mutually orthogonal, then these product functions are correct zeroth-order wave functions for this perturbation. This situation arises whenever the unperturbed wave equation can be separated in a set of variables in which x is included.

It may be pointed out that Equation 24-18 may also be written in the form

$$\begin{vmatrix} H_{11} - W & H_{12} & \cdots & H_{1\alpha} \\ H_{21} & H_{22} - W & \cdots & H_{2\alpha} \\ \cdots & \cdots & \cdots & \cdots \\ H_{\alpha 1} & H_{\alpha 2} & \cdots & H_{\alpha\alpha} - W \end{vmatrix} = 0,$$

in which $H_{ij} = H^0_{ij} + H'_{ij}$ and $W = W^0_k + W'_{kl}$, inasmuch as H^0_{ij} is equal to W^0_k for $i = j$ and to zero for $i \neq j$. This form is used in Section 30c.

24a. An Example: Application of a Perturbation to a Hydrogen Atom.—As an illustration of the application of perturbation theory to degenerate systems, let us consider a hydrogen atom to which a perturbation which is a function of x only has been applied. Since the lowest state of the hydrogen atom is non-degenerate, the treatment of Section 23 applies to it and we have the result that

$$W' = \int \psi^2_{100} f(x) d\tau$$

with $H' = f(x)$. For the second energy state, however, we need to use the treatment for degenerate systems, since for $W^0_2 = -\frac{1}{4} Rhc$ there are four wave functions,

$$\psi_{2s} = \psi^0_{200} = \sqrt{\frac{1}{32\pi a_0^3}} e^{-\frac{r}{2a_0}} \left(\frac{r}{a_0} - 2 \right),$$

$$\psi_{2p_0} = \psi^0_{210} = \sqrt{\frac{1}{32\pi a_0^3}} e^{-\frac{r}{2a_0}} \left(\frac{r}{a_0} \right) \cos \vartheta,$$

$$\psi_{2p_{-1}} = \psi_{21\bar{1}}^0 = \sqrt{\frac{1}{32\pi a_0^3}} e^{-\frac{r}{2a_0}} \left(\frac{r}{a_0}\right) \cdot \frac{1}{2}\sqrt{2} e^{-i\varphi} \sin \vartheta,$$

$$\psi_{2p_{+1}} = \psi_{211}^0 = \sqrt{\frac{1}{32\pi a_0^3}} e^{-\frac{r}{2a_0}} \left(\frac{r}{a_0}\right) \cdot \frac{1}{2}\sqrt{2} e^{+i\varphi} \sin \vartheta,$$

as given in Chapter V. In order to set up the secular equation for this system we need the integrals

$$H'_{2lm,2l'm'} = \int \psi_{2lm}^{0*} f(x) \psi_{2l'm'}^0 d\tau.$$

Even without specifying the form of the function $f(x)$ further, we can say certain things about these integrals. Since the complex conjugate of $e^{-i\varphi}$ is $e^{+i\varphi}$ and $e^{-i\varphi}e^{+i\varphi} = 1$, we see that

$$B = H'_{21\bar{1},21\bar{1}} = H'_{211,211}$$

regardless of the nature of H', so long as it is real. By expressing x in polar coordinates through the equation

$$x = r \sin \vartheta \cos \varphi,$$

we see that $f(x)$ is the same function of $\varphi' = 2\pi - \varphi$ as it is of φ, since $\cos (2\pi - \varphi) = \cos \varphi$. If we make this substitution in an integral over φ we get the result

$$\int_0^{2\pi} g(\varphi)d\varphi = -\int_{2\pi}^0 g(2\pi - \varphi')d\varphi' = \int_0^{2\pi} g(2\pi - \varphi')d\varphi' = \int_0^{2\pi} g(2\pi - \varphi)d\varphi, \quad (24\text{-}21)$$

since it is immaterial what symbol we use for the variable of integration in a definite integral. This substitution also changes $e^{-i\varphi}$ into $e^{-i(2\pi-\varphi')}$ or $e^{+i\varphi'}$, so that by its use we can prove the identity

$$D = H'_{200,21\bar{1}} = H'_{200,211}.$$

$f(x)$ is also unchanged in form by the substitution $\vartheta = \pi - \vartheta'$, since $\sin (\pi - \vartheta') = \sin \vartheta'$. Also, we have the relation

$$\int_0^\pi g(\vartheta) \sin \vartheta d\vartheta = \int_0^\pi g(\pi - \vartheta') \sin \vartheta' d\vartheta' = \int_0^\pi g(\pi - \vartheta) \sin \vartheta d\vartheta, \quad (24\text{-}22)$$

in which the factor $\sin \vartheta$ is introduced because it occurs in the volume element $d\tau$ of polar coordinates. The substitution $\vartheta = \pi - \vartheta'$ does not leave $\cos \vartheta$ unchanged, however, since $\cos (\pi - \vartheta') = -\cos \vartheta'$. By employing this substitution we can show that

$$H'_{210,200} = -H'_{210,200} \quad \text{or} \quad H'_{210,200} = 0,$$

since the integrand is unaltered by the substitution except for the cosine factor in ψ^0_{210} which changes sign. Similarly we find

$$H'_{210,211} = 0 \quad \text{and} \quad H'_{210,21\bar{1}} = 0.$$

Finally we have the general rule that

$$H'_{2lm,2l'm'} = H'^*_{2l'm',2lm}.$$

We are now in a position to write down the secular equation for this perturbation, using the relations we have obtained among the elements $H'_{2lm,2l'm'}$. It is (using the order 200, 211, 21$\bar{1}$, 210 for the rows and columns)

$$\begin{vmatrix} A - W' & D & D & 0 \\ D & B - W' & E & 0 \\ D & E & B - W' & 0 \\ 0 & 0 & 0 & C - W' \end{vmatrix} = 0. \quad (24\text{--}23)$$

The symbols A, B, etc., have the meanings: $A = H'_{200,200}$; $B = H'_{211,211}$; $C = H'_{210,210}$, $D = H'_{200,211}$, and $E = H'_{211,21\bar{1}}$. We may obtain one root of this equation at once. Since the other elements of the row and the column which contains $C - W'$ are all zero, $C - W'$ is a factor of the determinant and may be equated to zero to obtain the root $W' = C$. The other three roots may be obtained by solving the cubic equation which remains, but inspection of the secular equation suggests a simpler method. Determinants have the property of being unchanged in value when the members of any row are added to or subtracted from the corresponding members of any other row. The same is true of the columns. We therefore have

$$\begin{vmatrix} A - W' & D & D \\ D & B - W' & E \\ D & E & B - W' \end{vmatrix}$$

$$= -\frac{1}{2} \begin{vmatrix} A - W' & 2D & 0 \\ D & B - W' + E & B - W' - E \\ D & B - W' + E & E - B + W' \end{vmatrix}$$

$$= \frac{1}{4} \begin{vmatrix} A - W' & 2D & 0 \\ 2D & 2(B + E - W') & 0 \\ 0 & 0 & 2(B - E - W') \end{vmatrix} = 0, \quad (24\text{--}24)$$

in which we have first added the last column to the second column to form a new second column and subtracted the last column from the second column to form a new third column, and then repeated this process on the rows instead of the columns. The result shows that we have factored out another root, $W' = B - E$, leaving now the quadratic equation

$$(A - W')(B + E - W') - 2D^2 = 0$$

which determines the remaining two roots.

The process by which we have factored the secular equation into two linear factors and a quadratic corresponds to using the real functions ψ_{2s}, ψ_{2p_x}, ψ_{2p_y}, and ψ_{2p_z} for the ψ_{kl}^0's instead of the set ψ_{2s}, ψ_{2p_1}, $\psi_{2p_{-1}}$, and ψ_{2p_0} (see Sec. 18b). In terms of the real set the secular equation has the form

$$\begin{vmatrix} A - W' & \sqrt{2}D & 0 & 0 \\ \sqrt{2}D & B + E - W' & 0 & 0 \\ 0 & 0 & B - E - W' & 0 \\ 0 & 0 & 0 & C - W' \end{vmatrix} = 0,$$

$$(24\text{--}25)$$

which, aside from the last row and column, differs from the last determinant of Equation 24–24 only by a constant factor. The proper zeroth-order wave functions for this perturbation are therefore ψ_{2p_y}, ψ_{2p_z}, and two linear combinations $\alpha\psi_{2s} + \beta\psi_{2p_x}$ and $\beta\psi_{2s} - \alpha\psi_{2p_x}$, in which the constants α and β are determined by solving the quadratic factor of the secular equation, substituting the roots into the equations for the coefficients of the linear combinations, and solving for the ratio α/β. The normalization condition yields the necessary additional equation.

It is to be noted that in place of ψ_{2p_y} and ψ_{2p_z} any linear combinations of these might have been used in setting up the secular equation 24–25, without changing the factoring of that equation, so that these linear combinations would also be satisfactory zeroth-order wave functions for this perturbation.

Problem 24–1.   Prove the statement of the last paragraph.
Problem 24–2.   Discuss the effect of a perturbation $f(y)$ [in place of $f(x)$] on the system of Section 24a.

25. SECOND-ORDER PERTURBATION THEORY

In the discussion of Section 23 we obtained expressions for W' and ψ' in the series

$$W = W^0 + \lambda W' + \lambda^2 W'' + \cdots \qquad (25\text{–}1)$$

and

$$\psi = \psi^0 + \lambda\psi' + \lambda^2\psi'' + \qquad (25\text{–}2)$$

In most problems it is either unnecessary or impracticable to carry the approximation further, but in some cases the second-order calculation can be carried out and is large enough to be important. This is especially true in cases in which the first-order energy W' is zero, as it is for the Stark effect for a free rotator, a problem which is important in the theory of the measurement of dipole moments (Sec. 49f).

The expressions for W'' and ψ'' are obtained from the equation which results when the coefficient of λ^2 in Equation 23–8 is put equal to zero and a solution obtained in a manner similar to that of the first-order treatment. We shall not give the details of the derivation but only state the results, which are, for the energy correction,

$$W_k'' = \sum_l{}' \frac{H_{kl}' H_{lk}'}{W_k^0 - W_l^0} + H_{kk}'', \qquad (25\text{–}3)$$

in which

$$H_{kl}' = \int \psi_k^{0*} H' \psi_l^0 d\tau \qquad (25\text{–}4)$$

and

$$H_{kk}'' = \int \psi_k^{0*} H'' \psi_k^0 d\tau \qquad (25\text{–}5)$$

and the prime on Σ means that the term $l = k$ is omitted. All other values of l must be included in the sum, however, including those corresponding to the continuous spectrum, if there is one. If the state W_k^0 is degenerate and the first-order perturbation has removed the degeneracy, then the functions to be used in calculating H_{kl}', etc., are the correct zeroth-order functions found by solving the secular equation.

If the energy level for the unperturbed problem is degenerate and the first-order perturbation does not remove the degeneracy, the application of the second-order correction will also not remove the degeneracy unless the term $\lambda^2 H''$ is different from zero, in which case the degeneracy may or may not be removed. The treatment in this case is closely similar to that of Section 24.

25a. An Example: The Stark Effect of the Plane Rotator.—A rigid body with a moment of inertia I and electric moment[1] μ, constrained to rotate in a plane about an axis passing through its center of mass and under the influence of a uniform electric field E, is characterized by a wave equation of the form[2]

$$\frac{d^2\psi}{d\varphi^2} + \frac{8\pi^2 I}{h^2}(W + \mu E \cos\varphi)\psi = 0,$$

in which φ is the angle of rotation. If we call $-\mu E \cos\varphi$ the perturbation term, with E taking the place of the parameter λ, then the unperturbed equation which remains when $E = 0$ has the normalized solutions

$$\psi_m^0 = \frac{1}{\sqrt{2\pi}}e^{im\varphi}, \qquad m = 0, \pm 1, \pm 2, \pm 3, \cdots, \quad (25\text{--}6)$$

and the energy values

$$W_m^0 = \frac{m^2 h^2}{8\pi^2 I}. \quad = \frac{m^2 \hbar^2}{2I} \qquad (25\text{--}7)$$

In order to calculate the perturbation energy we shall need integrals of the type

$$
\begin{aligned}
H'_{mm'} &= -\mu \int_0^{2\pi} \psi_m^{0*}\psi_{m'}^0 \cos\varphi d\varphi = -\frac{\mu}{2\pi}\int_0^{2\pi} e^{i(m'-m)\varphi}\cos\varphi d\varphi \\
&= -\frac{\mu}{4\pi}\int_0^{2\pi} e^{i(m'-m+1)\varphi}d\varphi - \frac{\mu}{4\pi}\int_0^{2\pi} e^{i(m'-m-1)\varphi}d\varphi \\
&= 0 \text{ for } m' \neq m \pm 1, \\
&= -\frac{\mu}{2} \text{ for } m' = m \pm 1.
\end{aligned}
\left.\begin{aligned}\\\\\\\\\end{aligned}\right\} \qquad (25\text{--}8)
$$

Using this result we see at once that the first-order energy correction is zero, for

$$W_m' = EH'_{mm} = 0. \qquad (25\text{--}9)$$

---

[1] For a definition of μ see Equation 3–5.

[2] This equation can be obtained as the approximate wave equation for a system of two particles constrained by a potential function which restricts the particles to a plane and keeps them a fixed distance apart by an argument similar to that used in the discussion of the diatomic molecule mentioned in the footnote to Section 35c.

This problem is really a degenerate one, since W^0 depends only on $|m|$ and not on the sign of m, so that there are two wave functions for every energy level (other than the lowest). It is, however, not necessary to consider this circumstance in evaluating W'_m and W''_m because neither the first- nor the second-order perturbation removes the degeneracy, and either the exponential functions 25–6 or the corresponding sine and cosine functions are satisfactory zeroth-order wave functions.

The second-order energy, as given by Equation 25–3, is

$$W''_m = E^2\frac{(H'_{m,m-1})^2}{W^0_m - W^0_{m-1}} + E^2\frac{(H'_{m,m+1})^2}{W^0_m - W^0_{m+1}} = \frac{4\pi^2 I \mu^2 E^2}{h^2(4m^2 - 1)},$$

(25–10)

so that the total energy, to the second order, is

$$W = W^0 + \lambda W' + \lambda^2 W'' = \frac{m^2 h^2}{8\pi^2 I} + \frac{4\pi^2 I \mu^2 E^2}{h^2(4m^2 - 1)}. \quad (25–11)$$

It is interesting to point out the significance of this result in connection with the effect of the electric field on the *polarizability* of the rotator. The polarizability α is the proportionality factor between the induced dipole moment and the applied field E. The energy of an induced dipole in a field is then $-\frac{1}{2}\alpha E^2$. From this and a comparison with Equation 25–11 we obtain the relation

$$\alpha = -\frac{8\pi^2 I \mu^2}{h^2(4m^2 - 1)}, \quad (25–12)$$

which shows that α is positive for $m = 0$; the induced dipole (which in this case is due to the orienting effect of the field E on the permanent dipole μ of the rotator) is therefore in the direction of the field E. For $|m| > 0$, however, the opposite is true and the field tends to orient the dipole in the reverse direction.

This is similar to the classical-mechanical result, which is that a plane rotator with insufficient energy to make a complete rotation in the field tends to be oriented parallel to the field while a rotator with energy great enough to permit complete rotation is speeded up when parallel and slowed down when antiparallel to the field so that the resulting polarization is opposed to the field.[1]

[1] An interesting application of perturbation theory has been made to the Stark effect of the hydrogen atom, the first-order treatment having been

Problem 25–1. Carry out a treatment similar to the above treatment for the rigid rotator in space, using the wave equation and wave functions found in the footnote of Section 35c. Discuss the results from the viewpoint of the last paragraph above. Compute the average contribution to the polarizability of all the states with given l and with $m = -l, -l + 1, \cdots,$ $+ l$, assigning equal weights to the states in the averaging.

---

given independently by Schrödinger, *Ann. d. Phys.* **80**, 437 (1926), and P. S. Epstein, *Phys. Rev.* **28**, 695 (1926), the second order by Epstein, *loc. cit.*, G. Wentzel, *Z. f. Phys.* **38**, 518 (1926), and I. Waller, *ibid.* **38**, 635 (1926), and the third order by S. Doi, Y. Ishida, and S. Hiyama, *Sci. Papers Tokyo* **9**, 1 (1928), and M. A. El-Sherbini, *Phil. Mag.* **13**, 24 (1932). See also Sections 27a and 27e.

CHAPTER VII

THE VARIATION METHOD AND OTHER APPROXIMATE METHODS

There are many problems of wave mechanics which cannot be conveniently treated either by direct solution of the wave equation or by the use of perturbation theory. The helium atom, discussed in the next chapter, is such a system. No direct method of solving the wave equation has been found for this atom, and the application of perturbation theory is unsatisfactory because the first approximation is not accurate enough while the labor of calculating the higher approximations is extremely great.

In many applications, however, there are methods available which enable approximate values for the energy of certain of the states of the system to be computed. In this chapter we shall discuss some of these, paying particular attention to the *variation method*, inasmuch as this method is especially applicable to the lowest energy state of the system, which is the state of most interest in chemical problems.

26. THE VARIATION METHOD

26a. The Variational Integral and Its Properties.—We shall show[1] in this section that the integral

$$E = \int \phi^* H \phi d\tau \qquad (26\text{--}1)$$

is an upper limit to the energy W_0 of the lowest state of a system. In this equation, H is the complete Hamiltonian operator $H\left(\dfrac{h}{2\pi i}\dfrac{\partial}{\partial q}, q\right)$ for the system under discussion (Sec. 12a) and $\phi(q)$ is any normalized function of the coordinates of the system satisfying the auxiliary conditions of Section 9c for a satisfactory wave function. The function ϕ is otherwise completely unre-

---

[1] C. Eckart, *Phys. Rev.* **36**, 878 (1930).

stricted; its choice may be quite arbitrary, but the more wisely it is chosen the more closely will E approach the energy W_0.

If we used for our function ϕ, called the *variation function*, the true wave function ψ_0 of the lowest state, E would equal W_0; that is,

$$E = \int \psi_0^* H \psi_0 d\tau = W_0, \qquad (26\text{-}2)$$

since

$$H \psi_0 = W_0 \psi_0.$$

If ϕ is not equal to ψ_0 we may expand ϕ in terms of the complete set of normalized, orthogonal functions $\psi_0, \psi_1, \cdots, \psi_n, \cdots$, obtaining

$$\phi = \sum_n a_n \psi_n, \quad \text{with} \quad \sum_n a_n^* a_n = 1. \qquad (26\text{-}3)$$

Substitution of this expansion in the integral for E leads to the equation

$$E = \sum_n \sum_{n'} a_n^* a_{n'} \int \psi_n^* H \psi_{n'} d\tau = \sum_n a_n^* a_n W_n, \qquad (26\text{-}4)$$

inasmuch as the functions ψ_n satisfy the equations

$$H \psi_n = W_n \psi_n. \qquad (26\text{-}5)$$

Subtracting W_0, the lowest energy value, from both sides gives

$$E - W_0 = \sum_n a_n^* a_n (W_n - W_0). \qquad (26\text{-}6)$$

Since W_n is greater than or equal to W_0 for all values of n and the coefficients $a_n^* a_n$ are of course all positive or zero, the right side of Equation 26-6 is positive or zero. We have therefore proved that E is always an upper limit to W_0; that is,

$$E \geqslant W_0. \qquad (26\text{-}7)$$

This theorem is the basis of the variation method for the calculation of the approximate value of the lowest energy level of a system. If we choose a number of variation functions $\phi_1, \phi_2, \phi_3, \cdots$ and calculate the values $E_1, E_2, E_3, \cdots$ corresponding to them, then each of these values of E is greater than the energy W_0, so that the lowest one is the nearest to W_0. Often the functions $\phi_1, \phi_2, \phi_3, \cdots$ are only distinguished by having different values of some parameter. The process of minimizing E with respect to this parameter may then be carried out in order to obtain the best approximation to W_0 which the form of the trial function ϕ will allow.

If good judgment has been exercised in choosing the trial function ϕ, especially if a number of parameters have been introduced into ϕ in such a manner as to allow its form to be varied considerably, the value obtained for E may be very close to the true energy W_0. In the case of the helium atom, for example, this method has been applied with great success, as is discussed in the next chapter.

If E is equal to W_0 then ϕ is identical[1] with ψ_0 (as can be seen from Eq. 26–6), so that it is natural to assume that if E is nearly equal to W_0 the function ϕ will approximate closely to the true wave function ψ_0. The variation method is therefore very frequently used to obtain approximate wave functions as well as approximate energy values. From Equation 26–6 we see that the application of the variation method provides us with that function ϕ among those considered which approximates most closely to ψ_0 according to the following criterion: On expanding $\phi - \psi_0$ in terms of the correct wave functions ψ_n, the quantity $\sum_n a_n a_n(W_n - W_0)$ is minimized; that is, the sum of the squares of the absolute values of the coefficients of the wave functions for excited states with the weight factors $W_n - W_0$ is minimized. For some purposes (as of course for the calculation of the energy of the system) this is a good criterion to use; but for others the approximate wave function obtained in this way might not be the most satisfactory one.

Eckart[2] has devised the following way of estimating how closely a variation function approximates to the true solution ψ_0 by using E and the experimental values of W_0 and W_1. A very reasonable criterion of the degree of approximation of ϕ to ψ_0 (for real functions) is the smallness of the quantity

$$\epsilon = \int (\phi - \psi_0)^2 d\tau = \int (\phi^2 - 2\psi_0 \sum_n a_n \psi_n + \psi_0^2) d\tau = 2 - 2a_0,$$

$$(26–8)$$

[1] If the level W_0 is degenerate, the equality of E and W_0 requires that ϕ be identical with one of the wave functions corresponding to W_0.

[2] Reference on p. 180.

in which a_0 is the coefficient of ψ_0 in the expansion 26–3 of ϕ
From Equation 26–6 we can write

$$E - W_0 = \sum_{n=0}^{\infty} a_n^2 (W_n - W_0) \geqslant \sum_{n=1}^{\infty} a_n^2 (W_1 - W_0)$$

or

$$E - W_0 \geqslant (W_1 - W_0)(1 - a_0^2).$$

Therefore if ϵ^2 is small compared to ϵ, we may combine this
equation and Equation 26–8, obtaining

$$\epsilon < \frac{E - W_0}{W_1 - W_0} \quad \text{or} \quad 1 - a_0 < \frac{1}{2}\frac{E - W_0}{W_1 - W_0}. \quad (26\text{–}9)$$

Thus, from a knowledge of the correct energy values W_0 and W_1
for the two lowest levels of the systems and the energy integral E
for a variation function ϕ, we obtain an upper limit for the
deviation of a_0 from unity, that is, of the contribution to ϕ of
wave functions other than ψ_0.

The variation method has the great drawback of giving only an
upper limit to the energy, with no indication of how far from the
true energy that limit is. (In Section 26e we shall discuss a
closely related method, which is not, however, so easy to
apply, by means of which both an upper and a lower limit can be
obtained.) Nevertheless, it is very useful because there arise
many instances in which we have physical reasons for believing
that the wave function approximates to a certain form, and this
method enables these intuitions to be utilized in calculating a
better approximation to the energy than can be easily obtained
with the use of perturbation theory.

If we use for ϕ the zeroth-order approximation to the wave
function ψ_0^0 discussed under perturbation theory, Chapter VI,
and consider H as equal to $H^0 + H'$, this method gives for
E a value identical with the first-order perturbation energy
$W_0^0 + W_0'$. If therefore we use for ϕ a variation function containing parameters such that for certain values of the parameters
ϕ reduces to ψ_0^0, the value we obtain for E is always at least as
good as that given by the first-order perturbation treatment.
If ϕ is set equal to the first-order wave function, the energy value
E given by the variation method is the same, to the second
power in the parameter λ, as the second-order energy obtained
by the perturbation treatment.

In case that it is not convenient to normalize ϕ, the above considerations retain their validity provided that E is given by the expression

$$E = \frac{\int \phi^* H \phi d\tau}{\int \phi^* \phi d\tau}. \qquad (26\text{--}10)$$

26b. An Example: The Normal State of the Helium Atom.— In Section 23b we treated the normal state of the helium atom with the use of first-order perturbation theory. In this section we shall show that the calculation of the energy can be greatly increased in accuracy by considering the quantity Z which occurs in the exponent ($\rho = 2Zr/a_0$) of the zeroth-order function given in Equations 23–34 and 23–37 as a parameter Z' instead of as a constant equal to the atomic number. The value of Z' is determined by using the variation method with ϕ given by

$$\phi = \phi_1 \phi_2 = \left(\frac{Z'^3}{\pi a_0^3}\right) e^{-\frac{Z'r_1}{a_0}} e^{-\frac{Z'r_2}{a_0}}, \qquad (26\text{--}11)$$

in which Z', the effective atomic number, is a variable parameter. In this problem, the Hamiltonian operator is

$$H = -\frac{h^2}{8\pi^2 m_0}(\nabla_1^2 + \nabla_2^2) - Ze^2\left(\frac{1}{r_1} + \frac{1}{r_2}\right) + \frac{e^2}{r_{12}},$$

in which Z is the true atomic number. The factors ϕ_1 and ϕ_2 of ϕ are hydrogenlike wave functions for nuclear charge $Z'e$, so that ϕ_1 satisfies the equation

$$-\frac{h^2}{8\pi^2 m_0}\nabla_1^2 \phi_1 = \frac{Z'e^2}{r_1}\phi_1 - Z'^2 W_H \phi_1 \qquad (26\text{--}12)$$

(W_H being equal to $e^2/2a_0$), with a similar equation for ϕ_2. Using these and the expression for H, we obtain

$$E = -2Z'^2 W_H + (Z' - Z)e^2 \int \phi^*\left(\frac{1}{r_1} + \frac{1}{r_2}\right)\phi d\tau +$$

$$\int \phi^* \frac{e^2}{r_{12}}\phi d\tau. \qquad (26\text{--}13)$$

The first integral on the right has the value

$$e^2 \int \phi^* \left(\frac{1}{r_1} + \frac{1}{r_2} \right) \phi d\tau = 2e^2 \int \frac{\phi_1^2}{r_1} d\tau_1 =$$

$$\frac{2e^2 Z'^3}{\pi a_0^3} \int_0^\infty \int_0^\pi \int_0^{2\pi} \frac{1}{r_1} e^{-\frac{2Z'r_1}{a_0}} r_1^2 \sin\theta d\varphi d\vartheta dr_1 = \frac{8Z'^3 e^2}{a_0^3} \int_0^\infty r_1 e^{-\frac{2Z'r_1}{a_0}} dr_1 =$$

$$\frac{2Z' e^2}{a_0} = 4Z' W_H. \quad (26\text{--}14)$$

The second integral of Equation 26–13 is the same as that of Equation 23–38 if Z is replaced by Z'. It therefore has the value

$$\int \phi^* \frac{e^2}{r_{12}} \phi d\tau = \frac{5}{4} Z' W_H. \quad (26\text{--}15)$$

Combining these results, we obtain for E the expression

$$E = \{ -2Z'^2 + \tfrac{5}{4}Z' + 4Z'(Z' - Z) \} W_H. \quad (26\text{--}16)$$

Minimizing E with respect to Z' gives

$$\frac{\partial E}{\partial Z'} = 0 = \left(-4Z' + \frac{5}{4} + 8Z' - 4Z \right) W_H$$

or

$$Z' = Z - \tfrac{5}{16}, \quad (26\text{--}17)$$

which leads to

$$E = -2(Z - \tfrac{5}{16})^2 W_H. \quad (26\text{--}18)$$

As pointed out in Section 29c, this treatment cuts the error in the energy of helium to one-third of the error in the first-order perturbation treatment. In the same section, more elaborate variation functions are applied to this problem, with very accurate results.

Problem 26–1. Calculate the energy of a normal hydrogen atom in a uniform electric field of strength F along the z axis by the variation method, and hence evaluate the polarizability α, such that the field energy is $-\tfrac{1}{2}\alpha F^2$. Use for the variation function the expression[1]

[1] The correct value of α for the normal hydrogen atom, given by the second-order perturbation theory (footnote at end of preceding chapter) is

$$\alpha = \tfrac{9}{2} a_0^3 = 0.667 \cdot 10^{-24} \text{ cm}^3.$$

A value agreeing exactly with this has been obtained by the variation method by H. R. Hassé, *Proc. Cambridge Phil. Soc.* **26,** 542 (1930), using the variation function $\psi_{1s}(1 + Az + Bzr)$. Hassé also investigated the effects of additional terms (cubic and quartic) in the series, finding them to be negligible. The same result is given by the treatment of Section 27a.

$$\psi_{1s}(1 + Az),$$

minimizing the energy with respect to A, with neglect of powers of F higher than F^2.

$$\alpha = 4a_0^3 = 0.59 \cdot 10^{-24} \text{ cm}^3. \quad Ans.$$

26c. Application of the Variation Method to Other States.—
The theorem $E \geqslant W_0$, proved in Section 26a, may be extended in special cases to states of the system other than the lowest one. It is sometimes possible to choose ϕ so that the first few coefficients a_0, a_1, $\cdots$ of the expansion 26–3 are zero. If, for example, a_0, a_1, and a_2 are all zero, then by subtracting W_3 from both sides of Equation 26–4 we obtain

$$E - W_3 = \sum_n a_n a_n^*(W_n - W_3) \geqslant 0, \qquad (26\text{–}19)$$

since, although $W_0 - W_3$, $W_1 - W_3$, and $W_2 - W_3$ are negative, their coefficients are zero. In this case then we find the inequality $E \geqslant W_3$.

There are several cases in which such a situation may arise. The simplest illustration is a one-dimensional problem in which the independent variable x goes from $-\infty$ to $+\infty$ and the potential function V is an even function of x, so that

$$V(-x) = V(+x).$$

The wave function belonging to the lowest level of such a system is always an even function; i.e., $\psi_0(-x) = \psi_0(x)$; while ψ_1 is odd, with $\psi_1(-x) = -\psi_1(x)$ (see Sec. 9c). If we therefore use for ϕ an even function, we can only say that E is greater than or equal to W_0, but if ϕ is an odd function, a_0 will be zero (also all a_n's with n even) and the relation $E \geqslant W_1$ will hold. For such a problem the variation method may be used to obtain the two lowest energy levels.

The variation method may also be applied to the lowest state of given resultant angular momentum and of given electron-spin multiplicity, as will be discussed in the next chapter (Sec. 29d). Still another method of extending the variation method to levels other than the lowest is given in the following section.

26d. Linear Variation Functions.[1]—A very convenient type of variation function is one which is the sum of a number of linearly

[1] The generalized perturbation theory of Section 27a is closely related to the treatment discussed here.

independent functions χ_1, χ_2, $\cdots$, χ_m with undetermined coefficients c_1, c_2, $\cdots$, c_m. In other words the variation function ϕ has the form

$$\phi = c_1\chi_1 + c_2\chi_2 + \cdots + c_m\chi_m, \qquad (26\text{--}20)$$

in which c_1, c_2, $\cdots$, c_m are the parameters which are to be determined to give the lowest value of E and therefore the best approximation to W_0. It is assumed that the functions χ_1, χ_2, $\cdots$, χ_m satisfy the conditions of Section 9c. If we introduce the symbols

$$H_{nn'} = \int \chi_n H \chi_{n'} d\tau \qquad \text{and} \qquad \Delta_{nn'} = \int \chi_n \chi_{n'} d\tau, \qquad (26\text{--}21)$$

in which for simplicity we have assumed that ϕ is real, then the expression for E becomes

$$E = \frac{\int \phi H \phi \, d\tau}{\int \phi \phi \, d\tau} = \frac{\displaystyle\sum_{n=1}^{m} \sum_{n'=1}^{m} c_n c_{n'} H_{nn'}}{\displaystyle\sum_{n=1}^{m} \sum_{n'=1}^{m} c_n c_{n'} \Delta_{nn'}} \qquad (26\text{--}22)$$

or

$$E \sum_n \sum_{n'} c_n c_{n'} \Delta_{nn'} = \sum_n \sum_{n'} c_n c_{n'} H_{nn'}.$$

To find the values of c_1, c_2, $\cdots$, c_m which make E a minimum, we differentiate with respect to each c_k:

$$\frac{\partial E}{\partial c_k} \sum_n \sum_{n'} c_n c_{n'} \Delta_{nn'} + E \frac{\partial}{\partial c_k} \left(\sum_n \sum_{n'} c_n c_{n'} \Delta_{nn'} \right) =$$

$$\frac{\partial}{\partial c_k} \left(\sum_n \sum_{n'} c_n c_{n'} H_{nn'} \right).$$

The condition for a minimum is that $\dfrac{\partial E}{\partial c_k} = 0$ for $k = 1, 2, \cdots$, m, which leads to the set of equations

$$\sum_n c_n (H_{nk} - \Delta_{nk} E) = 0, \qquad k = 1, 2, \cdots, m. \quad (26\text{--}23)$$

This is a set of m simultaneous homogeneous linear equations in the m independent variables c_1, c_2, $\cdots$, c_m. For this set of equations to have a non-trivial solution it is necessary that the determinant of the coefficients vanish (cf. Sec. 24); i.e., that

$$\begin{vmatrix} H_{11} - \Delta_{11}E & H_{12} - \Delta_{12}E & \cdots & H_{1m} - \Delta_{1m}E \\ H_{21} - \Delta_{21}E & H_{22} - \Delta_{22}E & \cdots & H_{2m} - \Delta_{2m}E \\ \cdots & \cdots & \cdots & \cdots \\ H_{m1} - \Delta_{m1}E & H_{m2} - \Delta_{m2}E & \cdots & H_{mm} - \Delta_{mm}E \end{vmatrix} = 0.$$

$$(26\text{-}24)$$

This equation is closely similar to the secular equation 24–17 of perturbation theory. It may be solved by numerical methods,[1] or otherwise, and the lowest root $E = E_0$ is an upper limit to

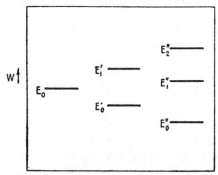

Fig. 26-1.—Figure showing the interleaving of energy values for linear variation functions with added terms.

the energy W_0. Substitution of this value of E_0 in Equations 26–23 and solution of these equations for $c_2, c_3, \cdots, c_m$ in terms of c_1 (which can be used as a normalizing factor) gives the variation function ϕ_0 corresponding to E_0.

The other roots $E_1, E_2, \cdots, E_{m-1}$ of Equation 26–24 are upper limits for $W_1, W_2, \cdots, W_{m-1}$, respectively.[2] Furthermore, it is possible to state how these roots will be changed when a new trial function ϕ' is used, containing one more function χ_{m+1},

$$\phi' = c_1\chi_1 + c_2\chi_2 + \cdots + c_m\chi_m + c_{m+1}\chi_{m+1}. \quad (26\text{-}25)$$

In this case the roots $E_0', E_1', E_2', \cdots, E_m'$ will be separated by the old ones $E_0, E_1, E_2, \cdots, E_m$ as shown in Figure 26–1.

---

[1] For a convenient numerical method see H. M. James and A. S. Coolidge, *J. Chem. Phys.* **1**, 825 (1933).

[2] J. K. L. MacDonald, *Phys. Rev.* **43**, 830 (1933).

In other words, the relations $E_0' \leqslant E_0$, $E_1' \leqslant E_1$, etc., and $E_0 \leqslant E_1'$, $E_1 \leqslant E_2'$, etc., are satisfied. This method has proved to be very useful in practice, as will be illustrated by examples discussed in Chapters VIII and XII.

The application of the variation method to wave mechanics grew from the work of Ritz, *J. f. reine u. angew. Math.* **135**, 1 (1909), who considered the solution of certain differential equations by discussing the equivalent variation problem. It can be shown that a general normalized function ϕ_1 which satisfies the boundary conditions of Section 9c and which makes the integral $E = \int \phi_1^* H \phi_1 d\tau$ a minimum relative to all variations in ϕ_1 is a solution of the differential equation $H\psi = W\psi$, E then being equal to the corresponding characteristic energy value. A similar minimization of E with respect to all variations in another general normalized function ϕ_2 with the added restriction that ϕ_2 is orthogonal to ϕ_1 leads to another solution ψ_2 of the wave equation. By the continuation of this process of minimization, all of the solutions can be found. Ritz proved that in certain cases a rigorous solution can be obtained by applying a limiting process to the integral $\int \phi^* H \phi d\tau$, in which ϕ is represented as the sum of m functions of a convenient set of normalized orthogonal functions $\psi_1, \psi_2, \cdots$ which satisfy the boundary conditions, taken with arbitrary coefficients $c_1, c_2, \cdots, c_m$. For each value of m the coefficients c_m are determined so that the integral $\int \phi^* H \phi d\tau$ is a minimum, keeping $\int \phi^* \phi d\tau = 1$. Ritz found that under certain restrictions the sequence of functions converges to a true solution of the wave equation and the sequence of values of the integral converges to the corresponding true characteristic value. The approximate method discussed in this section is very closely related to the Ritz method, differing from it in that the functions ψ are not necessarily members of a complete orthogonal set and the limiting process is not carried out.

Problem 26–2. Using a variation function of the form $\phi = A + B \cos \varphi + C \sin \varphi$, obtain an upper limit to the lowest energy level of the plane rotator in an electric field, for which the wave equation is

$$\frac{d^2\psi}{d\varphi^2} + \frac{8\pi^2 I}{h^2}(W + \mu E \cos \varphi)\psi = 0.$$

26e. A More General Variation Method.—A method has been devised[1] which gives both an upper and a lower limit for an energy level. If we represent by E and D the integrals

$$E = \int \phi^* H \phi d\tau \quad \text{and} \quad D = \int (H\phi)^*(H\phi)d\tau, \quad (26\text{–}26)$$

in which ϕ is a normalized trial variation function as before, then we shall show that some energy level W_k satisfies the relation

$$E + \sqrt{D - E^2} \geqslant W_k \geqslant E - \sqrt{D - E^2}. \quad (26\text{–}27)$$

[1] D. H. WEINSTEIN, *Proc. Nat. Acad. Sci.* **20**, 529 (1934); see also J. K. L. MACDONALD, *Phys. Rev.* **46**, 828 (1934).

To prove this we expand ϕ as before (Eq. 26-3), so that

$$E = \sum_n a_n^* a_n W_n, \qquad D = \sum_n a_n^* a_n W_n^2, \qquad \text{and} \qquad \sum_n a_n^* a_n = 1.$$

$$(26\text{-}28)$$

From this we obtain the result

$$\Delta = D - E^2 = \sum_n a_n^* a_n W_n^2 - 2E \sum_n a_n^* a_n W_n + E^2 \sum_n a_n^* a_n =$$

$$\sum_n a_n^* a_n (W_n - E)^2. \quad (26\text{-}29)$$

There will be some energy level W_k which lies at least as near E as any other, i.e., for which

$$(W_k - E)^2 \leqslant (W_n - E)^2.$$

Therefore Δ is related to $W_k - E$ by the inequality

$$\Delta \geqslant (W_k - E)^2 \sum_n a_n^* a_n$$

or

$$\Delta \geqslant (W_k - E)^2. \qquad (26\text{-}30)$$

There are now two possible cases,

$$W_k \geqslant E \qquad \text{and} \qquad W_k < E.$$

In the first case we have

$$\sqrt{\Delta} \geqslant W_k - E, \qquad \text{so that} \qquad E + \sqrt{\Delta} \geqslant W_k \geqslant E;$$

and in the second case

$$\sqrt{\Delta} \geqslant E - W_k, \qquad \text{and} \qquad E > W_k \geqslant E - \sqrt{\Delta}.$$

From this we see that the condition in Equation 26-27 applies to both cases.

The application of this method to actual problems of the usual type is more difficult than that of the simple variation method because, in addition to the integral E, it is necessary to evaluate D, which ordinarily is considerably more difficult than E.

It may be pointed out that by varying parameters in a function in such a way as to make Δ a minimum the function ϕ is made to approach some correct wave function ψ_k as closely as is permitted by the form of ϕ. This method consequently may be considered as another type of variation method applicable to any state of a system.

27. OTHER APPROXIMATE METHODS

There are a number of other methods which may be used to obtain approximate wave functions and energy levels. Five of these, a generalized perturbation method, the Wentzel-Kramers-Brillouin method, the method of numerical integration, the method of difference equations, and an approximate second-order perturbation treatment, are discussed in the following sections. Another method which has been of some importance is based on the polynomial method used in Section 11*a* to solve the harmonic oscillator equation. Only under special circumstances does the substitution of a series for ψ lead to a two-term recursion formula for the coefficients, but a technique has been developed which permits the computation of approximate energy levels for low-lying states even when a three-term recursion formula is obtained. We shall discuss this method briefly in Section 42*c*.

27a. A Generalized Perturbation Theory.—A method of approximate (and in some cases exact) solution of the wave equation which has been found useful in many problems was developed by Epstein[1] in 1926, immediately after the publication of Schrödinger's first papers, and applied by him in the complete treatment of the first-order and second-order Stark effects of the hydrogen atom. The principal feature of the method is the expansion of the wave function in terms of a complete set of orthogonal functions which are not necessarily solutions of the wave equation for any unperturbed system related to the system under treatment, nor even necessarily orthogonal functions in the same configuration space. Closely related discussions of perturbation problems have since been given by a number of authors, including Slater and Kirkwood[2] and Lennard-Jones.[3] In the following paragraphs we shall first discuss the method in general, then its application to perturbation problems and its relation to ordinary perturbation theory (Chap. VI), and finally as an illustration its application to the second-order Stark effect for the normal hydrogen atom.

In applying this method in the discussion of the wave equation

$$H\psi(x) = W\psi(x), \qquad (27-1)$$

[1] P. S. EPSTEIN, *Phys. Rev.* **28,** 695 (1926).

[2] J. C. SLATER and J. G. KIRKWOOD, *Phys. Rev.* **37,** 682 (1931).

[3] J. E. LENNARD-JONES, *Proc. Roy. Soc.* **A 129,** 598 (1930).

in which x is used to represent all of the independent variables for the system, we express $\psi(x)$ in terms of certain functions $F_n(x)$, writing

$$\psi(x) = \sum_n A_n F_n(x). \qquad (27\text{--}2)$$

The functions $F_n(x)$ are conveniently taken as the members of a complete set of orthogonal functions of the variables x; it is not necessary, however, that they be orthogonal in the same configuration space as that for the system under discussion. Instead, we assume that they satisfy the normalization and orthogonality conditions

with

$$\int F_m^*(x)F_n(x)\rho(x)dx = \delta_{mn}$$

$$\delta_{mn} = \begin{cases} 1 \text{ for } m = n, \\ 0 \text{ for } m \neq n, \end{cases} \qquad (27\text{--}3)$$

in which $\rho(x)dx$ may be different from the volume element $d\tau$ corresponding to the wave equation 27–1. $\rho(x)$ is called the *weight factor*[1] for the functions $F_n(x)$. On substituting the expression 27–2 in Equation 27–1, we obtain

$$\sum_n A_n(H - W)F_n(x) = 0, \qquad (27\text{--}4)$$

which on multiplication by $F_m^*(x)\rho(x)dx$ and integration becomes

$$\sum_n A_n(H_{mn} - W\delta_{mn}) = 0, \qquad m = 1, 2, \cdots, \quad (27\text{--}5)$$

in which

$$H_{mn} = \int F_m^*(x)HF_n(x)\rho(x)dx. \qquad (27\text{--}6)$$

[1] In case that the functions $F_n(x)$ satisfy the differential equation

$$\frac{d}{dx}\left\{p(x)\frac{dF}{dx}\right\} - q(x)F + \lambda\rho(x)F = 0,$$

in which λ is the characteristic-value parameter, they are known to form a complete set of functions which are orthogonal with respect to the weight factor $\rho(x)$. For a discussion of this point and other properties of differential equations of the Sturm-Liouville type see, for example, R. Courant and D. Hilbert, "Methoden der mathematischen Physik," Julius Springer, Berlin, 193.

For an arbitrary choice of the functions $F_n(x)$ Equation 27-5 represents an infinite number of equations in an infinite number of unknown coefficients A_n. Under these circumstances questions of convergence arise which are not always easily answered. In special cases, however, only a finite number of functions $F_n(x)$ will be needed to represent a given function $\psi(x)$; in these cases we know that the set of simultaneous homogeneous linear equations 27-5 has a non-trivial solution only when the determinant of the coefficients of the A_n's vanishes; that is, when the condition

$$\begin{vmatrix} H_{11} - W & H_{12} & H_{13} & \cdots \\ H_{21} & H_{22} - W & H_{23} & \cdots \\ H_{31} & H_{32} & H_{33} - W & \cdots \\ \cdots & \cdots & \cdots & \cdots \end{vmatrix} = 0 \qquad (27\text{-}7)$$

is satisfied. We shall assume that in the infinite case the mathematical questions of convergence have been settled, and that Equation 27-7, involving a convergent infinite determinant, is applicable.

Our problem is now in principle solved: We need only to evaluate the roots of Equation 27-7 to obtain the allowed energy values for the original wave equation, and substitute them in the set of equations 27-5 to evaluate the coefficients A_n and obtain the wave functions.

The relation of this treatment to the perturbation theory of Chapter VI can be seen from the following arguments. If the functions $F_n(x)$ were the true solutions $\psi_n(x)$ of the wave equation 27-1, the determinantal equation 27-7 would have the form

$$\begin{vmatrix} W_1 - W & 0 & 0 & \cdots \\ 0 & W_2 - W & 0 & \cdots \\ 0 & 0 & W_3 - W & \cdots \\ \cdots & \cdots & \cdots & \cdots \end{vmatrix} = 0, \qquad (27\text{-}8)$$

with roots $W = W_1$, $W = W_2$, etc. Now, if the functions $F_n(x)$ closely approximate the true solutions $\psi_n(x)$, the non-diagonal terms in Equation 27-7 will be small, and as an approximation we can neglect them. This gives

$$\left. \begin{aligned} W_1 &= H_{11}, \\ W_2 &= H_{22}, \\ W_3 &= H_{33}, \\ \text{etc.,} \end{aligned} \right\} \qquad (27\text{-}9)$$

which corresponds to ordinary first-order perturbation theory, inasmuch as, if H can be written as $H^0 + H'$, with

$$H^0 F_n(x) = W_n^0 F_n(x),$$

then $W_n = H_{nn}$ has the value $W_n = W_n^0 + \int F_n^*(x) H' F_n(x) \rho(x) dx$, which is identical with the result of ordinary first-order perturbation theory of Section 23 when $\rho(x) dx = d\tau$. Equation 27–9 is more general than the corresponding equation of first-order perturbation theory, since the functions $F_n(x)$ need not correspond to any unperturbed system. On the other hand, it may not be so reliable, in case that a poor choice of functions $F_n(x)$ is made; the first step of ordinary perturbation theory is essentially a procedure for finding suitable zeroth-order functions.

It may happen that some of the non-diagonal terms are large and others small; in this case neglect of the small terms leads to an equation such as

$$\begin{vmatrix} H_{11} - W & H_{12} & 0 & 0 & \cdots \\ H_{21} & H_{22} - W & 0 & 0 & \cdots \\ 0 & 0 & H_{33} - W & 0 & \cdots \\ 0 & 0 & 0 & H_{44} - W & \cdots \\ \cdots & \cdots & \cdots & \cdots & \cdots \end{vmatrix} = 0,$$

which can be factored into the equations

$$\left. \begin{aligned} \begin{vmatrix} H_{11} - W & H_{12} \\ H_{21} & H_{22} - W \end{vmatrix} &= 0, \\ H_{33} - W &= 0, \\ H_{44} - W &= 0, \\ \text{etc.} \end{aligned} \right\} \qquad (27\text{–}10)$$

It is seen that this treatment is analogous to the first-order perturbation treatment for degenerate states as given in Section 24. The more general treatment now under discussion is especially valuable in case that the unperturbed levels are not exactly equal, that is, in case of approximate degeneracy.

A second approximation to the solution of Equation 27–7 can be made in the following manner. Suppose that we are interested in the second energy level, for which the value H_{22} is found for the energy as a first approximation. We introduce this expression for W everywhere except in the term $H_{22} - W$, and neglect non-diagonal terms except H_{2n} and H_{n2}, thus obtaining the equation

$$
\begin{vmatrix}
H_{11} - H_{22} & H_{12} & 0 & 0 & \cdots \\
H_{21} & H_{22} - W & H_{23} & H_{24} & \cdots \\
0 & H_{32} & H_{33} - H_{22} & 0 & \cdots \\
0 & H_{42} & 0 & H_{44} - H_{22} & \cdots \\
\cdots & \cdots & \cdots & \cdots & \cdots
\end{vmatrix} = 0. \quad (27\text{-}11)
$$

On multiplying out the determinant, we convert this equation into the form

$$
\begin{aligned}
&(H_{22} - W)(H_{11} - H_{22})(H_{33} - H_{22})(H_{44} - H_{22}) \cdots \\
&- H_{12}H_{21}(H_{33} - H_{22})(H_{44} - H_{22}) \cdots \\
&- H_{32}H_{23}(H_{11} - H_{22})(H_{44} - H_{22}) \cdots - \cdots = 0,
\end{aligned}
$$

with the solution

$$
W = H_{22} - \sideset{}{'}\sum_{l} \frac{H_{2l}H_{l2}}{H_{ll} - H_{22}}, \quad (27\text{-}12)
$$

in which the prime indicates that the term with $l = 2$ is omitted. This is analogous to (and more general than) the second-order perturbation treatment of Section 25; Equation 27–12 becomes identical with Equation 25–3 when H_{ll} is replaced by W_l^0 and H_{2l} by H'_{kl}.

Higher approximations can be carried out by obvious extensions of this method. If Equation 27–7 can be factored into equations of finite degree, they can often be solved accurately by algebraic or numerical methods.

Let us now consider a simple example,[1] the second-order Stark effect of the normal hydrogen atom, using essentially the method of Epstein (mentioned above). This will also enable us to introduce and discuss a useful set of orthogonal functions.

The wave equation for a hydrogen atom in an electric field can be written as

$$
-\frac{h^2}{8\pi^2\mu}\nabla^2\psi - \frac{e^2}{r}\psi + eFz\psi = W\psi, \quad (27\text{-}13)
$$

in which eFz represents the interaction with an electric field of strength F along the z axis. In order to discuss this equation we shall make use of certain functions $F_{\nu\lambda\mu}(\xi, \vartheta, \varphi)$, defined in terms

---

[1] The study of this example can be omitted by the reader if desired.

of the associated Laguerre and Legendre functions (Secs. 19 and 20) as

$$F_{\nu\lambda\mu}(\xi, \vartheta, \varphi) = \Lambda_{\nu\lambda}(\xi)\Theta_{\lambda\mu}(\vartheta)\Phi_{\mu}(\varphi), \qquad (27\text{-}14)$$

in which

$$\Lambda_{\nu\lambda}(\xi) = \left[\frac{(\nu - \lambda - 1)!}{\{(\nu + \lambda)!\}^3}\right]^{\frac{1}{2}}\xi^{\lambda}L_{\nu+\lambda}^{2\lambda+1}(\xi)e^{-\frac{\xi}{2}}, \qquad (27\text{-}15)$$

$L_{\nu+\lambda}^{2\lambda+1}(\xi)$ being an associated Laguerre polynomial as defined in Section 20b. The functions $\Theta_{\lambda\mu}(\vartheta)$ and $\Phi_{\mu}(\varphi)$ are identical with the functions $\Theta_{lm}(\vartheta)$ and $\Phi_m(\varphi)$ of Equations 21–2 and 21–3 except for the replacement of l and m by λ and μ. It is found by the use of relations given in Sections 19 and 20 that $F_{\nu\lambda\mu}(\xi, \vartheta, \varphi)$ satisfies the differential equation

$$\frac{\partial^2 F}{\partial \xi^2} + \frac{2}{\xi}\frac{\partial F}{\partial \xi} + \left(\frac{\nu}{\xi} - \frac{1}{4}\right)F + \frac{1}{\xi^2 \sin^2 \vartheta}\frac{\partial^2 F}{\partial \varphi^2}$$
$$+ \frac{1}{\xi^2 \sin \vartheta}\frac{\partial}{\partial \vartheta}\left(\sin \vartheta \frac{\partial F}{\partial \vartheta}\right) = 0. \quad (27\text{-}16)$$

The functions are normalized and mutually orthogonal with weight factor ξ, satisfying the relations

$$\int_0^{2\pi}\int_0^{\pi}\int_0^{\infty} F_{\nu\lambda\mu}^* F_{\nu'\lambda'\mu'}\xi d\xi \sin \vartheta d\vartheta d\varphi = 1 \text{ for } \begin{Bmatrix} \nu = \nu' \\ \lambda = \lambda' \\ \mu = \mu' \end{Bmatrix} \quad (27\text{-}17)$$
$$= 0 \text{ otherwise.}$$

If we identify ξ with $2Zr/n'a_0$, where $a_0 = h^2/4\pi^2\mu e^2$, then the functions $F_{\nu\lambda\mu}$ become identical with the hydrogen-atom wave functions ψ_{nlm} for the value $n = n'$ of the principal quantum number n, but not for other values of n; the functions $F_{\nu\lambda\mu}$ all contain the same exponential function of r, whereas the hydrogen-atom wave functions for different values of n contain different exponential functions of r.

For the problem at hand we place n' equal to 1 and Z equal to 1, writing

$$\xi = \frac{2r}{a_0}, \qquad a_0 = \frac{h^2}{4\pi^2\mu e^2}. \qquad (27\text{-}18)$$

The functions $F_{\nu\lambda\mu}$ then satisfy the equation

$$\nabla^2 F_{\nu\lambda\mu} + \left(\frac{1}{\xi} - \frac{1}{4}\right)F_{\nu\lambda\mu} = -\frac{(\nu - 1)}{\xi}F_{\nu\lambda\mu}. \qquad (27\text{-}19)$$

Now let us write our wave equation 27–13 as

$$\nabla^2 \psi + \left(\frac{1}{\xi} - \frac{1}{4}\right)\psi - A\xi \cos \vartheta \psi = \beta \psi, \qquad (27\text{–}20)$$

in which

$$\left.\begin{array}{l} A = \dfrac{a_0^2 F}{4e}, \\[2mm] \beta = -\dfrac{W a_0}{2e^2} - \dfrac{1}{4}, \end{array}\right\} \qquad (27\text{–}21)$$

and the operation ∇^2 refers to the coordinate ξ rather than r, ξ being given by Equation 27–18. To obtain an approximate solution of this equation in terms of the functions $F_{\nu\lambda\mu}$ we shall set up the secular equation in the form corresponding to second-order perturbation theory for the normal state, as given in Equation 27–11; we thus obtain the equation

$$\begin{vmatrix} H_{11} - 2\beta & H_{12} & H_{13} & \cdots \\ H_{21} & H_{22} & 0 & \cdots \\ H_{31} & 0 & H_{33} & \cdots \\ \cdots & \cdots & \cdots & \cdots \end{vmatrix} = 0, \qquad (27\text{–}22)$$

in which

$$H_{ij} = \int\int\int F_i^* \left(\nabla^2 + \frac{1}{\xi} - \frac{1}{4} - A\xi \cos \vartheta\right) \\ F_j \xi^2 d\xi \sin \vartheta d\vartheta d\varphi, \qquad (27\text{–}23)$$

i and j being used to represent the three indices ν, λ, μ. The factor 2 before β arises from the fact that the functions $F_{\nu\lambda\mu}$ are not normalized to unity with respect to the volume element $\xi^2 d\xi \sin \vartheta d\vartheta d\varphi$.

It is found on setting up the secular equation 27–22 that only the three functions F_{100}, F_{210}, and F_{310} need be considered, inasmuch as the equation factors into a term involving these three functions only (to the degree of approximation considered) and terms involving other functions. The equations

$$\xi^2 \cos \vartheta F_{100} = 4\sqrt{2}F_{210} - 2\sqrt{2}F_{310} \qquad (27\text{–}24)$$

and

$$\xi F_{\nu\lambda\mu} = -\{(\nu - \lambda)(\nu + \lambda + 1)\}^{1/2} F_{\nu+1,\lambda\mu} + 2\nu F_{\nu\lambda\mu} - \\ \{(\nu + \lambda)(\nu - \lambda - 1)\}^{1/2} F_{\nu-1,\lambda\mu}, \qquad (27\text{–}25)$$

together with Equation 27-17, enable us to write as the secular equation for these three functions

$$\begin{vmatrix} -2\beta & -4\sqrt{2}A & 2\sqrt{2}A \\ -4\sqrt{2}A & -1 & 0 \\ 2\sqrt{2}A & 0 & -2 \end{vmatrix} = 0. \qquad (27\text{-}26)$$

The root of this is easily found to be $\beta = 18A^2$, which corresponds to

$$W = -\frac{e^2}{2a_0} - \frac{9}{4}a_0^3F^2,$$

or

$$W'' = W - W^0 = -\tfrac{9}{4}a_0^3F^2. \qquad (27\text{-}27)$$

This corresponds to the value

$$\alpha = \tfrac{9}{2}a_0^3 = 0.677 \cdot 10^{-24} \text{ cm}^3$$

for the polarizability of the normal hydrogen atom.

Problem 27-1. Derive the formulas 27-24 and 27-25.

Problem 27-2. Discuss the first-order and second-order Stark effects for the states $n = 2$ of the hydrogen atom by the use of the functions $F_{\nu\lambda\mu}$. Note that in this case the term in A can be neglected in calculating

$$H_{\nu'\lambda'\mu',\nu''\lambda''\mu''}$$

unless ν' or ν'' is equal to 2, and that the secular equation can be factored into terms for $\mu = +1$, $\mu = 0$, and $\mu = -1$, respectively.

27b. The Wentzel-Kramers-Brillouin Method.—For large values of the quantum numbers or of the masses of the particles in the system the quantum mechanics gives results closely similar to classical mechanics, as we have seen in several illustrations. For intermediate cases it is found that the old quantum theory often gives good results. It is therefore pleasing that there has been obtained[1] an approximate method of solution of the wave equation based on an expansion the first term of which leads to the classical result, the second term to the old-quantum-theory result, and the higher terms to corrections which bring in the effects characteristic of the new mechanics. This method is usually called the *Wentzel-Kramers-Brillouin method*. In our discussion we shall merely outline the principles involved in it.

[1] G. WENTZEL, *Z. f. Phys.* **38**, 518 (1926); H. A. KRAMERS, *Z. f. Phys.* **39**, 828 (1926); L. BRILLOUIN, *J. de phys.* **7**, 353 (1926); J. L. DUNHAM, *Phys. Rev.* **41**, 713 (1932).

For a one-dimensional problem, the wave equation is

$$\frac{d^2\psi}{dx^2} + \frac{8\pi^2 m}{h^2}(W - V)\psi = 0, \qquad -\infty < x < +\infty.$$

If we make the substitution

$$\psi = e^{\frac{2\pi i}{h}\int y\,dx}, \tag{27-28}$$

we obtain, as the equation for y,

$$\frac{h}{2\pi i}\frac{dy}{dx} = 2m(W - V) - y^2 = p^2 - y^2, \tag{27-29}$$

in which $p = \pm\sqrt{2m(W - V)}$ is the classical expression for the momentum of the particle. We may now expand y in powers of $h/2\pi i$, considering it as a function of h, obtaining

$$y = y_0 + \frac{h}{2\pi i}y_1 + \left(\frac{h}{2\pi i}\right)^2 y_2 + \cdots. \tag{27-30}$$

Substituting this expansion in Equation 27–29 and equating the coefficients of the successive powers of $h/2\pi i$ to zero, we obtain the equations

$$y_0 = p = \pm\sqrt{2m(W - V)}, \tag{27-31}$$

$$y_1 = -\frac{y_0'}{2y_0} = -\frac{p'}{2p} = \frac{V'}{4(W - V)}, \tag{27-32}$$

$$y_2 = -\tfrac{1}{32}\{5V'^2 + 4V''(W - V)\}(2m)^{-\frac{1}{2}}(W - V)^{-\frac{5}{2}}, \tag{27-33}$$

in which $V' = \dfrac{dV}{dx}$ and $V'' = \dfrac{d^2V}{dx^2}$.

The first two terms when substituted in Equation 27–28 lead to the expression

$$\psi \cong N(W - V)^{-\frac{1}{4}}e^{\frac{2\pi i}{h}\int\sqrt{2m(W-V)}\,dx} \tag{27-34}$$

as an approximate wave function, since

$$\int y_1\,dx = \frac{1}{4}\int\frac{V'}{W - V}dx = +\frac{1}{4}\int\frac{dV}{W - V} = -\frac{1}{4}\log(W - V)$$

so that

$$e^{\int y_1\,dx} = (W - V)^{-\frac{1}{4}}.$$

The probability distribution function to this degree of approximation is therefore

$$\psi^*\psi = N^2(W - V)^{-\frac{1}{2}} = \text{const.}\frac{1}{p}, \tag{27-35}$$

agreeing with the classical result, since p is proportional to the velocity and the probability of finding a particle in a range dx is inversely proportional to its velocity in the interval dx.

The approximation given in Equation 27–34 is obviously not valid near the classical turning points of the motion, at which $W = V$. This is related to the fact that the expansion in Equation 27–30 is not a convergent series but is only an asymptotic representation of y, accurate at a distance from the points at which $W = V$.

So far nothing corresponding to quantization has appeared. This occurs only when an attempt is made to extend the wave function beyond the points $W = V$ into the region with W less than V. It is found[1] that it is not possible to construct an approximate solution in this region satisfying the conditions of Section 9c and fitting smoothly on to the function of Equation 27–34, which holds for the classically allowed region, unless W is restricted to certain discrete values. The condition imposed on W corresponds to the restriction

$$\oint y\,dx = nh, \qquad n = 0, 1, 2, 3, \cdots, \qquad (27\text{–}36)$$

in which the integral is a phase integral of the type discussed in Section 5b. If we insert the first term of the series for y, $y = p$, we obtain the old-quantum-theory condition (Sec. 5b)

$$\oint p\,dx = nh, \qquad n = 0, 1, 2, 3 \cdots. \qquad (27\text{–}37)$$

For systems of the type under discussion, the second term introduces half-quantum numbers; i.e., with $y = y_0 + \frac{h}{2\pi i}y_1$,

$$\oint y\,dx = \oint p\,dx + \frac{h}{2\pi i}\oint y_1\,dx = \oint p\,dx - \frac{h}{2} = nh,$$

so that

$$\oint p\,dx = (n + \tfrac{1}{2})h \qquad (27\text{–}38)$$

to the second approximation. (The evaluation of integrals such as $\oint y_1\,dx$ is best carried out by using the methods of complex variable theory, which we shall not discuss here.[2])

This method has been applied to a number of problems and is a convenient one for many types of application. Its main

[1] Even in its simplest form the discussion of this point is too involved to be given in detail here.

[2] J. L. Dunham, *Phys. Rev.* **41**, 713 (1932).

drawback is the necessity of a knowledge of contour integration, but the labor involved in obtaining the energy levels is often considerably less than other methods require.

27c. Numerical Integration.—There exist well-developed methods[1] for the numerical integration of total differential equations which can be applied quite rapidly by a practiced investigator. The problem is not quite so simple when it is desired to find characteristic values such as the energy levels of the wave equation, but the method is still practicable.

Hartree,[2] whose method of treating complex atoms is discussed in Chapter IX, utilizes the following procedure. For some assumed value of W, the wave equation is integrated numerically, starting with a trial function which satisfies the boundary conditions at one end of the range of the independent variable x and carrying the solution into the middle of the range. Another solution is then computed for this same value of W, starting with a function which satisfies the boundary conditions at the other end of the range of x. For arbitrary values of W these two solutions will not in general join smoothly when they meet for some intermediate value of x. W is then changed by a small amount and the process repeated. After several trials a value of W is found such that the right-hand and left-hand solutions join together smoothly (i.e., with the same slope), giving a single wave function satisfying all the boundary conditions.

This method is a quantitative application of the qualitative ideas discussed in Section 9c. The process of numerical integration consists of starting with a given value and slope for ψ at a point A and then calculating the value of ψ at a near-by point B by the use of values of the slope and curvature $\frac{d^2\psi}{dx^2}$ at A, the latter being obtained from the wave equation.

This procedure is useful only for total differential equations in one independent variable, but there are many problems involving several independent variables which can be separated into total

[1] E. P. Adams, "Smithsonian Mathematical Formulae," Chap. X, The Smithsonian Institution, Washington, 1922; E. T. Whittaker and G. Robinson, "Calculus of Observations," Chap. XIV, Blackie and Son., Ltd., London, 1929.

[2] D. R. Hartree, *Proc. Cambridge Phil. Soc.* **24**, 105 (1928); *Mem. Manchester Phil. Soc.* **77**, 91 (1932–1933).

differential equations to which this method may then be applied. Hartree's method of treating complicated atoms (Sec. 32) and Burrau's calculation of the energy of H_2^+ (Sec. 42c) are illustrations of the use of numerical integration.

27d. Approximation by the Use of Difference Equations.—The wave equation

$$\frac{d^2\psi}{dx^2} + k^2(W - V)\psi = 0, \qquad k^2 = \frac{8\pi^2 m}{h^2}, \qquad (27\text{-}39)$$

may be approximated by a set of difference equations,[1]

$$\frac{1}{a^2}(\psi_{i-1} - 2\psi_i + \psi_{i+1}) + k^2\{W - V(x_i)\}\psi_i = 0, \qquad (27\text{-}40)$$

or

$$\sum_j b_{ij}\psi_j = W\psi_i, \qquad (27\text{-}41)$$

in which $\psi_1, \psi_2, \cdots, \psi_i, \cdots$ are numbers, the values of the function ψ at the points $x_1, x_2, \cdots, x_i, \cdots$ uniformly spaced

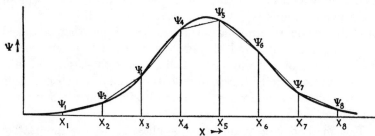

Fig. 27-1.—The approximation to a wave function Ψ by segments of straight lines.

along the x axis with a separation $x_i - x_{i-1} = a$. To prove this we consider the approximation to ψ formed by the polygon of straight lines joining the points $(x_1, \psi_1), (x_2, \psi_2), \cdots, (x_i, \psi_i), \cdots$ of Figure 27-1. The slope of ψ at the point halfway between x_{i-1} and x_i is approximately equal to the slope of the straight line connecting x_{i-1} and x_i, which is $(\psi_i - \psi_{i-1})/a$. The second derivative of ψ at $x = x_i$ is likewise approximated by $1/a$ times the change in slope from $(x_i + x_{i-1})/2$ to $(x_i + x_{i+1})/2$; that is,

$$\frac{d^2\psi}{dx^2} \cong \frac{\psi_{i-1} - 2\psi_i + \psi_{i+1}}{a^2}. \qquad (27\text{-}42)$$

[1] R. G. D. RICHARDSON, *Trans. Am. Math. Soc.* **18**, 489 (1917); R. COURANT, K. FRIEDRICHS, and H. LEWY, *Mathematische Annalen* **100**, 32 (1928).

The differential equation 27–39 is the relation between the curvature $\dfrac{d^2\psi}{dx^2}$ at a point and the function $k^2(W - V)\psi$ at that point, so that we may approximate to the differential equation by the set of equations 27–40, there being one such equation for each point x_i. The more closely we space the points x_i, the more accurately do Equations 27–40 correspond to Equation 27–39.

Just as the lowest energy W of the differential equation can be obtained by minimizing the energy integral $E = \int \phi^* H \phi \, d\tau$ with respect to the function ϕ, keeping $\int \phi^* \phi \, d\tau = 1$, so the lowest value of W giving a solution of Equations 27–40 may be obtained by minimizing the quadratic form

$$E = \frac{\displaystyle\sum_{ij} b_{ij}\phi_i\phi_j}{\displaystyle\sum_i \phi_i^2}, \qquad (27\text{–}43)$$

in which $\phi_1,\ \phi_2,\ \cdots,\ \phi_i,\ \cdots$ are numbers which are varied until E is a minimum. (Just as ϕ must obey the boundary conditions of Section 9c, so the numbers ϕ_i must likewise approximate a curve which is a satisfactory wave function.)

A convenient method[1] has been devised for carrying out this minimization. A set of trial values of ϕ_i is chosen and the value of E is calculated from them. The true solutions ψ_i, to which the values of ϕ_i will converge as we carry out the variation, satisfy Equations 27–40. Transposing one of these gives

$$\psi_i = \frac{\psi_{i-1} + \psi_{i+1}}{2 - a^2 k^2 \{ W - V(x_i) \}}. \qquad (27\text{–}44)$$

If the ϕ_i's we choose are near enough to the true values ψ_i, then it can be shown[1] that, by putting ϕ_{i-1} and ϕ_{i+1} in place of ψ_{i-1} and ψ_{i+1} and E in place of W in Equation 27–44, the resulting expression gives an improved value ϕ_i' for ϕ_i, namely,

$$\phi_i' = \frac{\phi_{i-1} + \phi_{i+1}}{2 - a^2 k^2 \{ E - V(x_i) \}}. \qquad (27\text{–}45)$$

In this way a new set $\phi_1',\ \phi_2',\ \cdots,\ \phi_i',\ \cdots$ can be built up from the initial set $\phi_1,\ \phi_2,\ \cdots,\ \phi_i,\ \cdots$, the new set giving a

[1] G. E. KIMBALL and G. H. SHORTLEY, *Phys. Rev.* **45**, 815 (1934).

lower and therefore a better value of E than the first set. This process may then be repeated until the best set ϕ_i is obtained and the best value of E.

This procedure may be modified by the use of unequal intervals, and it can be applied to problems in two or more dimensions, but the difficulty becomes much greater in the case of two dimensions.

Problem 27–3. Using the method of difference equations with an interval $a = \frac{1}{2}$, obtain an upper limit to the lowest energy W_0 and an approximation to ψ_0 for the harmonic oscillator, with wave equation

$$\frac{d^2\psi}{dx^2} + (\lambda - x^2)\psi = 0 \qquad \text{(see Eq. 11–1)}.$$

27e. An Approximate Second-order Perturbation Treatment. The equation for the second-order perturbation energy (Eq. 25–3) is

$$W_k'' = \sum_l{}' \frac{H_{kl}'H_{lk}'}{W_k^0 - W_l^0} + H_{kk}'', \qquad (27\text{–}46)$$

with

$$H_{kl}' = \int \psi_k^0 {}^* H' \psi_l^0 d\tau$$

and

$$H_{kk}'' = \int \psi_k^0 {}^* H'' \psi_k^0 d\tau.$$

The sum may be rearranged in such a manner as to permit an approximate value to be easily found. On multiplying by $1 - \dfrac{W_l^0}{W_k^0} + \dfrac{W_l^0}{W_k^0}$, it becomes

$$\frac{1}{W_k^0}\sum_l{}' H_{kl}'H_{lk}' + \sum_l{}' \frac{W_l^0}{W_k^0} \frac{H_{kl}'H_{lk}'}{(W_k^0 - W_l^0)}.$$

Now we can replace[1] $\sum_l{}' H_{kl}'H_{lk}'$ by $(H'^2)_{kk} - (H_{kk}')^2$, obtaining

[1] To prove this, we note that $H'\psi_k^0 = \sum_l H_{lk}'\psi_l^0$ (as is easily verified by multiplication by $\psi_l^0{}^*$ and integration). Hence

$$(H'^2)_{kk} = \int \psi_k^0 {}^* H'^2 \psi_k^0 d\tau = \int \psi_k^0 {}^* H'H' \psi_k^0 d\tau = \int \psi_k^0 {}^* H' \left(\sum_l H_{lk}'\psi_l^0\right) d\tau$$

$$= \sum_l H_{kl}'H_{lk}'.$$

The sum $\sum_l{}' = H_{kl}'H_{lk}'$ differs from this only by the term with $l = k$, $(H_{kk}')^2$.

for W_k'' the expression

$$W_k'' = \frac{(H'^2)_{kk}}{W_k^0} - \frac{(H_{kk}')^2}{W_k^0} + H_{kk}'' + \sum_l' \frac{W_l^0}{W_k^0} \frac{H_{kl}' H_{lk}'}{(W_k^0 - W_l^0)}, \quad (27\text{-}47)$$

in which

$$(H'^2)_{kk} = \int \psi_k^0 {}^* H'^2 \psi_k^0 d\tau. \quad (27\text{-}48)$$

This expression is of course as difficult to evaluate as the original expression 27–46. However, it may be that the sum is small compared with the other terms. For example, if k represents the normal state of the system, and the origin for energy measurements is such that W_k^0 is negative, the terms in the sum will be negative for W_l^0 negative, and positive for W_l^0 positive, and there may be considerable cancellation. It must be emphasized that the individual terms in this expression are dependent on the origin chosen for the measurement of energy (the necessity for an arbitrary choice of this origin being the main defect of the approximate treatment we are describing). If this origin were to be suitably chosen, this sum could be made to vanish, the second-order perturbation energy then being given by integrals involving only one unperturbed wave function, that for the state under consideration. The approximate treatment consists in omitting the sum.

As an example let us take the now familiar problem of the polarizability of the normal hydrogen atom, with $H' = eFz$. We know that $H_{1s,1s}'$ vanishes. The integral $(H'^2)_{1s,1s}$ is equal to $e^2 F^2 (z^2)_{1s,1s}$, and, inasmuch as $r^2 = x^2 + y^2 + z^2$ and the wave function for the normal state is spherically symmetrical, the value of $(z^2)_{1s,1s}$ is just one-third that of $(r^2)_{1s,1s}$, given in Section 21c as $3a_0^2$. Thus we obtain

$$W_{1s}'' = \frac{e^2 F^2 a_0^2}{W_{1s}^0}.$$

If we use the value $-e^2/2a_0$ for W_{1s}^0 (taking the ionized atom at zero energy), we obtain

$$W'' = -2F^2 a_0^3,$$

which corresponds to the value $\alpha = 4a_0^3$ for the polarizability. This is only 11 per cent less than the true value (Sec. 27a), being just equal to the value given by the simple treatment of Problem 26–1.

It is interesting to note that if, in discussing the normal state of a system, we take as the zero of energy the first unperturbed excited level, then the sum is necessarily positive and the approximate treatment gives a lower limit to W_0''. In the problem of the normal hydrogen atom this leads to

$$W_{1s}'' \geqslant -\tfrac{8}{3}F^2 a_0^3,$$

giving for α the upper limit $16\tfrac{2}{3}a_0^3$, which is 18 per cent larger than the correct value $\tfrac{9}{2}a_0^3$. Inasmuch as the value $\alpha = 4a_0^3$ given by the variation method is a lower limit, these two very simple calculations fix α to within a few per cent.

It was pointed out by Lennard-Jones[1] that this approximate treatment of W_k'' corresponds to taking as the first-order perturbed wave function the approximate expression (not normalized)

$$\psi_k = \psi_k^0(1 + AH' + \cdots) \tag{27-49}$$

in which $A = 1/W_k^0$.

This suggests that, when practicable, it may be desirable to introduce the perturbation function in the variation function in this way in carrying out a variation treatment. Examples of calculations in which this is done are given in Sections 29e and 47.

[1] J. E. LENNARD-JONES, *Proc. Roy. Soc.* **A129,** 598 (1930).

CHAPTER VIII

THE SPINNING ELECTRON AND THE PAULI EXCLUSION PRINCIPLE, WITH A DISCUSSION OF THE HELIUM ATOM

28. THE SPINNING ELECTRON[1]

The expression obtained in Chapter V for the energy levels of the hydrogen atom does not account completely for the lines observed in the hydrogen spectrum, inasmuch as many of the lines show a splitting into several components, corresponding to a fine structure of the energy levels not indicated by the simple theory. An apparently satisfactory quantitative explanation of this fine structure was given in 1916 by the brilliant work of Sommerfeld,[2] who showed that the consideration of the relativistic change in mass of the electron caused the energy levels given by the old quantum theory to depend to some extent on the azimuthal quantum number k as well as on the total quantum number n, the splitting being just that observed experimentally not only for hydrogen and ionized helium but also for x-ray lines of heavy atoms. This explanation was accepted for several years. Shortly before the development of the quantum mechanics, however, it became evident that there were troublesome features connected with it, relating in particular to the spectra of alkalilike atoms. A neutral alkali atom consists in its normal state of an alkali ion of particularly simple electronic structure (a completed outer group of two or eight electrons) and one valence electron. The interaction of the valence electron and the ion is such as to cause the energy of the atom in various quantum states to depend largely on the azimuthal quantum number for the valence electron as well as on its total quantum number, even neglecting the small relativistic effect, which is negligible compared with the electron-ion interaction. How-

---

[1] For a more detailed treatment of this subject see L. Pauling and S. Goudsmit, "The Structure of Line Spectra," Chap. IV.

[2] A. SOMMERFELD, *Ann. d. Phys.* **51**, 1 (1916).

ever, the levels corresponding to given values of these two quantum numbers were found to be often split into two levels, and it was found that the separations of these doublet levels are formally representable by the Sommerfeld relativistic equation. Millikan and Bowen[1] and Landé,[2] who made this discovery, pointed out that it was impossible to accept the relativistic mechanism in this case, inasmuch as the azimuthal quantum number is the same for the two components of a doublet level, and they posed the question as to the nature of the phenomenon involved.

The answer was soon given by Uhlenbeck and Goudsmit,[3] who showed that the difficulties were removed by attributing to the electron the new properties of angular momentum and magnetic moment, such as would be associated with the spinning motion of an electrically charged body about an axis through it. The magnitude of the total angular momentum of the electron is $\sqrt{s(s+1)}\dfrac{h}{2\pi}$, in which s, *the electron-spin quantum number*, is required by the experimental data to have the value $\tfrac{1}{2}$. The component of angular momentum which the electron spin can possess along any prescribed axis is either $+\dfrac{1}{2}\dfrac{h}{2\pi}$ or $-\dfrac{1}{2}\dfrac{h}{2\pi}$; that is, it is given by the expression $m_s\dfrac{h}{2\pi}$, in which the quantum number m_s can assume only the values $+\tfrac{1}{2}$ and $-\tfrac{1}{2}$. To account for the observed fine-structure splitting and Zeeman effects it is found that the magnetic moment associated with the electron spin is to be obtained from its angular momentum by multiplication not by the factor $e/2m_0c$, as in the case of orbital magnetic moment (Sec. 21d), but by twice this factor, the extra factor 2 being called the *Landé g factor for electron spin*. In consequence the total magnetic moment of the electron spin

[1] R. A. MILLIKAN and I. S. BOWEN, *Phys. Rev.* **24**, 223 (1924).

[2] A. LANDÉ, *Z. f. Phys.* **25**, 46 (1924).

[3] G. E. UHLENBECK and S. GOUDSMIT, *Naturwissenschaften* **13**, 953 (1925); *Nature* **117**, 264 (1926). The electron spin was independently postulated by R. Bichowsky and H. C. Urey, *Proc. Nat. Acad. Sci.* **12**, 80 (1926) (in whose calculations there was a numerical error) and had been previously suggested on the basis of unconvincing evidence by several people.

is $2\dfrac{e}{2m_0c}\dfrac{h}{2\pi}\sqrt{\dfrac{1}{2}\cdot\dfrac{3}{2}}$ or $\sqrt{3}$ Bohr magnetons, and the component along a prescribed axis is either $+1$ or -1 Bohr magneton.

It was shown by Uhlenbeck and Goudsmit and others[1] that the theory of the spinning electron resolves the previous difficulties, and the electron spin is now accepted as a property of the electron almost as well founded as its charge or mass. The doublet splitting for alkalilike atoms is due purely to the magnetic interaction of the spin of the electron and its orbital motion. The fine structure of the levels of hydrogenlike atoms is due to a particular combination of spin and relativity effects, resulting in an equation identical with Sommerfeld's original relativistic equation. The anomalous Zeeman effect shown by most atoms (the very complicated splitting of spectral lines by a magnetic field) results from the interaction of the field with both the orbital and the spin magnetic moments of the electrons, the complexity of the effect resulting from the anomalous value 2 for the g factor for electron spin.[2]

The theory of the spinning electron has been put on a particularly satisfactory basis by the work of Dirac. In striving to construct a quantum mechanics compatible with the requirements of the theory of relativity, Dirac[3] was led to a set of equations representing a one-electron system which is very different in form from the non-relativistic quantum-mechanical equations which we are discussing. On solving these, he found that the spin of the electron and the anomalous g factor 2 were obtained automatically, without the necessity of a separate postulate. The equations led to the complete expression for the energy levels for a hydrogenlike atom, with fine structure, and even to the foreshadowing of the positive electron or positron, discovered four years later by Anderson.

So far the Dirac theory has not been extended to systems containing several electrons. Various methods of introducing

[1] W. Pauli, Z. f. Phys. **36**, 336 (1926); W. Heisenberg and P. Jordan, Z. f. Phys. **37**, 266 (1926); W. Gordon, Z. f. Phys. **48**, 11 (1928); C. G. Darwin, Proc. Roy. Soc. **A 118**, 654 (1928); A. Sommerfeld and A. Unsöld, Z. f. Phys. **36**, 259; **38**, 237 (1926).

[2] For a fuller discussion see Pauling and Goudsmit, "The Structure of Line Spectra," Secs. 17 and 27.

[3] P. A. M. Dirac, Proc. Roy. Soc. **A117**, 610; **A118**, 351 (1928).

the spin in non-relativistic quantum mechanics have been devised. Of these we shall describe and use only the simplest one, which is satisfactory so long as magnetic interactions are neglected, as can be done in treating most chemical and physical problems. This method consists in introducing a spin variable ω, representing the orientation of the electron, and two spin wave functions, $\alpha(\omega)$ and $\beta(\omega)$, the former corresponding to the value $+\frac{1}{2}$ for the spin-component quantum number m_s (that is, to a component of spin angular momentum along a prescribed axis in space of $+\frac{1}{2}h/2\pi$) and the latter to the value $-\frac{1}{2}$ for m_s. The two wave functions are normalized and mutually orthogonal, so that they satisfy the equations

$$\left.\begin{array}{l} \int \alpha^2(\omega)d\omega = 1, \\ \int \beta^2(\omega)d\omega = 1, \\ \int \alpha(\omega)\beta(\omega)d\omega = 0. \end{array}\right\} \tag{28-1}$$

A wave function representing a one-electron system is then a function of four coordinates, three positional coordinates such as x, y, and z, and the spin coordinate ω. Thus we write $\psi(x, y, z)\alpha(\omega)$ and $\psi(x, y, z)\beta(\omega)$ as the two wave functions corresponding to the positional wave function $\psi(x, y, z)$, which is a solution of the Schrödinger wave equation. The introduction of the spin wave functions for systems containing several electrons will be discussed later.

Various other simplified methods of treating electron spin have been developed, such as those of Pauli,[1] Darwin,[2] and Dirac.[3] These are especially useful in the approximate evaluation of interaction energies involving electron spins in systems containing more than one electron.

29. THE HELIUM ATOM. THE PAULI EXCLUSION PRINCIPLE

29a. The Configurations 1s2s and 1s2p.—In Section 23b we applied the first-order perturbation theory to the normal helium atom. Let us now similarly discuss the first excited states of this atom,[4] arising from the unperturbed level for which one

---

[1] W. PAULI, *Z. f. Phys.* **43**, 601 (1927).

[2] C. G. DARWIN, *Proc. Roy. Soc.* **A116**, 227 (1927).

[3] P. A. M. DIRAC, *Proc. Roy. Soc.* **A123**, 714 (1929).

[4] This was first done by W. Heisenberg, *Z. f. Phys.* **39**, 499 (1926).

electron has the total quantum number $n = 1$ and the other $n = 2$. It was shown that, if the interelectronic interaction term e^2/r_{12} be considered as a perturbation, the solutions of the unperturbed wave equation are the products of two hydrogen-like wave functions

$$\psi_{n_1 l_1 m_1}(1)\ \psi_{n_2 l_2 m_2}(2),$$

in which the symbol (1) represents the coordinates $(r_1,\ \vartheta_1,\ \varphi_1)$ of the first electron, and (2) those of the second electron. The corresponding zeroth-order energy is

$$W^0_{n_1 n_2} = -4Rhc\left(\frac{1}{n_1^2} + \frac{1}{n_2^2}\right).$$

We shall ignore the contribution of electron spin to the wave function until the next section.

The first excited level, with the energy $W^0 = -5Rhc$, is that for $n_1 = 1,\ n_2 = 2$ and $n_1 = 2,\ n_2 = 1$. This is eight-fold degenerate, the eight corresponding zeroth-order wave functions being

$$\left.\begin{array}{ll}
1s(1) & 2s(2), \\
2s(1) & 1s(2), \\
1s(1) & 2p_x(2), \\
2p_x(1) & 1s(2), \\
1s(1) & 2p_y(2), \\
2p_y(1) & 1s(2), \\
1s(1) & 2p_z(2), \\
2p_z(1) & 1s(2),
\end{array}\right\} \qquad (29\text{-}1)$$

in which we have chosen to use the real φ functions and have represented $\psi_{100}(1)$ by $1s(1)$, and so on.

On setting up the secular equation, it is found to have the form

$$\begin{vmatrix}
J_s - W' & K_s & 0 & 0 & 0 & 0 & 0 & 0 \\
K_s & J_s - W' & 0 & 0 & 0 & 0 & 0 & 0 \\
0 & 0 & J_p - W' & K_p & 0 & 0 & 0 & 0 \\
0 & 0 & K_p & J_p - W' & 0 & 0 & 0 & 0 \\
0 & 0 & 0 & 0 & J_p - W' & K_p & 0 & 0 \\
0 & 0 & 0 & 0 & K_p & J_p - W' & 0 & 0 \\
0 & 0 & 0 & 0 & 0 & 0 & J_p - W' & K_p \\
0 & 0 & 0 & 0 & 0 & 0 & K_p & J_p - W'
\end{vmatrix} = 0.$$

$$(29\text{-}2)$$

Here the symbols J_s, K_s, J_p, and K_p represent the perturbation integrals

$$
\left.
\begin{aligned}
J_s &= \int\int 1s(1)\ 2s(2)\ \frac{e^2}{r_{12}}\ 1s(1)\ 2s(2)\ d\tau_1 d\tau_2, \\
K_s &= \int\int 1s(1)\ 2s(2)\ \frac{e^2}{r_{12}}\ 2s(1)\ 1s(2)\ d\tau_1 d\tau_2, \\
J_p &= \int\int 1s(1)\ 2p_x(2)\ \frac{e^2}{r_{12}}\ 1s(1)\ 2p_x(2)\ d\tau_1 d\tau_2, \\
K_p &= \int\int 1s(1)\ 2p_x(2)\ \frac{e^2}{r_{12}}\ 2p_x(1)\ 1s(2)\ d\tau_1 d\tau_2.
\end{aligned}
\right\} \quad (29\text{-}3)
$$

J_p and K_p also represent the integrals obtained by replacing $2p_x$ by $2p_y$ or $2p_z$, inasmuch as these three functions differ from one another only with regard to orientation in space. The integrals J_s and J_p are usually called *Coulomb integrals*; J_s, for example, may be considered to represent the average Coulomb interaction energy of two electrons whose probability distribution functions are $\{1s(1)\}^2$ and $\{2s(2)\}^2$. The integrals K_s and K_p are usually called *resonance integrals* (Sec. 41), and sometimes *exchange integrals* or *interchange integrals*, since the two wave functions involved differ from one another in the interchange of the electrons.

It can be seen from symmetry arguments that all the remaining perturbation integrals vanish; we shall discuss $\int\int 1s(1)\ 2s(2)\ \frac{e^2}{r_{12}} 1s(1)\ 2p_x(2)\ d\tau_1 d\tau_2$ as an example. In this integral the function $2p_x(2)$ is an odd function of the coordinate x_2, and inasmuch as all the other terms in the integrand are even functions of x_2, the integral will vanish, the contribution from a region with x_2 negative canceling that from the corresponding region with x_2 positive.

The solution of Equation 29-2 leads to the perturbation energy values

$$
\left.
\begin{aligned}
W' &= J_s + K_s, \\
& \quad J_s - K_s, \\
& \quad J_p + K_p, \quad \text{(triple root),} \\
& \quad J_p - K_p, \quad \text{(triple root).}
\end{aligned}
\right\} \quad (29\text{-}4)
$$

The splitting of the unperturbed level represented by these equations is shown in Figure 29-1.

One part of the splitting, due to the difference of the Coulomb integrals J_s and J_p, can be easily interpreted as resulting from the difference in the interaction of an inner $1s$ electron with an outer $2s$ electron or $2p$ electron. This effect was recognized in the days of the old quantum theory, being described as resulting from greater penetration of the core of the atom (the nucleus plus the inner electrons) by the more eccentric orbits of the

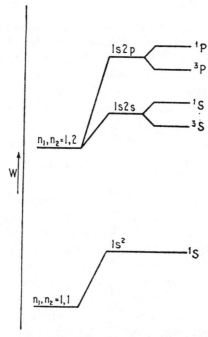

Fig. 29-1.—The splitting of energy levels for the helium atom.

outer electron, with a consequent increase in stability, an s orbit being more stable than a p orbit with the same value of n, and so on.[1] (It is this dependence of the energy of an electron on l as well as n which causes the energy levels of an atom to depend largely on the *electronic configuration*, this expression meaning the n and l values of all electrons. These values are usually indicated by writing ns, np, etc., with the number of similar electrons indicated by a superscript. Thus $1s^2$ indicates two $1s$ electrons, $1s^2 2p$ these plus a $2p$ electron, and so on.)

[1] Pauling and Goudsmit, "The Structure of Line Spectra," Chap. III.

On the other hand, the further splitting due to the integrals K_s and K_p was not satisfactorily interpreted before the development of the quantum mechanics. It will be shown in Section 41 that we may describe it as resulting from the *resonance phenomenon* of the quantum mechanics. The zeroth-order wave function for the state with $W' = J_s + K_s$, for example, is

$$\frac{1}{\sqrt{2}}\{1s(1)\ 2s(2) + 2s(1)\ 1s(2)\};$$

the atom in this state may be described as resonating between the structure in which the first electron is in the $1s$ orbit and the second in the $2s$ orbit and that in which the electrons have been interchanged.

A wave function of the type just mentioned is said to be *symmetric in the positional coordinates of the two electrons*, inasmuch as the interchange of the coordinates of the two electrons leaves the function unchanged. On the other hand, the wave function

$$\frac{1}{\sqrt{2}}\{1s(1)\ 2s(2) - 2s(1)\ 1s(2)\}$$

is *antisymmetric* in the positional coordinates of the electrons, their interchange causing the function to change sign. It is found that all wave functions for a system containing two identical particles are either symmetric or antisymmetric in the coordinates of the particles.

For reasons discussed in the next section, the stationary states of two-electron atoms represented by symmetric and by antisymmetric positional wave functions are called *singlet states* and *triplet states*, respectively. The triplet state from a given configuration is in general more stable than the singlet state.

29b. The Consideration of Electron Spin. The Pauli Exclusion Principle.—In reconsidering the above perturbation problem, taking cognizance of the spin of the electrons, we must deal with thirty-two initial spin-orbit wave functions instead of the eight orbital functions $1s(1)\ 2s(2)$, $1s(1)\ 2p_x(2)$, etc. These thirty-two functions are obtained by multiplying each of the eight orbital functions by each one of the four spin functions

$$\alpha(1)\ \alpha(2),$$
$$\alpha(1)\ \beta(2),$$
$$\beta(1)\ \alpha(2),$$
$$\beta(1)\ \beta(2).$$

Instead of using the second and third of these, it is convenient to use certain linear combinations of them, taking as the four spin functions for two electrons

$$\left.\begin{array}{c} \alpha(1)\ \alpha(2), \\ \dfrac{1}{\sqrt{2}}\{\alpha(1)\ \beta(2) + \beta(1)\ \alpha(2)\}, \\ \beta(1)\ \beta(2), \\ \dfrac{1}{\sqrt{2}}\{\alpha(1)\ \beta(2) - \beta(1)\ \alpha(2)\}. \end{array}\right\} \tag{29-5}$$

These are normalized and mutually orthogonal. The first three of them are symmetric in the spin coordinates of the two electrons, and the fourth is antisymmetric. It can be shown that these are correct zeroth-order spin functions for a perturbation involving the spins of the two electrons.

Taking the thirty-two orbit functions in the order

$$1s(1)\ 2s(2)\ \alpha(1)\ \alpha(2),$$
$$2s(1)\ 1s(2)\ \alpha(1)\ \alpha(2),$$
$$1s(1)\ 2p_x(2)\ \alpha(1)\ \alpha(2),$$
$$\cdots \cdots \cdots \cdots \cdots,$$
$$1s(1)\ 2s(2) \cdot \dfrac{1}{\sqrt{2}}\{\alpha(1)\ \beta(2) + \beta(1)\ \alpha(2)\},$$
$$\cdots \cdots \cdots \cdots \cdots \cdots \cdots \cdots$$

obtained by multiplying the eight orbital functions by the first spin function, then by the second spin function, and so on, we find that the secular equation has the form

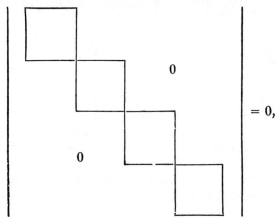

$$= 0,$$

in which each of the small squares is an eight-rowed determinant identical with that of Equation 29–2. The integrals outside of these squares vanish because of the orthogonality of the spin functions and the non-occurrence of the spin coordinates in the perturbation function e^2/r_{12}. The roots of this equation are the same as those of Equation 29–2, each occurring four times, however, because of the four spin functions.

The correct zeroth-order wave functions are obtained by multiplying the correct positional wave functions obtained in the preceding section by the four spin functions. For the configuration $1s2s$ alone they are

$$\frac{1}{\sqrt{2}}\{1s(1)\ 2s(2) + 2s(1)\ 1s(2)\} \cdot \alpha(1)\ \alpha(2),$$

$$\frac{1}{\sqrt{2}}\{1s(1)\ 2s(2) + 2s(1)\ 1s(2)\} \cdot \frac{1}{\sqrt{2}}\{\alpha(1)\ \beta(2) + \beta(1)\ \alpha(2)\}$$

$$\frac{1}{\sqrt{2}}\{1s(1)\ 2s(2) + 2s(1)\ 1s(2)\} \cdot \beta(1)\ \beta(2),$$

$$\frac{1}{\sqrt{2}}\{1s(1)\ 2s(2) - 2s(1)\ 1s(2)\} \cdot \frac{1}{\sqrt{2}}\{\alpha(1)\ \beta(2) - \beta(1)\ \alpha(2)\},$$

$$\text{Triplet}\begin{cases} \dfrac{1}{\sqrt{2}}\{1s(1)\ 2s(2) - 2s(1)\ 1s(2)\} \cdot \alpha(1)\ \alpha(2), \\[2mm] \dfrac{1}{\sqrt{2}}\{1s(1)\ 2s(2) - 2s(1)\ 1s(2)\} \cdot \dfrac{1}{\sqrt{2}}\{\alpha(1)\ \beta(2) + \\[4mm] \hspace{5cm} \beta(1)\ \alpha(2)\}, \\[2mm] \dfrac{1}{\sqrt{2}}\{1s(1)\ 2s(2) - 2s(1)\ 1s(2)\} \cdot \beta(1)\ \beta(2), \end{cases}$$

$$\text{Singlet}\ \ \frac{1}{\sqrt{2}}\{1s(1)\ 2s(2) + 2s(1)\ 1s(2)\} \cdot \frac{1}{\sqrt{2}}\{\alpha(1)\ \beta(2) - \\ \hspace{6cm} \beta(1)\ \alpha(2)\}.$$

Of these eight functions, the first four are symmetric in the coordinates of the two electrons, the functions being unchanged on interchanging these coordinates. This symmetric character results for the first three functions from the symmetric character of the orbital part and of the spin part of each function. For the fourth function it results from the antisymmetric character of the two parts of the function, each of which changes sign on interchanging the two electrons.

The remaining four functions are antisymmetric in the two electrons, either the orbital part being antisymmetric and the spin part symmetric, or the orbital part symmetric and the spin part antisymmetric.

Just as for $1s2s$, so each configuration leads to some symmetric and some antisymmetric wave functions. For $1s^2$, for example, there are three of the former and one of the latter, obtained by combining the symmetric orbital wave function of Section 23b with the four spin functions. For $1s2p$ there are twelve of each

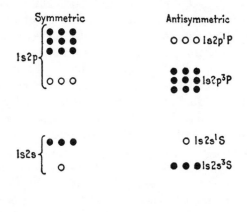

FIG. 29-2.—Levels for configurations $1s^2$, $1s2s$, and $1s2p$ of the helium atom. ●, spin-symmetric wave functions; ○, spin-antisymmetric wave functions

type, nine spin-symmetric and orbital-symmetric, three spin-antisymmetric and orbital-antisymmetric; nine spin-symmetric and orbital-antisymmetric, and three spin-antisymmetric and orbital-symmetric. The levels thus obtained for the helium atom by solution of the wave equation are shown in Figure 29–2, the completely symmetric wave functions being represented on the left and the completely antisymmetric ones on the right.

Now it can be shown that if a helium atom is initially in a symmetric state no perturbation whatever will suffice to cause it to change to any except symmetric states (the two electrons being considered to be identical). Similarly, if it is initially in an antisymmetric state it will remain in an antisymmetric state. The solution of the wave equation has provided us with two completely independent sets of wave functions. To show that no perturbation will cause the system in a state represented by the symmetric wave function ψ_S to change to a state represented by the antisymmetric wave function ψ_A we need only show that the integral

$$\int \psi_A^* H' \psi_S d\tau$$

vanishes (H' being the perturbation function, involving the spin as well as the positional coordinates of the electrons), inasmuch as it is shown in Chapter XI that the probability of transition from one stationary state to another as a result of a perturbation is determined by this integral. Now, if the electrons are identical, the expression $H'\psi_S$ is a symmetrical function of the coordinates, whereas ψ_A^* is antisymmetric; hence the integrand will change sign on interchanging the coordinates of the two electrons, and since the region of integration is symmetrical in these coordinates, the contribution of one element of configuration space is just balanced by that of the element corresponding to the interchange of the electrons, and the integral vanishes.[1]

The question as to which types of wave functions actually occur in nature can at present be answered only by recourse to experiment. So far all observations which have been made on helium atoms have shown them to be in antisymmetric states.[2] We accordingly make the additional postulate that *the wave function representing an actual state of a system containing two or more electrons must be completely antisymmetric in the coordinates of the electrons; that is, on interchanging the coordinates of any two electrons it must change its sign.* This is the statement of the *Pauli exclusion principle* in wave-mechanical language.

This is a principle of the greatest importance. A universe based on some other principle, that is, represented by wave functions of different symmetry character, would be completely different in nature from our own universe. The chemical properties in particular of substances are determined by this principle, which, for example, restricts the population of the K shell of an atom to two electrons, and thus makes lithium

---

[1] The same conclusion is reached from the following argument: On interchanging the subscripts 1 and 2 the entire integral is converted into itself with the negative sign, and hence its value must be zero.

[2] The states are identified through the splitting due to spin-orbit interactions neglected in our treatment.

an alkali metal, the third electron being forced into an outer
shell where it is only loosely bound.

To show this, we may mention that if A represents a spin-orbit
function for one electron (such that $A(1) = 1s(1) \; \alpha(1)$, for
example) and B, C, $\cdots$, E others, then the determinantal
function

$$\begin{vmatrix} A(1) & B(1) & \ldots & E(1) \\ A(2) & B(2) & \ldots & E(2) \\ \ldots & \ldots & \ldots & \ldots \\ A(N) & B(N) & \ldots & E(N) \end{vmatrix}$$

is completely antisymmetric in the N electrons, and hence a
wave function of this form for the N-electron system satisfies
Pauli's principle, since from the properties of determinants the
interchange of two rows changes the sign of the determinant.
Moreover, no two of the functions A, B, $\cdots$, E can be equal,
as then the determinant would vanish. Since the only spin-orbit
functions based on a given one-electron orbital function are the
two obtained by multiplying by the two spin functions α and β,
we see that *no more than two electrons can occupy the same orbital
in an atom, and these two must have their spins opposed;* in other
words, no two electrons in an atom can have the same values of
the four quantum numbers n, l, m, and m_s. Pauli's original
statement[1] of his exclusion principle was in nearly this language;
its name is due to its limitation of the number of electrons in an
orbit.

The equations of quantum statistical mechanics for a system of
non-identical particles, for which all solutions of the wave
equations are accepted, are closely analogous to the equations
of classical statistical mechanics (Boltzmann statistics). The
quantum statistics resulting from the acceptance of only antisymmetric wave functions is considerably different. This
statistics, called *Fermi-Dirac statistics,* applies to many problems,
such as the Pauli-Sommerfeld treatment of metallic electrons
and the Thomas-Fermi treatment of many-electron atoms.
The statistics corresponding to the acceptance of only the
completely symmetric wave functions is called the *Bose-Einstein
statistics.* These statistics will be briefly discussed in Section 49.

[1] W. PAULI, *Z. f. Phys.* **31**, 765 (1925).

It has been found that for protons as well as electrons the wave functions representing states occurring in nature are antisymmetric in the coordinates of the particles, whereas for deuterons they are symmetric (Sec. 43*f*).

The stationary states of the helium atom, represented on the right side of Figure 29-2, are conveniently divided into two sets, shown by open and closed circles, respectively. The wave functions for the former, called *singlet states*, are obtained by multiplying the symmetric orbital wave functions by the single antisymmetric spin function $\dfrac{1}{\sqrt{2}}\{\alpha(1) \quad \beta(2) - \beta(1) \quad \alpha(2)\}$. Those for the latter, called *triplet states*, are obtained by multiplying the antisymmetric orbital wave functions by the three symmetric spin functions.[1] The spin-orbit interactions which we have neglected cause some of the triplet levels to be split into three adjacent levels. Transitions from a triplet to a singlet level can result only from a perturbation involving the electron spins, and since interaction of electron spins is small for light atoms, these transitions are infrequent; no spectral line resulting from such a transition has been observed for helium.

It is customary to represent the spectral state of an atom by a term symbol such as ^1S, ^3S, ^3P, etc. Here the superscript on the left gives the multiplicity, 1 signifying singlet and 3 triplet. The letters S, P, etc., represent the resultant of the orbital angular-momentum vectors of all the electrons in the atom. This is also given by a resultant azimuthal quantum number L, the symbols $S, P, D, F, \cdots$ corresponding to $L = 0, 1, 2, 3, \cdots$. If all the electrons but one occupy s orbitals, the value of L is the same as that of l for the odd electron, so that for helium the configurations $1s^2$, $1s2s$, and $1s2p$ lead to the states ^1S, ^1S and ^3S, and ^1P and ^3P. Use is also made of a resultant spin quantum number S (not to be confused with the symbol S for $L = 0$),

---

[1] The electrons are often said to have their spins opposed or antiparallel in singlet states and parallel in triplet states, the spin function

$$\frac{1}{\sqrt{2}}\{\alpha\,(1)\,\beta(2) + \beta(1)\,\alpha(2)\}$$

in the latter case representing orientation of the resultant spin with zero component along the z axis.

which has the value 0 for singlet states and 1 for triplet states, the multiplicity being equal to $2S + 1$.[1]

The results which we have obtained regarding the stationary states of two-electron atoms may be summarized in the following way. The main factors determining the term values are the values of the principal quantum numbers n_1 and n_2 for the two electrons and of the azimuthal quantum numbers l_1 and l_2, smaller values of these numbers leading to greater stability. These numbers determine the configuration of the atom. The configuration $1s^2$ leads to the normal state, $1s2s$ to the next most stable states, then $1s2p$, and so on. For configurations with n_1l_1 different from n_2l_2 there is a further splitting of the levels for a given configuration, due to the resonance integrals, leading to singlet and triplet levels, and to levels with different values of the resultant azimuthal quantum number L in case that both l_1 and l_2 are greater than zero. The triplet levels may be further split into their fine-structure components by the spin-orbit interaction, which we have neglected in our treatment. It is interesting to notice that these interactions completely remove the degeneracy for some states, such as $1s2s$ ${}^1S$, but not for others, such as $1s2s$ ${}^3S$, which then show a further splitting (Zeeman effect) on the application of a magnetic field to the atom.

Problem 29–1. Evaluate the integrals J and K for $1s2s$ and $1s2p$ of helium, and calculate by the first-order perturbation theory the term values for the levels obtained from these configurations. Observed term values (relative to He$^+$) are $1s2s$ ${}^1S$ 32033, $1s2s$ ${}^3S$ 38455, $1s2p$ ${}^1P$ 27176, and $1s2p$ ${}^3P$ 29233 cm^{-1}.

29c. The Accurate Treatment of the Normal Helium Atom.— The theoretical calculation of the energy of the normal helium atom proved to be an effective stumbling block for the old quantum theory. On the other hand, we have already seen that even the first attack on the problem by wave-mechanical methods,

---

[1] For a detailed discussion of spectroscopic nomenclature and the vector model of the atom see Pauling and Goudsmit, "The Structure of Line Spectra." The triplet levels of helium were long called doublets, complete resolution being difficult. Their triplet character was first suggested by J. C. Slater, *Proc. Nat. Acad. Sci.* **11**, 732 (1925), and was soon verified experimentally by W. V. Houston, *Phys. Rev.* **29**, 749 (1927). The names parhelium and orthohelium were ascribed to the singlet and triplet levels, respectively, before their nature was understood.

the first-order perturbation treatment given in Section 23b, led to a promising result, the discrepancy of about 4 v.e. (accepting the experimental value as correct) being small compared with the discrepancies shown by the old-quantum-theory calculations. It is of interest to see whether or not more and more refined wave-mechanical treatments continue to diminish the discrepancy with experiment and ultimately to provide a theoretical value of the ionization potential agreeing exactly with the experimental (spectroscopic) value 24.463 v.[1] The success of this program would strengthen our confidence in our wave-mechanical equations, and permit us to proceed to the discussion of many-electron atoms and molecules.

No exact solution of the wave equation has been made, and all investigators have used the variation method.[2] The simplest extension of the zeroth-order wave function e^{-2s}, with $s = (r_1 + r_2)/a_0$, is to introduce an effective nuclear charge $Z'e$ in place of the true nuclear charge $2e$ in the wave function. This function, $e^{-Z's}$, minimizes the energy when the atomic number Z' has the value $27\!/\!16$, corresponding to a screening constant of value $5\!/\!16$ (Sec. 26b). The discrepancy with the observed energy[3] (Table 29-1) is reduced by this simple change to 1.5 v.e., which is one-third the discrepancy for Unsöld's treatment. This wave function corresponds to assuming that each electron screens the other

---

[1] Calculated from Lyman's term value 198298 cm^{-1} corrected by Paschen to 198307.9 cm^{-1}; T. Lyman, *Astrophys. J.* **60**, 1 (1924); F. Paschen, *Sitzber. preuss. Akad. Wiss.* 1929, p. 662.

[2] The principal papers dealing with the normal helium atom are A. Unsöld, *Ann. d. Phys.* **82**, 355 (1927); G. W. Kellner, *Z. f. Phys.* **44**, 91, 110 (1927); J. C. Slater, *Proc. Nat. Acad. Sci.* **13**, 423 (1927); *Phys. Rev.* **32**, 349 (1928); C. Eckart, *Phys. Rev.* **36**, 878 (1930); E. A. Hylleraas, *Z. f. Phys.* **48**, 469 (1928); **54**, 347 (1929); **65**, 209 (1930). A summary of his work is given by Hylleraas in *Skrifter det Norske Vid.-Ak. Oslo*, I. Mat.-Naturv. Klasse 1932, pp. 5–141. For the special methods of evaluating and minimizing the energy integral, the reader is referred to these papers.

[3] The experimental value -78.605 v.e. $= -5.8074\ R_{He}hc$ for the energy of the normal helium atom is obtained by adding to the observed first ionization energy 24.463 v.e. (with the minus sign) the energy

$$4R_{He}hc = -54.1416 \text{ v.e.}$$

of the helium ion. Hylleraas has shown that the correction for motion of the nucleus in the neutral helium atom is to be made approximately by using R_{He}; that is, by assigning to each electron the reduced mass with the helium nucleus.

from the nucleus in the same way as a charge $-\frac{5}{16}e$ on the nucleus.

Problem 29-2. (a) Calculate approximately the energy the normal lithium atom would have if the allowed wave functions were completely symmetric in the electrons, using for the positional wave function the product function $1s(1)\ 1s(2)\ 1s(3)$, in which $1s$ contains the effective nuclear charge $Z' = 3 - S$, and minimizing the energy relative to Z' or S. From this and a similar treatment of Li$^+$ obtain the first ionization energy. The observed value is 5.368 v.e. (b) Obtain a general formula for the Nth ionization energy of an atom with atomic number Z in such a Bose-Einstein universe, using screening-constant wave functions. Note the absence of periodicity in the dependence on Z.

We might now consider other functions of the type $F(r_1)F(r_2)$, introducing other parameters. This has been done in a general way by Hartree, in applying his theory of the self-consistent field (Chap. IX), the function $F(r_1)$ being evaluated by special numerical and graphical methods. The resulting energy value, as given in Table 29-1, is still 0.81 v.e. from the experimental value. Even the simple algebraic function

$$e^{-\frac{Z_1 r_1}{a_0}} e^{-\frac{Z_2 r_2}{a_0}} + e^{-\frac{Z_1 r_1}{a_0}} e^{-\frac{Z_2 r_2}{a_0}}$$

leads to as good a value of the energy. (This is function 4 of the table, there expressed in terms of the hyperbolic cosine.) This variation function we may interpret as representing one electron in an inner orbit and the other in an outer orbit, the values of the constants, $Z_1 = 2.15$ and $Z_2 = 1.19$, corresponding to no shielding (or, rather, a small negative shielding) for the inner electron by the outer, and nearly complete shielding for the outer electron by the inner. By taking the sum of two terms the orbital wave function is made symmetric in the two electrons. It is interesting that the still simpler function 5 leads to a slightly better value for the energy. Various more complicated functions of s and t were also tried by Kellner and Hylleraas, with considerable improvement of the energy value. Then a major advance was made by Hylleraas by introducing in the wave function the coordinate $u = r_{12}/a_0$, which occurs in the interaction term for the two electrons. The simple two-parameter functions 6 and 7 provide values of the energy of the atom accurate to $\frac{1}{2}$ per cent. Here again the polynomial in u is more satisfactory than the more complicated exponential function, suggesting that a polynomial factor containing further powers of u, t, and s be used. The functions 8, 9, and 10 show that this procedure leads quickly to a value which is only slightly changed by further terms, the last three terms of 10 being reported by Hylleraas as making negligible contributions. The final theoretical value for the energy of the helium atom is 0.0016 v.e. *below* the experimental value. Inasmuch as this theoretical value, obtained by the variation method, should be an upper limit, the discrepancy is to be attributed to a numerical error in the calculations or to experimental error in the ionization energy, or possibly to some small effects such as electron-spin interactions, motion of the nucleus, etc. At any rate the agreement to within 0.0016 v.e. may be considered as a triumph for wave mechanics when applied to many-electron atoms.

TABLE 29–1.—VARIATION FUNCTIONS FOR THE NORMAL HELIUM ATOM[1]

Symbols: $s = \dfrac{r_1 + r_2}{a_0}$, $t = \dfrac{r_1 - r_2}{a_0}$, $u = \dfrac{r_{12}}{a_0}$

Experimental value of $W = -5.80736 R_{He}hc$

Variation function, with best values of constants[2]	Energy, in units $-R_{He}hc$	Difference with experiment	
		Units $-R_{He}hc$	V.e.
1. e^{-2s}	5.50	0.31	4.19
2. $e^{-Z's}$, $Z' = 2\frac{7}{16} = 1.6875$	5.6953	0.1120	1.53
3. $F(r_1)F(r_2)$	5.75	0.06	0.81
4. $e^{-Z's} \cosh ct$, $Z' = 1.67$, $c = 0.48$	5.7508	0.0565	0.764
5. $e^{-Z's}(1 + c_2 t^2)$, $Z' = 1.69$, $c_2 = 0.142$ ...	5.7536	0.0537	0.726
6. $e^{-Z's}e^{cu}$, $Z' = 1.86$, $c = 0.26$	5.7792	0.0281	0.380
7. $e^{-Z's}(1 + c_1 u)$, $Z' = 1.849$, $c_1 = 0.364$ ..	5.7824	0.0249	0.337
8. $e^{-Z's}(1 + c_1 u + c_2 t^2)$ .... $Z' = 1.816$, $c_1 = 0.30$, $c_2 = 0.13$.	5.80488	0.00245	0.0332
9. $e^{-Z's}(1 + c_1 u + c_2 t^2 + c_3 s + c_4 s^2 + c_5 u^2)$ $Z' = 1.818$, $c_1 = 0.353$, $c_2 = 0.128$, $c_3 = -0.101$, $c_4 = 0.033$, $c_5 = -0.032$	5.80648	0.00085	0.0115
10. $e^{-Z's}$(polynomial with fourteen terms)..	5.80748	-0.00012	-0.0016

[1] A few variation functions which have been tried are not included in the table because they are only slightly better than simpler ones; for example, the function $e^{-z's}(1 - c_1 e^{-c_2 u})$, which is scarcely better than function 6. (D. R. HARTREE and A. L. INGMAN, *Mem. Manchester Phil. Soc.* **77**, 69 (1932).)

[2] The normalization factor is omitted. Of these functions, 1 is due to Unsöld, 2 to Kellner, 3 to Hartree and Gaunt, 4 to Eckart and Hylleraas, and the remainder to Hylleraas.

Hylleraas's masterly attack on the problem of the energy of normal helium and heliumlike ions culminated in his derivation of a general formula for the first ionization energy I of these atoms and ions.[1] This formula, obtained by purely theoretical considerations, is

$$I = \frac{R_\infty hc}{1 + \frac{m_0}{M}}\left(Z^2 - \frac{5}{4}Z + 0.31488 - \frac{0.01752}{Z} + \frac{0.00548}{Z^2}\right), \quad (29\text{-}6)$$

in which M is the mass and Z the atomic number of the atom. Values calculated by this formula[2] are given in Table 29-2, together with experimental values obtained spectroscopically, mainly by Edlén[3] and coworkers. It is seen that there is agree-

TABLE 29-2.—IONIZATION ENERGIES OF TWO-ELECTRON ATOMS

Atom	I calculated, v.e.	I observed, v.e.
H^-	0.7149	
He	24.465	24.463
Li^+	75.257	75.279 ± 0.012
Be^{++}	153.109	153.09 ± 0.10
B^{+++}	258.029	258.1 ± 0.2
C^{4+}	390.020	389.9 ± 0.4
N^{5+}	549.085	
O^{6+}	735.222	

ment to within the experimental error. Indeed, the calculated values are now accepted as reliable by spectroscopists.[4]

Included in the table is the value 0.7149 v.e. for the ionization energy of the negative hydrogen ion H^-. This shows that the hydrogen atom has a positive electron affinity, amounting to 16480 cal/mole. The consideration of the crystal energy of the alkali hydrides has provided a rough verification of this value.

29d. Excited States of the Helium Atom.—The variation method can be applied to the lowest triplet state of helium as well as to the lowest singlet state, inasmuch as (neglecting

---

[1] E. A. HYLLERAAS, *Z. f. Phys.* **65**, 209 (1930).

[2] Using 1 v.e. = 8106.31 cm^{-1} and R_∞ = 109737.42 cm^{-1}.

[3] A. ERICSON and B. EDLÉN, *Nature* **124**, 688 (1929); *Z. f. Phys.* **59**, 656 (1930); B. EDLÉN, *Nature* **127**, 405 (1930).

[4] B. EDLÉN, *Z. f. Phys.* **84**, 746 (1933).

spin-orbit interactions) the triplet wave functions are anti-symmetric in the positional coordinates of the two electrons, and contain no contribution from singlet functions (Sec. 26c).
A simple and reasonable variation function is

$$1s_{Z'}(1)\ 2s_{Z''}(2)\ -\ 2s_{Z''}(1)\ 1s_{Z'}(2),$$

in which $1s_{Z'}$ and $2s_{Z''}$ signify hydrogenlike wave functions with the indicated effective nuclear charges as parameters. We would expect the energy to be minimized for $Z' = 2$ and $Z'' = 1$. Calculations for this function have not been made. However, Hylleraas[1] has discussed the function

$$se^{-Z's}\sinh ct, \qquad (29\text{--}7)$$

obtaining the energy value $-4.3420R_{He}hc$, not far above the observed value $-4.3504R_{He}hc$. This function is similar to the hydrogenlike function (containing some additional terms), and the parameter values found, $Z' = 1.374$ and $c = 0.825$, correspond to the reasonable values $Z' = 2.198$ and $Z'' = 1.099$. Hylleraas has also replaced s in 29–7 by $s + c_1u$, obtaining the energy $-4.3448R_{He}hc$, and by $s + c_2t^2$, obtaining the energy $-4.3484R_{He}hc$. It is probable that the series $s + c_1u + c_2t^2$ would lead to very close agreement with experiment.

Numerous investigations by Hylleraas and others[2] have shown that wave mechanics can be applied in the treatment of other states of the helium atom. We shall not discuss further the rather complicated calculations.

29e. The Polarizability of the Normal Helium Atom.—A quantity of importance for many physical and chemical considerations (indices of refraction, electric dipole moments, term values of non-penetrating orbits, van der Waals forces, etc.) is the *polarizability* of atoms and molecules, mentioned in Problem 26–1 and Sections 27a and 27e. We may write as the energy of a system in an electric field of strength F the expression

$$W = W^0 - \tfrac{1}{2}\alpha F^2 + \cdots \qquad (29\text{--}8)$$

---

[1] E. A. HYLLERAAS, Z. f. Phys. **54**, 347 (1929).

[2] W. HEISENBERG, Z. f. Phys. **39**, 499 (1926); A. UNSÖLD, Ann. d. Phys. **82**, 355 (1927); E. A. HYLLERAAS and B. UNDHEIM, Z. f. Phys. **65**, 759 (1930); E. A. HYLLERAAS, ibid. **66**, 453 (1930); **83**, 739 (1933); J. P. SMITH, Phys. Rev. **42**, 176 (1932); etc.

in case that the term linear in F vanishes, the permanent electric moment of the system being zero. The electric moment induced in the system by the field is αF, the factor of proportionality α being called the polarizability. The polarizability of the molecules in a gas determines its index of refraction n (for light of very large wave length) and its dielectric constant D, according to the equation

$$\frac{N}{V}\alpha = \frac{3}{4\pi}\frac{n^2-1}{n^2+2} = \frac{3}{4\pi}\frac{D-1}{D+2}, \tag{29-9}$$

in which N is Avogadro's number and V is the molal volume of the substance. The *mole refraction* R is defined as

$$R = \frac{4\pi N}{3}\alpha = 2.54 \cdot 10^{24}\alpha. \tag{29-10}$$

The dimensions of R and α are those of volume, and their magnitudes are roughly those of molal volumes and molecular volumes, respectively; for example, for monatomic hydrogen $R = 1.69$ cm^3 and $\alpha = 0.667 \cdot 10^{-24}$ cm^3 (Sec. 27a). Values of R and α are determined experimentally mainly by measurement of indices of refraction and of dielectric constants,[1] rough values being also obtainable from spectral data.[2]

The value of the polarizability α of an atom or molecule can be calculated by evaluating the second-order Stark effect energy $-\frac{1}{2}\alpha F^2$ by the methods of perturbation theory or by other approximate methods. A discussion of the hydrogen atom has been given in Sections 27a and 27e (and Problem 26-1). The helium atom has been treated by various investigators by the variation method, and an extensive approximate treatment of many-electron atoms and ions based on the use of screening constants (Sec. 33a) has also been given.[3] We shall discuss the variational treatments of the helium atom in detail.

The additional term in the Hamiltonian due to the electric

---

[1] The total polarization of a gas may be due to polarization of the electrons in the gas molecules (for fixed nuclear positions), polarization of the nuclei (with change in the relative positions of the nuclei in the molecules), and orientation of molecules with permanent electric dipole moments. We are here discussing only the first of these mechanisms; the second is usually unimportant, and the third is treated briefly in Section 49f.

[2] See PAULING and GOUDSMIT, "The Structure of Line Spectra," Sec. 11.

[3] L. PAULING, *Proc. Roy. Soc.* **A114**, 181 (1927).

field (assumed to lie along the z axis) is $eF(z_1 + z_2)$, z_1 and z_1 being the z coordinates of the two electrons relative to the nucleus. The argument of Section 27e suggests that the variation function be of the form

$$\psi = \psi^0\{1 + (z_1 + z_2)f(x_1, y_1, z_1, x_2, y_2, z_2)\}, \quad (29\text{--}11)$$

in which ψ^0 is an approximate wave function for zero field. Variation functions of this form (or approximating it) have been discussed by Hassé, Atanasoff, and Slater and Kirkwood,[1] whose results are given in Table 29–3.

TABLE 29–3.—VARIATION FUNCTIONS FOR THE CALCULATION OF THE POLARIZABILITY OF THE NORMAL HELIUM ATOM

Experimental value: $\alpha = 0.205 \cdot 10^{-24}\,\text{cm}^3$

$$s = \frac{r_1 + r_2}{a_0}$$

Variation function	α	References[1]
1. $e^{-Z's}\{1 + A(z_1 + z_2)\}$	$0.150 \cdot 10^{-24}\,\text{cm}^3$	H
2. $e^{-Z's}\{1 + A(z_1 e^{-Z''r_1} + z_2 e^{-Z''r_2})\}$	.164	SK
3. $(r_1 r_2)^{0.255} e^{-Z's}\{1 + A(z_1 e^{-Z''r_1} + z_2 e^{-Z''r_2})\}$	.222	SK
4. $e^{-Z's}\{1 + A(z_1 + z_2) + B(z_1 r_1 + z_2 r_2)\}$	.182	H
5. $e^{-Z's}\{1 + A(z_1 + z_2) + \text{terms to quartic})\}$	.183	H
6. $e^{-Z's}(1 + c_1 u)\{1 + A(z_1 + z_2) + B(z_1 r_1 + z_2 r_2)\}$	.201	H
7. $e^{-Z's}\{1 + c_1 u + c_2 l^2 + A(z_1 + z_2)\}$	.127	A
8. $e^{-Z's}\{1 + c_1 u + c_2 l^2 + (A + Bs)(z_1 + z_2) + Ct(z_1 - z_2)\}$	.182	A
9. $e^{-Z's}\{1 + c_1 u + c_2 l^2 + c_3 s + c_4 s^2 + c_5 u^2 + (A + Bs)(z_1 + z_2) + Ct(z_1 - z_2) + Du(z_1 + z_2)\}$	.194	A
10. $e^{-Z's}(1 + c_1 u + c_2 l^2)\{1 + A(z_1 + z_2) + B(z_1 r_1 + z_2 r_2)\}$	.231	H
11. A non-algebraic function	.210	SK

[1] H = Hassé, A = Atanasoff, SK = Slater and Kirkwood.

Of these functions, 1, 2, 4, and 5 are based on the simple screening-constant function 2 of Table 29–1; these give low values of α, the experimental value (from indices of refraction extrapolated to large wave length of light and from dielectric

[1] H. R. HASSÉ, *Proc. Cambridge Phil. Soc.* **26**, 542 (1930), **27**, 66 (1931); J. V. ATANASOFF, *Phys. Rev.* **36**, 1232 (1930); J. C. SLATER and J G. KIRKWOOD, *Phys. Rev.* **37**, 682 (1931).

constant) being about $0.205 \cdot 10^{-24}$ cm³. The third function, supposed to provide a better approximation to the correct wave function for large values of r_1 and r_2 (that is, in the region of the atom in which most of the polarization presumably occurs), overshoots the mark somewhat. (The fundamental theorem of the variation method (Sec. 26a) does not require that a calculation such as these give a lower limit for α, inasmuch as the wave function and energy value for the unperturbed system as well as for the perturbed system are only approximate.) Function 6 is based on 7 of Table 29–1, 7, 8, and 10 on 8, and 9 on 9. It is seen that functions of the form 29–11 (6, 10) seem to be somewhat superior to functions of the same complexity not of this form (7, 8, 9). Function 11 is based on a helium-atom function (not given by a single algebraic expression) due to Slater.[1]

It is seen that the values of α given by these calculations in the main lie within about 10 per cent of the experimental value[2] $0.205 \cdot 10^{-24}$ cm³. For Li⁺, Hassé, using function 6, found the value $\alpha = 0.0313 \cdot 10^{-24}$ cm³; the only other values with which this can be compared are the spectroscopic value[3] 0.025 and the screening-constant value[2] $0.0291 \cdot 10^{-24}$ cm³.

Problem 29–3. Using the method of Section 27e and the screening-constant wave function 2 of Table 29–1, evaluate the polarizability of the helium atom, taking as the zero point for energy the singly ionized atom.

[1] J. C. SLATER, *Phys. Rev.* **32**, 349 (1928).

[2] The rough screening-constant treatment mentioned above gives the values $0.199 \cdot 10^{-24}$ cm³ for He and $0.0291 \cdot 10^{-24}$ cm³ for Li⁺.

[3] J. E. MAYER and M. G. MAYER, *Phys. Rev.* **43**, 605 (1933).

CHAPTER IX

MANY-ELECTRON ATOMS

Up to the present time no method has been applied to atoms with more than two electrons which makes possible the computation of wave functions or energy levels as accurate as those for helium discussed in Section 29c. With the increasing complexity of the atom, the labor of making calculations similar to those used for the ground state of helium increases tremendously. Nevertheless, many calculations of an approximate nature have been carried out for larger atoms with results which have been of considerable value. We shall discuss some of these in this chapter.[1]

30. SLATER'S TREATMENT OF COMPLEX ATOMS

30a. Exchange Degeneracy.—All of the methods which we shall consider are based on a first approximation in which the interaction of the electrons with each other has either been omitted or been replaced by a centrally symmetric field approximately representing the average effect of all the other electrons on the one under consideration. We may first think of the problem as a perturbation problem. The wave equation for an atom with N electrons and a stationary nucleus is

$$\sum_{i=1}^{N}\nabla_i^2\psi + \frac{8\pi^2 m_0}{h^2}\left(W + \sum_{i=1}^{N}\frac{Ze^2}{r_i} - \sum_{i,j>i}\frac{e^2}{r_{ij}}\right)\psi = 0, \quad (30\text{--}1)$$

in which r_i is the distance of the ith electron from the nucleus, r_{ij} is the distance between the ith and jth electrons, and Z is the atomic number.

If the terms in r_{ij} are omitted, this equation is separable into N three-dimensional equations, one for each electron, just as was found to be the case for helium in Section 23b. To this

---

[1] This chapter can be omitted by readers not interested in atomic spectra and related subjects; however, the treatment is closely related to that for molecules given in Chapter XIII.

degree of approximation the wave function for the atom may be
built up out of single-electron wave functions; that is, a solution
of the equation for the atom with $\Sigma e^2/r_{ij}$ omitted is

$$\psi_1^0 = u_\alpha(1)\, u_\beta(2) \,\cdots\, u_\nu(N), \qquad (30\text{-}2)$$

in which $u_\alpha(1)$, etc., are the solutions of the separated single-
electron equations with the three quantum numbers[1] symbolized
by α, β, $\cdots$, ν and the three coordinates symbolized by
$1, 2, \cdots, N$. With this form for ψ^0 the individual electrons
retain their identity and their own quantum numbers. However, an equally good solution of the unperturbed equation corresponding to the same energy as Equation 30-2 is

$$\psi_2^0 = u_\alpha(2)\, u_\beta(1) \,\cdots\, u_\nu(N), \qquad (30\text{-}3)$$

in which electrons 1 and 2 have been interchanged. In general,
the function

$$\psi_P^0 = P u_\alpha(1)\, u_\beta(2) \,\cdots\, u_\nu(N), \qquad (30\text{-}4)$$

in which P is any permutation of the electron coordinates, is an
unperturbed solution for this energy level.

The meaning of the operator P may be illustrated by a simple example.
Let us consider the permutations of the three symbols x_1, x_2, x_3. These are
$x_1, x_2, x_3; x_2, x_3, x_1; x_3, x_1, x_2; x_2, x_1, x_3; x_1, x_3, x_2; x_3, x_2, x_1$. Any one of these
six may be represented by $P x_1, x_2, x_3$, in which P represents the operation of
permuting the symbols x_1, x_2, x_3 in one of the above ways. The operation P
which yields x_1, x_2, x_3 is called the *identity operation*.

Any of the above permutations can be formed from x_1, x_2, x_3 by successive
interchanges of pairs of symbols. This can be done in more than one way,
but the number of interchanges necessary is either always even or always
odd, regardless of the manner in which it is carried out. A permutation
is said to be *even* if it is equivalent to an even number of interchanges, and
odd if it is equivalent to an odd number. We shall find it convenient to
use the symbol $(-1)^P$ to represent $+1$ when P is an even permutation and
-1 when P is an odd permutation.

Multiplication of the operators P and P' means that P and P' are to be
applied successively. The set of all the permutations of N symbols has the
property that the product PP' of any two of them is equal to some other
permutation of the set. A set of operators with this property is called a
group, if in addition the set possesses an identity operation and if every
operation P possesses an *inverse operation* P^{-1} such that PP^{-1} is equivalent
to the identity operation. There are $N!$ permutations of N different
symbols.

---

[1] The symbols α, β, $\cdots$, ν are of course not related to the spin functions
α and β.

At this point we may introduce the spin of the electrons into the wave function (in the same manner as for helium) by multiplying each single-electron orbital function by either $\alpha(\omega)$ or $\beta(\omega)$. For convenience we shall include these spin factors in the functions $u_\alpha(1)$, etc., so that hereafter $\alpha, \beta, \gamma, \cdots$ represent four quantum numbers n, l, m_l, and m_s for each electron and 1, 2, $\cdots$ represent four coordinates $r_i, \vartheta_i, \varphi_i$, and ω_i. As discussed in Section 29a for the two-electron case, treatment of this degenerate energy level by perturbation theory (the electron interactions being the perturbation) leads to certain combinations

$$\psi^0 = \frac{1}{\sqrt{N!}} \sum_P c_P P u_\alpha(1)\, u_\beta(2)\, \cdots\, u_\nu(N) \qquad (30\text{-}5)$$

for the correct zeroth-order normalized wave functions. One of these combinations will have the value $+1$ for each of the coefficients c_P. Interchange of any pair of electrons in this function leaves the function unchanged; i.e., it is completely symmetric in the electron coordinates. For another combination the coefficients c_P are equal to $+1$ or to -1, according as P is an even or an odd permutation. This combination is completely antisymmetric in the electrons; i.e., the interchange of any two electrons changes the sign of the function without otherwise altering it. Besides these two combinations, which were the only ones which occurred in helium, there are for many-electron atoms others which have intermediate symmetries. However, this complexity is entirely eliminated by the application of the Pauli exclusion principle (Sec. 29b) which says that only the completely antisymmetric combination

$$\psi^0 = \frac{1}{\sqrt{N!}} \sum_P (-1)^P P u_\alpha(1)\, u_\beta(2)\, \cdots\, u_\nu(N) \qquad (30\text{-}6)$$

has physical significance. This solution may also be written as a determinant.

$$\psi^0 = \frac{1}{\sqrt{N!}} \begin{vmatrix} u_\alpha(1) & u_\beta(1) & \cdots & u_\nu(1) \\ u_\alpha(2) & u_\beta(2) & \cdots & u_\nu(2) \\ \cdots & \cdots & \cdots & \cdots \\ u_\alpha(N) & u_\beta(N) & \cdots & u_\nu(N) \end{vmatrix}, \qquad (30\text{-}7)$$

as was done in Section 29b. The two forms are identical.

30b. Spatial Degeneracy.—In the previous section we have taken care of the degeneracy due to the $N!$ possible distributions of the N electrons in a fixed set of N functions u. There still remains another type of degeneracy, due to the possibility of there being more than one set of spin-orbit functions corresponding to the same unperturbed energy. In particular there may be other sets of u's differing from the first in that one or more of the quantum numbers m_l or m_s have been changed. These quantum numbers, which represent the z components of orbital and spin angular momentum of the individual electrons, do not affect the unperturbed energy. It is therefore necessary for us to construct the secular equation for all these possible functions in order to find the correct combinations and first approximation to the energy levels.[1]

Before doing this, however, we should ask if there are any more unperturbed wave functions belonging to this level. If, in setting up the perturbation problem, we had called the term $\Sigma e^2/r_{ij}$ the perturbation, then the single-electron functions would have been hydrogenlike functions with quantum numbers n, l, m_l, and m_s. The energy of these solutions depends only on n, as we have seen. However, a better starting point is to add and subtract a term $\sum_i v(x_i)$ representing approximately the average effect of the electrons on each other. If this term is added to H^0 and subtracted from H', the true Hamiltonian $H = H^0 + H'$ is of course unaltered and the unperturbed equation is still separable. The single-electron functions are, however, no longer hydrogenlike functions and their energies are no longer independent of the quantum number l, because it is only with a Coulomb field that such a degeneracy exists (see Sec. 29a). Therefore, in considering the wave functions to be combined we do not ordinarily include any but those involving a single set of values of n and l; i.e., those belonging to a single configuration.

The consideration of a simple example, the configuration $1s^2 2p$ of lithium, may make clearer what the different unperturbed functions are. Table 30–1 gives the sets of quantum num-

---

[1] The treatment of atoms which we are giving is due to J. C. Slater, *Phys. Rev.* **34**, 1293 (1929), who showed that this method was very much simpler and more powerful than the complicated group-theory methods previously used.

bers possible for this configuration. The notation $\{100\frac{1}{2}\}$ means $n = 1, l = 0, m_l = 0, m_s = +\frac{1}{2}$. Each line of the table n corresponds to a set of functions $u_\alpha \cdots u_\nu$ which when substituted into the determinant of Equation 30–7 gives a satisfactory antisymmetrical wave function ψ_n^0 corresponding to the

TABLE 30–1.—SETS OF QUANTUM NUMBERS FOR THE CONFIGURATION $1s^2 2p$

1. $(100\frac{1}{2})$ $(100 -\frac{1}{2})$ $(211\frac{1}{2})$; $\Sigma m_l = +1,$ $\Sigma m_s = +\frac{1}{2},$
2. $(100\frac{1}{2})$ $(100 -\frac{1}{2})$ $(211 -\frac{1}{2})$; $\Sigma m_l = +1,$ $\Sigma m_s = -\frac{1}{2},$
3. $(100\frac{1}{2})$ $(100 -\frac{1}{2})$ $(210\frac{1}{2})$; $\Sigma m_l = 0,$ $\Sigma m_s = +\frac{1}{2},$
4. $(100\frac{1}{2})$ $(100 -\frac{1}{2})$ $(210 -\frac{1}{2})$; $\Sigma m_l = 0,$ $\Sigma m_s = -\frac{1}{2},$
5. $(100\frac{1}{2})$ $(100 -\frac{1}{2})$ $(21 -1\frac{1}{2})$; $\Sigma m_l = -1,$ $\Sigma m_s = +\frac{1}{2},$
6. $(100\frac{1}{2})$ $(100 -\frac{1}{2})$ $(21 -1 -\frac{1}{2})$; $\Sigma m_l = -1,$ $\Sigma m_s = -\frac{1}{2}.$

same unperturbed energy level. No other sets satisfying the Pauli exclusion principle can be written for this configuration. The order of the expressions n, l, m_l, m_s in a given row is unimportant.

This simple case illustrates the idea of *completed shells* of electrons. The first two sets of quantum numbers remain the

TABLE 30–2.—SETS OF QUANTUM NUMBERS FOR THE CONFIGURATION np^2

		Σm_l	Σm_s
1. $(n11\frac{1}{2})$	$(n11 -\frac{1}{2})$	2	0
2. $(n11\frac{1}{2})$	$(n10\frac{1}{2})$	1	+1
3. $(n11\frac{1}{2})$	$(n10 -\frac{1}{2})$	1	0
4. $(n11 -\frac{1}{2})$	$(n10\frac{1}{2})$	1	0
5. $(n11 -\frac{1}{2})$	$(n10 -\frac{1}{2})$	1	-1
6. $(n11\frac{1}{2})$	$(n1 -1\frac{1}{2})$	0	+1
7. $(n11\frac{1}{2})$	$(n1 -1-\frac{1}{2})$	0	0
8. $(n11 -\frac{1}{2})$	$(n1 -1\frac{1}{2})$	0	0
9. $(n10\frac{1}{2})$	$(n10 -\frac{1}{2})$	0	0
10. $(n11 -\frac{1}{2})$	$(n1 -1 -\frac{1}{2})$	0	-1
11. $(n1 -1\frac{1}{2})$	$(n10\frac{1}{2})$	-1	+1
12. $(n1 -1 -\frac{1}{2})$	$(n10\frac{1}{2})$	-1	0
13. $(n1 -1\frac{1}{2})$	$(n10 -\frac{1}{2})$	-1	0
14. $(n1 -1 -\frac{1}{2})$	$(n10 -\frac{1}{2})$	-1	-1
15. $(n1 -1\frac{1}{2})$	$(n1 -1 -\frac{1}{2})$	-2	0

same throughout this table because $1s^2$ is a completed shell; i.e., it contains as many electrons as there are possible sets of quantum numbers. The shell ns can contain two electrons, np six electrons, nd ten electrons, etc. In determining the number of wave functions which must be combined, it is only necessary to consider electrons outside of completed shells, because there can be only one set of functions $u_\alpha \cdots u_\nu$ for the completed shells.

Table 30–2 gives the allowed sets of quantum numbers for two equivalent p electrons, i.e., two electrons with the same value of n and with $l = 1$.

Problem 30–1. Construct tables similar to Table 30–2 for the configurations np^3 and nd^2.

30c. Factorization and Solution of the Secular Equation.—We have now determined the unperturbed wave functions which must be combined in order to get the correct zeroth-order wave functions for the atom. The next step is to set up the secular equation for these functions as required by perturbation theory, the form given at the end of Section 24 being the most convenient. This equation has the form

$$\begin{vmatrix} H_{11} - W & H_{12} & \cdots & H_{1k} \\ H_{21} & H_{22} - W & \cdots & H_{2k} \\ \cdots & \cdots & \cdots & \cdots \\ H_{k1} & H_{k2} & \cdots & H_{kk} - W \end{vmatrix} = 0, \quad (30\text{–}8)$$

in which

$$H_{nm} = \int \psi_n^* H \psi_m d\tau. \quad (30\text{–}9)$$

ψ_n is an antisymmetric normalized wave function of the form of Equation 30–6 or 30–7, the functions u composing it corresponding to the nth row of a table such as Table 30–1 or 30–2. H is the true Hamiltonian for the atom, including the interactions of the electrons.

This equation is of the kth degree, k being the number of allowed sets of functions $u_\alpha \cdots u_\nu$. Thus for the configuration $1s^2 2p$ k is equal to 6, as is seen from Table 30–1. However, there is a theorem which greatly simplifies the solution of this equation: *the integral H_{mn} is zero unless ψ_m and ψ_n have the same value of Σm_s and the same value of Σm_l, these quantities being the sums of quantum numbers m_s and m_l of the functions u making up ψ_m and ψ_n.* We shall prove this theorem in Section 30d in connection with the evaluation of the integrals H_{mn}, and in the meantime we shall employ the result to factor the secular equation.

Examining Table 30–1, we see that the secular equation for $1s^2 2p$ factors into six linear factors; i.e., no two functions ψ_n and ψ_m have the same values of Σm_s and Σm_l. The equation for np^2, as seen from Table 30–2, has the factors indicated by Figure 30–1, the shaded squares being the only non-zero elements. A fifteenth-degree equation has, therefore, by the use of this theorem been reduced to a cubic, two quadratic, and eight linear factors.

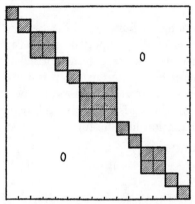

Fig. 30–1.—The secular determinant for the configuration np^2, represented diagrammatically.

By evaluating the integrals H_{mn} and solving these equations, the approximate energy levels W corresponding to this configuration could be obtained; but a still simpler method is available, based on the fact that the roots W of the equations of lower degree will coincide with some of the roots of the equations of higher degree. The reason for this may be made clear by the following argument. The wave functions ψ_1, ψ_2, $\cdots$, ψ_k, which we are combining, differ from one another only in the quantum numbers m_s and m_l of the single electrons, these quantum numbers representing the z components of the spin and orbital angular momenta of the electrons. The energy of a single electron in a central field does not depend on m_l or m_s (neglecting magnetic effects), since these quantum numbers refer essentially to orientation in space. The energy of an atom with several electrons does depend on these quantum numbers, because the mutual interaction of the electrons is influenced by the relative orientations of the angular-momentum vectors of the individual electrons. Just as for one-electron atoms, however, the orientation of the whole atom in space does not affect its energy and we expect to find a number of states having the same energy but corresponding to different values of the *z* components of the total orbital angular momentum and of the total spin angular momentum; i.e., to different values of Σm_l and Σm_s.

This type of argument is the basis of the *vector model*[1] for atoms, a very convenient method of illustrating and remembering the results of quantum-mechanical discussions such as the one we are giving here. In the vector model of the atom the orbital and spin angular momenta of the individual electrons are considered as vectors (see Section 1e) which may be combined to give resultant vectors for the whole atom, the manner in which these vectors are allowed to combine being restricted by certain rules in such a way as to duplicate the results of quantum mechanics. The vector picture is especially useful in classifying and naming the energy levels of an atom, the values of the resultant vectors being used to specify the different levels.

In Chapter XV we shall show that not only is the energy of a stationary state of a free atom a quantity which has a definite value (and not a probability distribution of values) but also the total angular momentum and the component of angular momentum in any one chosen direction (say the *z* direction) are similar quantities. Whereas it is not possible to specify exactly both the energy and the positions of the electrons in an atom, it is possible to specify the above three quantities simultaneously. If the magnetic effects are neglected we may go further and specify the total spin and total orbital angular momenta separately, and likewise their *z* components. However, we may not give the angular momenta of the individual electrons separately, these being quantities which fluctuate because of the electron interactions.

It will likewise be shown that when magnetic effects are neglected the square of the total orbital angular momentum must assume only the quantized values $L(L + 1)(h/2\pi)^2$ where L is an integer, while the square of the total spin angular momentum can take on only the values $S(S + 1)(h/2\pi)^2$ where S is integral or half-integral. (The letter L is usually used for the

---

[1] See PAULING and GOUDSMIT, "The Structure of Line Spectra."

total resultant orbital angular momentum of the atom, and the letter S for the total spin angular momentum; see Section 29b.) In the approximation[1] we are using, the states of an atom may be labeled by giving the configuration and the quantum numbers L, S, $M_L = \Sigma m_l$, and $M_S = \Sigma m_s$, the last two having no effect on the energy. Just as for one electron, the allowed values of M_L are L, $L - 1$, $\cdots$, $-L + 1$, $-L$; M_S is similarly restricted to S, $S - 1$, $\cdots$, $-S + 1$, $-S$, all of these values of M_S and M_L belonging to the same degenerate energy level and corresponding to different orientations in space of the vectors **L** and **S**.

We shall now apply these ideas to the solution of the secular equation, taking the configuration np^2 as an example. From Table 30–2 we see that $H_{11} - W$ is a linear factor of the equation, since ψ_1 alone has $\Sigma m_l = 2$ and $\Sigma m_s = 0$. A state with $M_L = 2$ must from the above considerations have $L \geqslant 2$. Since 2 is the highest value of M_L in the table, it must correspond to $L = 2$. Furthermore the state must have $S = 0$, because otherwise there would appear entries in the table with $M_L = 2$ and $M_S \neq 0$. This same root W must appear five times in the secular equation, corresponding to the degenerate states $L = 2$, $S = 0$, $M_S = 0$, $M_L = 2, 1, 0, -1, -2$. From this it is seen that this root (which can be obtained from the linear factors) must occur in two of the linear factors ($M_L = 2, -2$; $M_S = 0$), in two of the quadratic factors ($M_L = 1, -1$; $M_S = 0$), and in the cubic factor ($M_L = 0$, $M_S = 0$). The linear factor $H_{22} - W$ with $M_L = 1$, $M_S = 1$ must belong to the level $L = 1$, $S = 1$, because no terms with higher values of M_L and M_S appear in the table except those already accounted for. This level will correspond to the nine states with $M_L = 1, 0, -1$, and $M_S = 1$, 0, -1. Six of these are roots of linear factors ($M_L = \pm 1$, $M_S = \pm 1$; $M_L = 0$; $M_S = \pm 1$), two of them are roots of the quadratic factors ($M_L = \pm 1$, $M_S = 0$), and one is a root of the cubic factor ($M_L = 0$, $M_S = 0$).

Without actually solving the quadratic equations or evaluating the integrals involved in them, we have determined their roots, since all the roots of the quadratics occur also in linear factors.

---

[1] This approximation, called (LS) or Russell-Saunders coupling, is valid for light atoms. Other approximations must be made for heavy atoms in which the magnetic effects are more important.

Likewise we have obtained two of the three roots of the cubic. The third root of the cubic can be evaluated without solving the cubic or calculating the non-diagonal elements of the equation, by appealing to the theorem that the sum of the roots of a secular equation (or of one of its factors) is equal to the sum of the diagonal elements of the equation[1] (or of the factor). Since two of the roots of the cubic have been found and the sum of the roots is given by the theorem, the third may be found. It corresponds to the state $L = 0$, $S = 0$, since this is the only possibility left giving one state with $M_L = 0$ and $M_S = 0$.

The three energy levels for np^2 which we have found are

$$\left.\begin{aligned}
W &= H_{11}, \ ^1D \ (L = 2, S = 0, M_S = 0, M_L = 2, 1, 0, -1, -2); \\
W &= H_{22}, \ ^3P \ (L = 1, \ S = 1, \ M_S = \pm 1, \ 0; \ M_L = \pm 1, \ 0); \\
W &= H_{77} + H_{88} + H_{99} - H_{11} - H_{22}, \ ^1S \ (L = 0, \ S = 0, \\
&\qquad\qquad\qquad\qquad\qquad\qquad\qquad\qquad M_S = 0, \ M_L = 0).
\end{aligned}\right\}$$

$$(30\text{--}10)$$

The term symbols ^1D, ^3P, ^1S have been explained in Section **29b**.

Problem 30-2. Investigate the factorization of the secular equation for np^3, using the results of Problem 30-1, and list terms which belong to this configuration.

30d. Evaluation of Integrals.—We need to obtain expressions for integrals of the type

$$H_{mn} = \int \psi_m^* H \psi_n d\tau = \frac{1}{N!} \sum_P \sum_{P'} (-1)^{P+P'} \int P' u_\alpha^*(1) \cdots$$
$$u_\nu^*(N) H P u_\alpha(1) \cdots u_\nu(N) d\tau. \quad (30\text{--}11)$$

---

[1] To prove this theorem, we expand the secular equation 30–8 and arrange according to powers of W. The resulting algebraic equation in W will have k roots, $W_1, W_2, \cdots, W_k$ and can therefore be factored into k factors

$$(W - W_1)(W - W_2) \cdots (W - W_k) = 0.$$

The coefficient of $-W^{k-1}$ in this form of the equation is seen to be

$$W_1 + W_2 + \cdots + W_k;$$

the coefficient of $-W^{k-1}$ in Equation 30–8 is seen to be

$$H_{11} + H_{22} + \cdots + H_{kk}.$$

These two expressions must therefore be equal, which **proves the theorem.**

We may eliminate one of the summations by the following device:

$$\int P'u_\alpha^*(1) \cdots u_\nu^*(N)HPu_\alpha(1) \cdots u_\nu(N)d\tau =$$
$$P''\int P'u_\alpha^*(1) \cdots u_\nu^*(N)HPu_\alpha(1) \cdots u_\nu(N)d\tau =$$
$$\int P''P'u_\alpha^*(1) \cdots u_\nu^*(N)HP''Pu_\alpha(1) \cdots u_\nu(N)d\tau,$$

the first step being allowed because P'' only interchanges the names of the variables of a definite integral. If we choose P'' to be P'^{-1}, the inverse permutation to P', then $P''P' = 1$; i.e., $P''P'$ is the identity operation, while $P''P$ is still some member of the set of permutations, all members of which are summed over. The integral therefore no longer involves P' and the sum over P' reduces to multiplication by $N!$, the number of permutations. We thus obtain the equation

$$H_{mn} = \sum_P (-1)^P \int u_\alpha^*(1) \cdots u_\nu^*(N)HPu_\alpha(1) \cdots u_\nu(N)d\tau.$$

$$(30\text{–}12)$$

We shall now prove the theorem that $H_{mn} = 0$ unless Σm_s is the same for ψ_m and ψ_n. H does not involve the spin coordinates so that integration over these coordinates yields a product of orthogonality integrals for the spin functions of the various electrons. Unless the spins of corresponding electrons in the two functions $u_\nu^*(1) \cdots u_\nu^*(N)$ and $Pu_\alpha(1) \cdots u_\nu(N)$ are the same, the integral is zero. If Σm_s is not the same for ψ_m and ψ_n there can be no permutation P which will make such a matching of the spins possible, because the number of positive and negative spins is different in the two functions.

To prove the theorem concerning Σm_l it is necessary to specify further the nature of H. We write

$$H = \sum_i f_i + \sum_{i,j>i} g_{ij},$$

where

$$f_i = -\frac{h^2}{8\pi^2 m_0}\nabla_i^2 - \frac{Ze^2}{r_i} \text{ and } g_{ij} = \frac{e^2}{r_{ij}}.$$

The functions $u_\alpha(1) \cdots$ are solutions of

$$\left\{-\frac{h^2}{8\pi^2 m_0}\nabla_i^2 - \frac{Ze^2}{r_i} - v(r_i)\right\}u_\zeta(i) = \epsilon_i u_\zeta(i).$$

From this we see that

$$f_i u_\zeta(i) = f(r_i)u_\zeta(i),$$

where $f(r_i)$ is a function of r_i alone. The integral of the first term in H thus reduces to

$$\sum_P (-1)^P \sum_i \int u_\alpha^*(1)Pu_\alpha(1)d\tau_1 \cdots \int u_\zeta^*(i)f(r_i)Pu_\zeta(i)d\tau_i \cdots$$
$$\int u_\nu^*(N)Pu_\nu(N)d\tau_N, \quad (30\text{--}13)$$

in which $Pu_\zeta(i)$ is used as a symbol for $u_\zeta(j)$ in which electron j has replaced i as a result of the permutation P. Because of the orthogonality of the u's, this is zero unless $Pu_\zeta(i) = u_\zeta(j)$ except perhaps for j equal to the one value i. In addition, since

$$u_\zeta(i) = R_{nl}(r_i) \cdot \Theta_{lm}(\vartheta_i) \cdot e^{im\,\varphi_i}, \quad (30\text{--}14)$$

the factor $\int u_\zeta^*(i)f(r_i)Pu_\zeta(i)d\tau_i$ will be zero unless $u_\zeta(i)$ and $Pu_\zeta(i)$ have the same quantum number m_l. We thus see that this integral will vanish, unless all the u's but one pair match and the members of that pair have the same value of m_l.

Similar treatment of the term Σg_{ij} shows that all but perhaps two pairs must match. The factor containing these unmatched functions is

$$\int u_\zeta^*(i)u_\xi^*(j)\frac{e^2}{r_{ij}}Pu_\zeta(i)Pu_\xi(j)d\tau_i d\tau_j. \quad (30\text{--}15)$$

It can be shown[1] that

$$\frac{1}{r_{ij}} = \sum_{k,m} \frac{(k-|m|)!}{(k+|m|)!}\frac{r_a^k}{r_b^{k+1}}P_k^{|m|}(\cos\vartheta_i)P_k^{|m|}(\cos\vartheta_j)e^{im\,(\varphi_i-\varphi_j)}, \quad (30\text{--}16)$$

in which r_a is the smaller of r_i and r_j, and r_b is the greater. $P_k^{|m|}(\cos\vartheta)$ is an associated Legendre function, discussed in Section 19b. Using this expansion we obtain for the φ part of the above integral $\int_0^{2\pi}\int_0^{2\pi} e^{i(Pm_l-m_l+m)\varphi_i}e^{i(Pm_l'-m_l'-m)\varphi_j}d\varphi_i d\varphi_j$, in which m_l is associated with $u_\zeta(i)$, m_l' with $u_\xi(j)$, Pm_l with $Pu_\zeta(i)$, and Pm_l' with $Pu_\xi(j)$. This vanishes unless $Pm_l - m_l + m = 0$ and $Pm_l' - m_l' - m = 0$; i.e., unless $Pm_l + Pm_l' = m_l + m_l'$.

[1] For a proof of this see J. H. Jeans, "Electricity and Magnetism," 5th ed., Equations 152 and 196, Cambridge University Press, 1927.

This completes the proof of the theorem that $H_{mn} = 0$ unless Σm_l is the same for ψ_n and ψ_m.

Of the non-vanishing elements H_{mn} only certain of the diagonal ones need to be evaluated in order to calculate the energy levels, as we have seen in the last section. Because of the orthogonality of the u's, Equation 30–13 vanishes unless $P = 1$ (the identity operation) when a diagonal element H_{mm} is being considered. Since the u's are also normalized, this expression reduces to

$$\sum_i \int u_\zeta^*(i)f(r_i)u_\zeta(i)d\tau_i = \sum_i I_i, \qquad (30\text{--}17)$$

a relation which defines the quantities I_i.

Similarly, the orthogonality of the u's restricts P in Equation 30–15 to $P = 1$ and $P = (ij)$, the identity operation and the interchange of i and j, respectively. The first choice of P contributes the terms

$$\sum_{i,j>i} \int u_\zeta^*(i)u_\xi^*(j)\frac{e^2}{r_{ij}}u_\zeta(i)u_\xi(j)d\tau_i d\tau_j = \sum_{i,j>i} J_{ij}, \qquad (30\text{--}18)$$

while the second yields

$$-\sum_{i,j>i} \int u_\zeta^*(i)u_\xi^*(j)\frac{e^2}{r_{ij}}u_\xi(i)u_\zeta(j)d\tau_i d\tau_j = -\sum_{i,j>i} K_{ij}. \qquad (30\text{--}19)$$

The integral K_{ij} vanishes unless the spins of $u_\zeta(i)$ and $u_\xi(j)$ are parallel, i.e., unless $m_{s_i} = m_{s_j}$.

The functions I_i reduce to integrals over the radial part of $u(i)$,

$$I_i = \int R_{nl}^*(r_i)f(r_i)R_{nl}(r_i)d\tau_i. \qquad (30\text{--}20)$$

We shall not evaluate these further.

The functions J_{ij} and K_{ij} may be evaluated by using the expansion for $1/r_{ij}$ given in Equation 30–16. For J_{ij} the φ_i part of the integral has the form $\int_0^{2\pi} e^{im\varphi_i}d\varphi_i$, which vanishes unless $m = 0$. The double sum in the expansion 30–16 thus reduces to a single sum over k, which can be written

$$J_{ij} = \sum_k a^k(lm_l; l'm_l')F^k(nl; n'l'), \qquad (30\text{--}21)$$

in which nlm_l and $n'l'm_l'$ are the quantum numbers previously represented by n_i and n_j, respectively. a^k and F^k are given by

$$a^k(lm_l; l'm_l') = \frac{(2l+1)(l-|m_l|)!}{2(l+|m_l|)!} \cdot \frac{(2l'+1)(l'-|m_l'|)!}{2(l'+|m_l'|)!}$$

$$\int_0^\pi \{P_l^{|m_l|}(\cos\vartheta_i)\}^2 P_k^0(\cos\vartheta_i) \sin\vartheta_i d\vartheta_i$$

$$\int_0^\pi \{P_{l'}^{|m_l'|}(\cos\vartheta_j)\}^2 P_k^0(\cos\vartheta_j) \sin\vartheta_j d\vartheta_j, \qquad (30\text{--}22)$$

and

$$F^k(nl; n'l') = (4\pi)^2 e^2 \int_0^\infty \int_0^\infty R_{nl}^2(r_i) R_{n'l'}^2(r_j) \frac{r_a^k}{r_b^{k+1}} r_i^2 r_j^2 dr_i dr_j. \qquad (30\text{--}23)$$

The a's are obtained from the angular parts of the wave functions, which are the same as for the hydrogen atom (Tables 21–1 and 21–2, Chap. V). Some of these are given in Table 30–3, taken

TABLE 30–3.—VALUES OF $a^k(lm_l; l'm_l')$
(In cases with two $\pm$ signs, the two can be combined in any of the four possible ways)

Electrons	l	m_l	l'	m_l'	a^0	a^2	a^4
ss	0	0	0	0	1	0	0
sp	0	0	1	± 1	1	0	0
	0	0	1	0	1	0	0
pp	1	± 1	1	± 1	1	$\tfrac{1}{25}$	0
	1	± 1	1	0	1	$-\tfrac{2}{25}$	0
	1	0	1	0	1	$\tfrac{4}{25}$	0
sd	0	0	2	± 2	1	0	0
	0	0	2	± 1	1	0	0
	0	0	2	0	1	0	0
pd	1	± 1	2	± 2	1	$\tfrac{2}{35}$	0
	1	± 1	2	± 1	1	$-\tfrac{1}{35}$	0
	1	± 1	2	0	1	$-\tfrac{2}{35}$	0
	1	0	2	± 2	1	$-\tfrac{4}{35}$	0
	1	0	2	± 1	1	$\tfrac{2}{35}$	0
	1	0	2	0	1	$\tfrac{4}{35}$	0
dd	2	± 2	2	± 2	1	$\tfrac{4}{49}$	$\tfrac{1}{441}$
	2	± 2	2	± 1	1	$-\tfrac{2}{49}$	$-\tfrac{4}{441}$
	2	± 2	2	0	1	$-\tfrac{4}{49}$	$\tfrac{6}{441}$
	2	± 1	2	± 1	1	$\tfrac{1}{49}$	$\tfrac{16}{441}$
	2	± 1	2	0	1	$\tfrac{2}{49}$	$-\tfrac{24}{441}$
	2	0	2	0	1	$\tfrac{4}{49}$	$\tfrac{36}{441}$

from Slater's paper. The F's, on the other hand, depend on the radial parts of the wave functions, which for the best approximation are not hydrogenlike.

K_{ij} may be similarly expressed as

$$K_{ij} = \sum_k b^k(lm_l; l'm_l')G^k(nl; n'l'), \qquad (30\text{-}24)$$

in which

$$b^k(lm_l; l'm_l') =$$
$$\frac{(k - |m_l - m_l'|)!(2l + 1)(l - |m_l|)!(2l' + 1)(l' - |m_l'|)!}{4(k + |m_l - m_l'|)!(l + |m_l|)\ (l' + |m_l'|)!}$$
$$\left\{ \int_0^\pi P_l^{|m_l|}(\cos \vartheta)P_{l'}^{|m_l'|}(\cos \vartheta)P_k^{|m_l - m_l'|}(\cos \vartheta) \sin \vartheta d\vartheta \right\}^2, \qquad (30\text{-}25)$$

and

$$G^k(nl; n'l') = e^2(4\pi)^2 \int_0^\infty \int_0^\infty R_{nl}(r_i)R_{n'l'}(r_i)R_{nl}(r_j)R_{n'l'}(r_j)$$
$$\frac{r_a^k}{r_b^{k+1}}r_i^2 r_j^2 dr_i dr_j. \qquad (30\text{-}26)$$

The functions b^k are given in Table 30–4. The functions G^k are characteristic of the atom.

30e. Empirical Evaluation of Integrals. Applications.—We have now carried the computations to a stage at which the energy levels may be expressed in terms of certain integrals I_i, F^k, and G^k which involve the radial factors of the wave functions. One method of proceeding further would be to assume some form for the central field $v(r_i)$, determine the functions $R_{ml}(r_i)$, and use them to evaluate the integrals. However, another and simpler method is available for testing the validity of this approximation, consisting in the use of the empirically determined energy levels to evaluate the integrals, a check on the theory resulting from the fact that there are more known energy levels than integrals to be determined.

For example, if we substitute for H_{11}, etc., for the configuration np^2 the expression in terms of I_i, F^k, and G^k, using the results of the previous section and Equation 30–10, we obtain for the energies of the terms ^1D, ^3P, and ^1S the quantities

$$^1D: W = 2I(n, 1) + F^0 + \tfrac{1}{25}F^2,$$
$$^3P: W = 2I(n, 1) + F^0 - \tfrac{2}{25}F^2 - \tfrac{3}{25}G^2,$$
$$^1S: W = 2I(n, 1) + F^0 + \tfrac{7}{25}F^2 + \tfrac{3}{25}G^2.$$

TABLE 30–4.—VALUES OF $b^k(lm_l; l'm_l')$

(In cases where there are two ± signs, the two upper, or the two lower, signs must be taken together)

Electrons	l	m_l	l'	m_l'	b^0	b^1	b^2	b^3	b^4
ss	0	0	0	0	1	0	0	0	0
sp	0	0	1	±1	0	$\frac{1}{3}$	0	0	0
	0	0	1	0	0	$\frac{1}{3}$	0	0	0
pp	1	±1	1	±1	1	0	$\frac{1}{25}$	0	0
	1	±1	1	0	0	0	$\frac{3}{25}$	0	0
	1	±1	1	∓1	0	0	$\frac{6}{25}$	0	0
	1	0	1	0	1	0	$\frac{4}{25}$	0	0
sd	0	0	2	±2	0	0	$\frac{1}{5}$	0	0
	0	0	2	±1	0	0	$\frac{1}{5}$	0	0
	0	0	2	0	0	0	$\frac{1}{5}$	0	0
pd	1	±1	2	±2	0	$\frac{2}{5}$	0	$\frac{3}{245}$	0
	1	±1	2	±1	0	$\frac{1}{5}$	0	$\frac{9}{245}$	0
	1	±1	2	0	0	$\frac{1}{15}$	0	$\frac{18}{245}$	0
	1	±1	2	∓1	0	0	0	$\frac{30}{245}$	0
	1	±1	2	∓2	0	0	0	$\frac{45}{245}$	0
	1	0	2	±2	0	0	0	$\frac{15}{245}$	0
	1	0	2	±1	0	$\frac{1}{5}$	0	$\frac{24}{245}$	0
	1	0	2	0	0	$\frac{4}{15}$	0	$\frac{27}{245}$	0
dd	2	±2	2	±2	1	0	$\frac{4}{49}$	0	$\frac{1}{441}$
	2	±2	2	±1	0	0	$\frac{6}{49}$	0	$\frac{5}{441}$
	2	±2	2	0	0	0	$\frac{4}{49}$	0	$\frac{15}{441}$
	2	±2	2	∓1	0	0	0	0	$\frac{35}{441}$
	2	±2	2	∓2	0	0	0	0	$\frac{70}{441}$
	2	±1	2	±1	1	0	$\frac{1}{49}$	0	$\frac{16}{441}$
	2	±1	2	0	0	0	$\frac{1}{49}$	0	$\frac{30}{441}$
	2	±1	2	∓1	0	0	$\frac{6}{49}$	0	$\frac{40}{441}$
	2	0	2	0	1	0	$\frac{4}{49}$	0	$\frac{36}{441}$

Examination of Equations 30–18 and 30–19 shows that for equivalent electrons F is equal to G (with the same index). We therefore have for the separations of the levels for np^2

$$^1D - {}^3P = \tfrac{6}{25}F^2(n1; n1),$$
$$^1S - {}^1D = \tfrac{9}{25}F^2(n1; n1).$$

The theory therefore indicates that, if the approximations which have been made are valid, the ratio of these intervals should be 2:3, a result which is obtained without the evaluation of any radial integrals at all. In addition, since F^2 is necessarily positive, this theory gives the order of the terms, ^3P lying lowest, ^1D next, and ^1S highest. This result is in agreement with Hund's empirical rules, that terms with largest multiplicity usually lie lowest, and that, for a given multiplicity, terms with largest L values usually lie lowest.[1]

Slater gives the example of the configuration[2] $1s^22s^22p^6$ $3s^23p^2$ of silicon, for which the observed term values[3] are

$$^3P = 65615 \text{ cm}^{-1},$$
$$^1D = 59466 \text{ cm}^{-1},$$
$$^1S = 50370 \text{ cm}^{-1},$$

so that the ratio $^1D - {}^3P$ to $^1S - {}^1D$ is 2:2.96, in excellent agreement with the theory. In other applications, however, large deviations have been found, most of which have been explained by considering higher approximations based on the same general principles.[4]

31. VARIATION TREATMENTS FOR SIMPLE ATOMS

The general discussion of Section 30, which is essentially a perturbation calculation, is not capable of very high accuracy, especially since it is not ordinarily practicable to utilize any central field except the coulombic one leading to hydrogenlike orbital functions. In this section we shall consider the application of the variation method (Sec. 26) to low-lying states of simple atoms such as lithium and beryllium. This type of treatment is much more limited than that of the previous section, but for the few states of simple atoms to which it has been applied it is more accurate.

[1] PAULING and GOUDSMIT, "The Structure of Line Spectra," p. 166.

[2] This configuration gives the same interval ratios as np^2, only the absolute energy being changed by the presence of the closed shells.

[3] As mentioned in Section 5a, term values are usually given in cm^{-1} and are measured *downward* from the lowest state of the ionized atom. Hence the largest term value represents the lowest energy level.

[4] There have been many papers on this subject; a few are: C. W. Ufford, *Phys. Rev.* **44**, 732 (1933); G. H. Shortley, *Phys. Rev.* **43**, 451 (1933); M. H. Johnson, Jr., *Phys. Rev.* **43**, 632 (1933); D. R. Inglis and N. Ginsburg, *Phys. Rev.* **43**, 194 (1933). A thorough treatment is given by E. U. Condon and G. H. Shortley, "The Theory of Atomic Spectra," Cambridge, 1935.

The principles involved are exactly the same as those discussed in Section 26 and applied to helium in Section 29c, so we shall not discuss them further but instead study the different types of variation functions used and the results achieved.

31a. The Lithium Atom and Three-electron Ions.—Table 31–1 lists the variation functions which have been tried for the lowest state of lithium, which has the configuration $1s^2 2s$. All these functions are of the determinant type given in Equation 30–7 and in all of them the orbital part of $u_{1s}(i)$ is of the form $e^{-Z'\frac{r_i}{a_0}}$, in which Z', the effective atomic number for the K shell, is one of the parameters determined by the variation method. The table gives the expressions for b, the orbital part of $u_{2s}(i)$, the function for the $2s$ electron. In addition, the upper limit to the total energy of the atom is given, and also the value of the first ionization potential calculated by subtracting the value of the energy calculated for Li^+ from the total energy calculated for Li. The Li^+ calculation was made with the use of the same type of $1s$ function used in Li for the K shell, in order to cancel part of the error introduced by this rather poor K function. The table also gives the differences between these calculated quantities and the experimental values.

TABLE 31–1.—VARIATION FUNCTIONS FOR THE NORMAL LITHIUM ATOM
Units: $R_\infty hc$
Experimental total energy: -14.9674; experimental ionization potential: 0.3966

2s function[1]	Total energy	Difference	Ionization potential	Difference
1. $b = e^{-\eta\frac{r}{a_0}}\left(\eta\frac{r}{a_0} - 1\right)$	-14.7844	0.1830	0.3392	0.0574
2. $b = re^{-\eta\frac{r}{a_0}}$	-14.8358	.1316	.3906	.0060
3. $b = e^{-\eta\frac{r}{a_0}}\left(\alpha\frac{r}{a_0} - 1\right)$	-14.8366	.1308	.3912	.0054
4. $b = \alpha\frac{r}{a_0}e^{-\eta\frac{r}{a_0}} - e^{-\zeta\frac{r}{a_0}}$	-14.8384	.1290	.3930	.0036

[1] The function 1 was used by C. Eckart, *Phys. Rev.* **36,** 878 (1930), 2 and 3 by V. G·ille-min and C. Zener, *Z. f. Phys.* **61,** 199 (1930), and 4 by E. B. Wilson, Jr., *J. Chem. Phys.* **1,** 211 (1933). The last paper includes similar tables for the ions Be⁺, B⁺⁺, and C⁺⁺⁺.

Table 31–2 lists the best values of the parameters for these lithium variation functions. Figure 31–1 shows the total electron distribution function $4\pi r^2 \rho$ for lithium, calculated using

TABLE 31–2.—PARAMETER VALUES FOR LITHIUM VARIATION FUNCTIONS

$$u_{1s} = e^{-\frac{Z'r}{a_0}}$$

Function	Z'	η	α	ζ
1. $b = e^{-\eta\frac{r}{a_0}}\left(\eta\frac{r}{a_0} - 1\right)$	2.686	0.888		
2. $b = re^{-\eta\frac{r}{a_0}}$	2.688	.630		
3. $b = e^{-\eta\frac{r}{a_0}}\left(\alpha\frac{r}{a_0} - 1\right)$	2.688	.630	5.56	
4. $b = \alpha\frac{r}{a_0}e^{-\eta\frac{r}{a_0}} - e^{-\zeta\frac{r}{a_0}}$	2.69	.665	1.34	1.5

the best of these functions. ρ is the electron density, which can be calculated from ψ in the following manner:

$$\rho = 3\int\psi^*\psi d\tau_1 d\tau_2. \tag{31–1}$$

$\psi^*\psi d\tau_1 d\tau_2 d\tau_3$ gives the probability of finding electron 1 in the volume element $d\tau_1$, electron 2 in $d\tau_2$, and electron 3 in $d\tau_3$. Integration over the coordinates of electrons 1 and 2 gives the probability of finding electron 3 in $d\tau_3$, regardless of the positions of 1 and 2. Since $\psi^*\psi$ is symmetric in the three electrons, the probability of finding one electron in a volume element $dxdydz$ in ordinary three-dimensional space is three times the probability of finding a particular one. Figure 31–1 shows clearly the two shells of electrons in lithium, the well-marked K shell and the more diffuse L shell. Due to the equivalence of the three electrons, we cannot say that a certain two occupy the K shell and the remaining one the L shell, but we can say that on the average there are two electrons in the K shell and one in the L shell.

The next step to be taken is to apply a variation function to lithium which recognizes explicitly the instantaneous, instead of just the average, influence of the electrons on each other. Such functions were found necessary to secure really accurate results for helium (Sec. 29c), but their application to lithium involves extremely great complications. This work has been begun (by James and Coolidge at Harvard[1]).

31b. Variation Treatments of Other Atoms.—Few efforts have been made to treat more complicated atoms by this method. Beryllium has been studied by several investigators but the functions which give good results for lithium are not nearly so accurate for heavier atoms. Hydrogenlike functions with variable effective nuclear charges (function 1 of Table 31–1 is

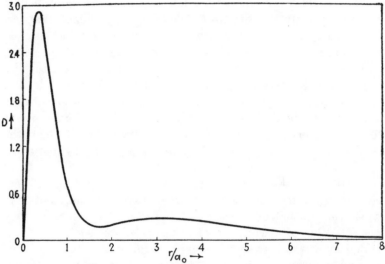

Fig. 31–1.—The electron distribution function $D = 4\pi r^2 \rho$ for the normal lithium atom.

such a function for $n = 2$, $l = 0$) have been applied to the case of the carbon atom,[2] the results being in approximate agreement with experiment. Functions of the types 2 and 3 of Table 31–1 have also been tried[3] for Be, B, C, N, O, F, and Ne. A more satisfactory attack has been begun by Morse and Young,[4] who have prepared numerical tables of integrals for wave functions dependent on four parameters (one for 1s, two for 2s, and one

---

[1] Private communication to the authors; see H. M. James and A. S. Coolidge, *Phys. Rev.* **47,** 700 (1935), for a preliminary report.

[2] N. F. BEARDSLEY, *Phys. Rev.* **39,** 913 (1932).

[3] C. ZENER, *Phys. Rev.* **36,** 51 (1930).

[4] P. M. MORSE and L. A. YOUNG, unpublished calculations (available at the Massachusetts Institute of Technology).

for $2p$ one-electron functions) for the treatment of the K and L shells of atoms.

The analytical treatment of complicated atoms by this method is at present too laborious for the accuracy obtained, but it may be possible to find new forms for the variation function which will enable further progress to be made.

32. THE METHOD OF THE SELF-CONSISTENT FIELD

The previous sections give some indication of the difficulty of treating many-electron atoms in even an approximate manner. In this section we shall discuss what is probably the most successful effort which has yet been made in attacking this problem, at least for those atoms which are too complicated to treat by any satisfactory variation function. Both the principle and the difficult technique involved are due to Hartree,[1] who, with the aid of his students, has now made the numerical computations for a number of atoms. In Section 32b we shall show the connection between this method and those previously discussed.

32a. Principle of the Method.—In Section 30b we have pointed out that the wave equation for a many-electron atom can be separated into single-electron wave functions not only when the mutual interactions of the electrons are completely neglected but also when a central field $v(x_i)$ for each electron is added to the unperturbed equation and subtracted from the perturbation term. Each of the resulting separated unperturbed wave equations describes the motion of an electron in a central field which is independent of the coordinates of the other electrons. The perturbation treatment considered in Section 30 was based on the idea that a suitable choice could be made of these central fields for the individual electrons so that they would represent as closely as possible the average effect upon one electron of all the other electrons in the atom.

The important step in the application of such a method of treatment is the choice of the potential-energy functions representing the central fields. The assumption made by Hartree is that the potential-energy function for one electron due to a second electron is determined approximately by the wave function for the second electron, $u_\beta(2)$, say, being given by the

[1] D. R. Hartree, *Proc. Cambridge Phil. Soc.* **24**, 89, 111, 426 (1928).

potential corresponding to the distribution of electricity determined by the probability distribution function $u_\beta^*(2)\ u_\beta(2)$. This is equivalent to assuming that the wave function for the second electron is independent of the coordinates of the first electron. The complete central-field potential-energy function for the first electron is then obtained by adding to the potential-energy function due to the nucleus those potential-energy functions due to all the other electrons, calculated in the way just described. The wave function for the first electron can then be found by solving the wave equation containing this complete potential-energy function.

It is seen, however, that in formulating a method of calculating the functions $u_\zeta(k)$ for an atom we have assumed them to be known. In practice there is adopted a method of successive approximations, each cycle of which involves the following steps:

1. A potential-energy function due to the nucleus and all of the electrons is estimated.

2. From this there is subtracted the estimated contribution of the kth electron, leaving the effective potential-energy function for this electron.

3. The resulting wave equation for the kth electron is then solved, to give the wave function $u_\zeta(k)$. Steps 2 and 3 are carried out for all of the electrons in the atom.

4. Using the functions $u_\zeta(k)$ obtained by step 3, the potential-energy functions due to the various electrons are calculated, and compared with those initially assumed in steps 1 and 2.

In general the final potential-energy functions are not identical with those chosen initially. The cycle is then repeated, using the results of step 4 as an aid in the estimation of new potential-energy functions. Ultimately a cycle may be carried through for which the final potential-energy functions are identical (to within the desired accuracy) with the initial ones. The field corresponding to this cycle is called a *self-consistent field* for the atom.

It may be mentioned that the potential-energy function due to an s electron is spherically symmetrical, inasmuch as the probability distribution function $u_{ns}^* u_{ns}$ is independent of φ and ϑ. Moreover, as a result of the theorem of Equation 21–16 the potential-energy function due to a completed shell of electrons is also spherically symmetrical. Spherical symmetry of the potential function greatly increases the ease of solution of the wave equation.

Hartree employs the method of numerical integration sketched in Section 27c to solve the single-electron wave equations. In addition he makes the approximation of considering all contributions to the field as spherically symmetrical. Thus if some electron (such as a p electron) gives rise to a charge distribution which is not spherically symmetrical, this is averaged over all directions. Finally, the simple product of Equation 30–2 is used for the wave function for the whole atom. As we have seen, this does not have the correct symmetry required by Pauli's principle. The error due to this involves the interchange energies of the electrons (Sec. 32c).

32b. Relation of the Self-consistent Field Method to the Variation Principle.—If we choose a variation function of the form

$$\phi = u_\alpha(1) \, u_\beta(2) \, \cdots \, u_\nu(N) \qquad (32\text{–}1)$$

and determine the functions $u_\zeta(i)$ by varying them individually until the variational integral in Equation 26–1 is a minimum, then, as shown in Section 26a, these are the best forms for the functions $u_\zeta(i)$ to use in a wave function of this product type for the lowest state. Neglecting the fact that Hartree averages all fields to make them spherically symmetrical, we shall now show[1] that the variation-principle criterion is identical, for this type of ϕ, with the criterion of the self-consistent field. If we keep each $u_\zeta(i)$ normalized, then $\int \phi^* \phi d\tau = 1$ and

$$E = \int \phi^* H \phi d\tau. \qquad (32\text{–}2)$$

The operator H may be written as

$$H = \sum_i H_i + \sum_{i,j>i} \frac{e^2}{r_{ij}}, \qquad (32\text{–}3)$$

with

$$H_i = -\frac{h^2}{8\pi^2 m_0} \nabla_i^2 - \frac{Ze^2}{r_i}. \qquad (32\text{–}4)$$

---

[1] J. C. SLATER, *Phys. Rev.* **35**, 210 (1930); V. FOCK, *Z. f. Phys.* **61**, 126 (1930).

Using this and the expression for ϕ in Equation 32–1, we obtain

$$E = \sum_i \int u_\varsigma^*(i) H_i u_\varsigma(i) d\tau_i +$$

$$\sum_{i,j>i} \int \int u_\varsigma^*(i) u_\xi^*(j) \frac{e^2}{r_{ij}} u_\varsigma(i) u_\xi(j) d\tau_i d\tau_j. \quad (32\text{–}5)$$

The variation principle can now be applied. This states that the best form for any function $u_\varsigma(i)$ is the one which makes E a minimum (keeping the function normalized). For this minimum, a small change $\delta u_\varsigma(i)$ in the form of $u_\varsigma(i)$ will produce no change in E; that is $\delta E = 0$.

The relation between $\delta u_\varsigma(i)$ and δE is

$$\delta E = \delta \int u_\varsigma^*(i) H_i u_\varsigma(i) d\tau_i +$$

$$\sum_j{}' \delta \int \int u_\varsigma^*(i) u_\xi^*(j) \frac{e^2}{r_{ij}} u_\varsigma(i) u_\xi(j) d\tau_i d\tau_j, \quad (32\text{–}6)$$

in which the prime on the summation sign indicates that the term with $j = i$ is not included. Let us now introduce the new symbol F_i, defined by the equation

$$F_i = H_i + \sum_j{}' \int u_\xi^*(j) \frac{e^2}{r_{ij}} u_\xi(j) d\tau_j, \quad (32\text{–}7)$$

or

$$F_i = H_i + V_i,$$

in which

$$V_i = \sum_j{}' \int u_\xi^*(j) \frac{e^2}{r_{ij}} u_\xi(j) d\tau_j. \quad (32\text{–}8)$$

F_i is an effective Hamiltonian function for the ith electron, and V_i the effective potential-energy function for the ith electron due to its interaction with the other electrons in the atom. Using the symbol F_i, we obtain as the condition that E be stationary with respect to variation in $u_\varsigma(i)$ the expression (Eq. 32–6)

$$\delta E = \delta \int u_\varsigma^*(i) F_i u_\varsigma(i) d\tau_i = 0. \quad (32\text{–}9)$$

A similar condition holds for each of the N one-electron functions $u_\alpha(1), \cdots, u_\nu(N)$.

Let us now examine the criterion used in the method of the self-consistent field. In this treatment the wave function $u_\zeta(i)$ is obtained as the solution of the wave equation

$$\nabla_i^2 u_\zeta(i) + \frac{8\pi^2 m_0}{h^2}\left(\epsilon_i + \frac{Ze^2}{r_i} - V_i\right)u_\zeta(i) = 0, \quad (32\text{--}10)$$

or, introducing the symbol F_i,

$$F_i u_\zeta(i) = \epsilon_i u_\zeta(i). \quad (32\text{--}11)$$

We know, however, that a normalized function $u_\zeta(i)$ satisfying this equation also satisfies the corresponding variational equation

$$\delta \int u_\zeta^*(i) F_i u_\zeta(i) d\tau_i = 0. \quad (32\text{--}12)$$

Equations 32–9 and 32–12 are identical, so that by using the variation method with a product-type variation function we obtain the same single-electron functions as by applying the criterion of the self-consistent field.

32c. Results of the Self-consistent Field Method.—Hartree and others have applied the method of the self-consistent field to a number of atoms and ions. In one series of papers[1] Hartree has published tables of values of single-electron wave functions for Cl^-, Cu^+, K^+, and Rb^+. These wave functions, as given, are not normalized or mutually orthogonal, but values of the normalizing factors are reported. For these atoms the total energy has not been calculated, although values of the individual ϵ_i's are tabulated. (The sum of these is not equal to the total energy, even if interchange is neglected.) For O, O^+, O^{++}, and O^{+++}, Hartree and Black[2] have given not only the wave functions but also the total energies calculated by inserting these single-electron wave functions into a determinant such as Equation 30–7 and evaluating the integral $E = \int \psi^* H \psi d\tau$.

Several other applications[3] have been made of this method and a considerable number are now in progress. Slater[4] has taken Hartree's results for certain atoms and has found analytic expres-

[1] D. R. HARTREE, *Proc. Roy. Soc.* **A 141**, 282 (1933); **A 143**, 506 (1933).

[2] D. R. HARTREE and M. M. BLACK, *Proc. Roy. Soc.* **A 139**, 311 (1933).

[3] F. W. BROWN, *Phys. Rev.* **44**, 214 (1933); F. W. BROWN, J. H. BARTLETT, JR., and C. G. DUNN, *Phys. Rev.* **44**, 296 (1933); J. McDOUGALL, *Proc. Roy. Soc.* **A 138**, 550 (1932); C. C. TORRANCE, *Phys. Rev.* **46**, 388 (1934).

[4] J. C. SLATER, *Phys. Rev.* **42**, 33 (1932).

sions for the single-electron wave functions which fit these results fairly accurately. Such functions are of course easier to use than numerical data.

The most serious drawback to Hartree's method is probably the neglect of interchange effects, i.e., the use of a simple product-

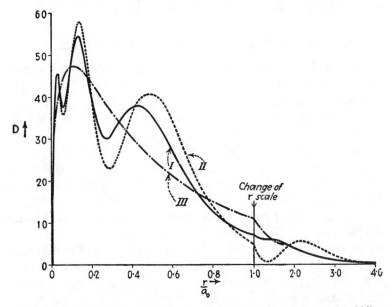

FIG. 32-1.—The electron distribution function D for the normal rubidium atom, as calculated: I, by Hartree's method of the self-consistent field; II, by the screening-constant method; and III, by the Thomas-Fermi statistical method.

type wave function instead of a properly antisymmetric one. This error is partially eliminated by the procedure of Hartree and Black described above, but, although in that way the energy corresponding to a given set of functions $u_{\zeta}(k)$ is properly calculated, the functions $u_{\zeta}(k)$ themselves are not the best obtainable because of the lack of antisymmetry of ψ. Fock[1] has considered this question and has given equations which may be numerically solved by methods similar to Hartree's, but which include interchange. So far no applications have been made of these, but several computations are in progress.[2]

Figures 32-1, from Hartree, shows the electron distribution function for Rb⁺ calculated by this method, together with those given by other methods for comparison.

[1] V. FOCK, *Z. f. Phys.* **61**, 126 (1930).
[2] See D. R. HARTREE and W. HARTREE, *Proc. Roy. Soc.* A **150**, 9 (1935).

Problem 32-1. (a) Obtain an expression for the potential due to an electron in a hydrogenlike $1s$ orbital with effective atomic number $Z' = 2\,7\!/\!16$. (b) Using this result, set up the wave equation for one electron in a helium atom in the field due to the nucleus and the other electron (assumed to be represented by the wave function mentioned above). Solve the wave equation by the method of difference equations (Sec. 27d), and compare the resultant wave function with that chosen initially.

33. OTHER METHODS FOR MANY-ELECTRON ATOMS

Besides the methods discussed in the previous sections there are others yielding useful results, some of which will be briefly outlined in the following sections. Several methods have been proposed which are beyond the scope of this book, notably the Dirac[1]-Van Vleck[2] vector model, which yields results similar to those given by the method of Slater of Section 30.

33a. Semi-empirical Sets of Screening Constants.—One of the methods mentioned in Section 31b consists in building up an approximate wave function for an atom by the use of hydrogenlike single-electron functions with effective nuclear charges determined by the variation method. Instead of giving the effective atomic number Z', it is convenient to use the difference between the true atomic number and the effective atomic number, this difference being called the *screening constant*. Pauling[3] has obtained sets of screening constants for all atoms, not by the application of the variation method (which is too laborious), but by several types of reasoning based in part on empirical considerations, involving such quantities as x-ray term values and molecular refraction values. It is not to be expected that wave functions formed in this manner will be of very great accuracy, but for many purposes they are sufficient and for many atoms they are the best available. The results obtained for Rb^+ are shown in Figure 32–1.

Slater[4] has constructed a similar table, based, however, on Zener's variation-method calculations for the first ten elements (Sec. 31b). His screening constants are meant to be used in

[1] P. A. M. DIRAC, "The Principles of Quantum Mechanics," Chap. XI.
[2] J. H. VAN VLECK, *Phys. Rev.* **45**, 405 (1934).
[3] L. PAULING, *Proc. Roy. Soc.* **A 114**, 181 (1927); L. PAULING and J SHERMAN, *Z. f. Krist.* **81**, 1 (1932).
[4] J. C. SLATER, *Phys. Rev.* **36**, 57 (1930).

functions of the type $r_i^{n'} e^{-z'r}$ instead of in hydrogenlike functions, the exponent n' being an effective quantum number.

A discussion of an approximate expression for the wave function in the outer regions of atoms and ions and its use in the treatment of various physical properties (polarizability, ionization potentials, ionic radii, etc.) has been given by Wasastjerna.[1]

33b. The Thomas-Fermi Statistical Atom.—In treating a system containing a large number of particles statistical methods are frequently applicable, so that it is natural to see if such methods will give approximate results when applied to the collection of electrons which surround the nucleus of a heavy atom. Thomas[2] and Fermi[3] have published such a treatment. In applying statistical mechanics to an electron cloud, it was recognized that it is necessary to use the Fermi-Dirac quantum statistics, based on the Pauli exclusion principle, rather than classical statistics, which is not even approximately correct for an electron gas. The distinctions between these have been mentioned in Section 29b and will be further discussed in Section 49.

The statistical treatment of atoms yields electron distributions that are surprisingly good in view of the small number of electrons involved. These results have been widely used for calculating the scattering power of an atom for x-rays and for obtaining an initial field for carrying out the self-consistent-field computations described in the previous section. However, the Thomas-Fermi electron distribution does not show the finer features, such as the concentration of the electrons into shells, which are characteristic of the more refined treatments. Figure 32–1 shows how the Thomas-Fermi results compare with Hartree's and Pauling's calculations for Rb^+.

General References on Line Spectra

Introductory treatments:

L. Pauling and S. Goudsmit: "The Structure of Line Spectra," McGraw-Hill Book Company, Inc., New York, 1930.

H. E. White: "Introduction to Atomic Spectra," McGraw-Hill Book Company, Inc., New York, 1934.

[1] J. A. Wasastjerna, *Soc. Scient. Fennica Comm. Phys.-Math.*, vol. 6, Numbers 18–22 (1932).

[2] L. H. Thomas, *Proc. Cambridge Phil. Soc.* **23**, 542 (1927).

[3] E. Fermi, *Z. f. Phys.* **48**, 73; **49**, 550 (1928).

A. E. RUARK and H. C. UREY: "Atoms, Molecules and Quanta," McGraw-Hill Book Company, Inc., New York, 1930.

A thorough quantum-mechanical treatment:

E. U. CONDON and G. H. SHORTLEY: "The Theory of Atomic Spectra," Cambridge University Press, 1935.

Tabulation of term values:

R. F. BACHER and S. GOUDSMIT: "Atomic Energy States," McGraw-Hill Book Company, Inc., New York, 1932.

CHAPTER X

THE ROTATION AND VIBRATION OF MOLECULES

The solution of the wave equation for any but the simplest molecules (some of which are discussed in Chap. XII) is a very difficult problem. However, the empirical results of molecular spectroscopy show that in many cases the energy values bear a simple relation to one another, such that the energy of the molecule (aside from translational energy) can be conveniently considered to be made up of several parts, called the *electronic energy*, the *vibrational energy*, and the *rotational energy*. This is indicated in Figure 34–1, showing some of the energy levels for a molecule of carbon monoxide, as calculated from the observed spectral lines by the Bohr frequency rule (Sec. 5a). It is seen that the energy levels fall into widely separated groups, which are said to correspond to different electronic states of the molecule. For a given electronic state the levels are again divided into groups, which follow one another at nearly equal intervals. These are said to correspond to successive states of vibration of the nuclei. Superimposed on this is the fine structure due to the different states of rotation of the molecule, the successive rotational energy levels being separated by larger and larger intervals with increasing rotational energy. This simplicity of structure of the energy levels suggests that it should be possible to devise a method of approximate solution of the wave equation involving its separation into three equations, one dealing with the motion of the electrons, one with the vibrational motion of the nuclei, and one with the rotational motion of the nuclei. A method of this character has been developed and is discussed in the following section. The remaining sections of this chapter are devoted to the detailed treatment of the vibrational and rotational motion of molecules of various types.

34. THE SEPARATION OF ELECTRONIC AND NUCLEAR MOTION

By making use of the fact that the mass of every atomic nucleus is several thousand times as great as the mass of an electron,

259

Born and Oppenheimer[1] were able to show that an approximate solution of the complete wave equation for a molecule can be obtained by first solving the wave equation for the electrons

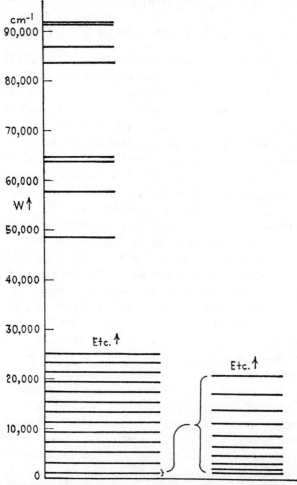

Fig. 34–1.—Energy levels for the carbon monoxide molecule.   On the left are shown various electronic levels, with vibrational fine structure for the normal state, and on the right, with one hundred fold increase of scale, the rotational fine structure for the lowest vibrational level.

alone, with the nuclei in a fixed configuration, and then solving a wave equation for the nuclei alone, in which a characteristic energy value of the electronic wave equation, regarded as a

[1] M. Born and J. R. Oppenheimer, Ann. d. Phys. 84, 457 (1927).

function of the internuclear distances, occurs as a potential function. Even in its simplest form the argument of Born and Oppenheimer is very long and complicated. On the other hand, the results of their treatment can be very simply and briefly described. Because of these facts, we shall content ourselves with describing their conclusions in detail.

The complete wave equation for a molecule consisting of r nuclei and s electrons is

$$\sum_{j=1}^{r}\frac{1}{M_j}\nabla_j^2\psi + \frac{1}{m_0}\sum_{i=1}^{s}\nabla_i^2\psi + \frac{8\pi^2}{h^2}(W - V)\psi = 0, \quad (34\text{-}1)$$

in which M_j is the mass of the jth nucleus, m_0 the mass of each electron, ∇_j^2 the Laplace operator in terms of the coordinates of the jth nucleus, and ∇_i^2 the same operator for the ith electron. V is the potential energy of the system, of the form

$$V = \sum_{i,i'}\frac{e^2}{r_{ii'}} + \sum_{j,j'}\frac{Z_jZ_{j'}e^2}{r_{jj'}} - \sum_{i,j}\frac{Z_je^2}{r_{ij}},$$

the sums including each pair of particles once. Here Z_j is the atomic number of the jth nucleus.

Let us use the letter ξ to represent the $3r$ coordinates of the r nuclei, relative to axes fixed in space, and the letter x to represent the $3s$ coordinates of the s electrons, relative to axes determined by the coordinates of the nuclei (for example, as described in Section 48). Let us also use the letter ν to represent the quantum numbers associated with the motion of the nuclei, and n to represent those associated with the motion of the electrons. The principal result of Born and Oppenheimer's treatment is that an approximate solution $\psi_{n,\nu}(x, \xi)$ of Equation 34-1 can be obtained of the form

$$\psi_{n,\nu}(x, \xi) = \psi_n(x, \xi)\psi_{n,\nu}(\xi). \quad (34\text{-}2)$$

The different functions $\psi_n(x, \xi)$, which may be called the *electronic wave functions*, correspond to different sets of values of the electronic quantum numbers n only, being independent of the nuclear quantum numbers ν. On the other hand, each of these functions is a function of the nuclear coordinates ξ as well as the electronic coordinates x. These functions are obtained by *solving a wave equation for the electrons alone, the nuclei being restricted to a fixed configuration.* This wave equation is

$$\sum_{i=1}^{s} \nabla_i^2 \psi_n(x, \xi) + \frac{8\pi^2 m_0}{h^2} \{U_n(\xi) - V(x, \xi)\} \psi_n(x, \xi) = 0. \quad (34\text{-}3)$$

It is obtained from the complete wave equation 34–1 by omitting the terms involving ∇_j^2, replacing ψ by $\psi_n(x, \xi)$, and writing $U_n(\xi)$ in place of W. The potential function $V(x, \xi)$ is the complete potential function of Equation 34–1. It is seen that for any fixed set of values of the s nuclear coordinates ξ this equation 34–3, which we may call the *electronic wave equation,* is an ordinary wave equation for the s electrons, the potential-energy function V being dependent on the values selected for the nuclear coordinates ξ. In consequence the characteristic electronic energy values U_n and the electronic wave functions ψ_n will also be dependent on the values selected for the nuclear coordinates; we accordingly write them as $U_n(\xi)$ and $\psi_n(x, \xi)$. The first step in the treatment of a molecule is to solve this electronic wave equation for all configurations of the nuclei. It is found that the characteristic values $U_n(\xi)$ of the electronic energy are continuous functions of the nuclear coordinates ξ. For example, for a free diatomic molecule the electronic energy function for the most stable electronic state ($n = 0$) is a function only of the distance r between the two nuclei, and it is a continuous function of r, such as shown in Figure 34–2.

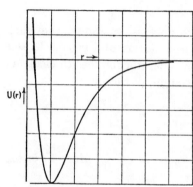

U(r)↑

FIG. 34–2.—A typical function $U(r)$ for a diatomic molecule (Morse function).

Having evaluated the characteristic electronic energy $U_n(\xi)$ as a function of the nuclear coordinates ξ for a given set of values of the electronic quantum numbers n by solving the wave equation 34–3 for various nuclear configurations, we next obtain expressions for the *nuclear wave functions* $\psi_{n,\nu}(\xi)$. It was shown by Born and Oppenheimer that these functions are the acceptable solutions of a wave equation in the nuclear coordinates ξ in which the characteristic electronic energy function $U_n(\xi)$ plays the role of the potential energy; that is, the *nuclear wave equation* is

$$\sum_{j=1}^{r} \frac{1}{M_j} \nabla_j^2 \psi_{n,\nu}(\xi) + \frac{8\pi^2}{h^2} \{W_{n,\nu} - U_n(\xi)\} \psi_{n,\nu}(\xi) = 0. \quad (34\text{-}4)$$

There is one such equation for each set of values of the electronic quantum numbers n, and each of these equations possesses an extensive set of solutions, corresponding to the allowed values of the nuclear quantum numbers ν. The values of $W_{n,\nu}$ are the characteristic energy values for the entire molecule; they depend on the electronic and nuclear quantum numbers n and ν.

The foregoing treatment can be formally justified by a procedure involving the expansion of the wave functions and other quantities entering in the complete wave equation 34-1 as power series in $(m_0/M)^{1/4}$, in which M is an average nuclear mass. The physical argument supporting the treatment is that on account of the disparity of masses of electrons and nuclei the electrons carry out many cycles of their motion in the time required for the nuclear configuration to change appreciably, and that in consequence we are allowed to quantize their motion for fixed configurations (by solving the electronic wave equation), and then to use the electronic energy functions as potential energy functions determining the motion of the nuclei.

When great accuracy is desired, and in certain cases when only ordinary accuracy is required, it is necessary to consider the coupling between electronic and nuclear motions, and especially between the electronic angular momentum (either spin or orbital) and the rotation of the molecule. We shall not discuss these questions,[1] but shall treat only the simplest problems in the complex field of molecular structure and molecular spectra in the following sections. Some further discussion is also given in Chapter XII and in Section 48 of Chapter XIV.

35. THE ROTATION AND VIBRATION OF DIATOMIC MOLECULES

In the previous section we have stated that an approximate wave function for a molecule can be written as a product of two

---

[1] See the references at the end of the chapter.

factors, one a function of the electronic coordinates relative to the nuclei and the other a function of the nuclear coordinates. In this section we shall consider the nuclear function and the corresponding energy levels for the simplest case, the diatomic molecule, assuming the electronic energy function $U_n(r)$ to be known.

35a. The Separation of Variables and Solution of the Angular Equations.—The wave equation for the rotation and vibration of a diatomic molecule (Eq. 34–4) has the form

$$\frac{1}{M_1}\nabla_1^2\psi_{n,\nu} + \frac{1}{M_2}\nabla_2^2\psi_{n,\nu} + \frac{8\pi^2}{h^2}\{W_{n,\nu} - U_n(r)\}\psi_{n,\nu} = 0, \quad (35\text{--}1)$$

in which $\psi_{n,\nu} = \psi_{n,\nu}(x_1, y_1, z_1, x_2, y_2, z_2)$ is the wave function for the nuclear motion, M_1 and M_2 are the masses of the two nuclei, and

$$\nabla_i^2 = \frac{\partial^2}{\partial x_i^2} + \frac{\partial^2}{\partial y_i^2} + \frac{\partial^2}{\partial z_i^2}, \quad i = 1, 2, \quad (35\text{--}2)$$

x_i, y_i, and z_i being the Cartesian coordinates of the ith nucleus relative to axes fixed in space. Equation 35–1 is identical with the wave equation for the hydrogen atom, the two particles here being the two nuclei instead of an electron and a proton. We may therefore refer to the treatment which has already been given of this equation in connection with hydrogen. All the steps are the same until the form for $U_n(r)$ is inserted into the radial equation.

In Section 18a we have seen that Equation 35–1, expressed in terms of the Cartesian coordinates of the two particles, can be separated into two equations, one describing the translational motion of the molecule and the other its internal motion. The latter has the form

$$\frac{1}{r^2}\frac{\partial}{\partial r}\left(r^2\frac{\partial\psi}{\partial r}\right) + \frac{1}{r^2\sin\vartheta}\frac{\partial}{\partial\vartheta}\left(\sin\vartheta\frac{\partial\psi}{\partial\vartheta}\right) + \frac{1}{r^2\sin^2\vartheta}\frac{\partial^2\psi}{\partial\varphi^2} +$$
$$\frac{8\pi^2\mu}{h^2}\{W - U(r)\}\psi = 0, \quad (35\text{--}3)$$

in which μ, the reduced mass, is given by the equation

$$\mu = \frac{M_1 M_2}{M_1 + M_2} \quad (35\text{--}4)$$

and r, ϑ, φ are polar coordinates of the second nucleus relative to the first as origin. In Section 18a it was also shown that this equation can be separated into three equations in the three variables φ, ϑ, and r, respectively. The solutions of the φ and ϑ equations, which are obtained in Sections 18b, 18c, and 19, are

$$\Phi_M(\varphi) = \frac{1}{\sqrt{2\pi}}e^{iM\varphi} \qquad (35\text{--}5)$$

and

$$\Theta_{KM}(\vartheta) = \left\{ \frac{(2K+1)(K-|M|)!}{2(K+|M|)!} \right\}^{\frac{1}{2}} P_K^{|M|}(\cos\vartheta), \quad (35\text{--}6)$$

in which $P_K^{|M|}(\cos\vartheta)$ is an associated Legendre function (Sec. 19b). Φ and Θ are the φ and ϑ factors, respectively, in the product function

$$\psi(r, \vartheta, \varphi) = R(r)\Theta(\vartheta)\Phi(\varphi). \qquad (35\text{--}7)$$

Instead of the azimuthal quantum number l, used for the hydrogen atom, we have here adopted the letter K, and for the magnetic quantum number m we here use M, in agreement with the usual notation for molecular spectra. Both M and K must be integers, for the reasons discussed in Sections 18b and 18c, and, as there shown, their allowed values are

$$K = 0, 1, 2, \cdots ; M = -K, -K+1, \cdots, K-1, K. \qquad (35\text{--}8)$$

Just as in the case of hydrogen, the quantum numbers M and K represent angular momenta (see also Sec. 52), the square of the total angular momentum due to the rotation of the molecule[1] being

$$K(K+1)\frac{h^2}{4\pi^2}, \qquad (35\text{--}9)$$

while the component of this angular momentum in any specially chosen direction (taken as the z direction) is

$$M\frac{h}{2\pi}. \qquad (35\text{--}10)$$

In Section 40d it will be shown that dipole radiation is emitted or absorbed only for transitions in which the quantum number

---

[1] There may be additional angular momentum due to the electrons.

K changes by one unit; i.e., the selection rule for K is

$$\Delta K = \pm 1.$$

Likewise, the selection rule for M is

$$\Delta M = 0 \text{ or } \pm 1.$$

The energy of the molecule does not depend on M (unless there is a magnetic field present), so that this rule is not ordinarily of importance in the interpretation of molecular spectra.

The equation for $R(r)$ (Eq. 18–26) is

$$\frac{1}{r^2}\frac{d}{dr}\left(r^2\frac{dR}{dr}\right) + \left[-\frac{K(K+1)}{r^2} + \frac{8\pi^2\mu}{h^2}\{W - U(r)\}\right]R = 0,$$

(35–11)

in which for simplicity we have omitted the subscripts n and ν This may be simplified by the substitution

$$R(r) = \frac{1}{r}S(r),$$ (35–12)

which leads to the equation

$$\frac{d^2S}{dr^2} + \left[-\frac{K(K+1)}{r^2} + \frac{8\pi^2\mu}{h^2}\{W - U(r)\}\right]S = 0.$$ (35–13)

35b. The Nature of the Electronic Energy Function.—The solution of the radial equation 35–13 involves a knowledge of the electronic energy function $U(r)$ discussed in Section 34. The theoretical calculation of $U(r)$ requires the solution of the wave equation for the motion of the electrons, a formidable problem which has been satisfactorily treated only for the very simplest molecules, such as the hydrogen molecule (Sec. 43). It is therefore customary to determine $U(r)$ empirically by assuming some reasonable form for it involving adjustable parameters which are determined by a comparison of the observed and calculated energy levels.

From the calculations on such simple molecules as the hydrogen molecule and from the experimental results, we know that $U(r)$ for a stable diatomic molecule is similar to the function plotted in Figure 34–2. When the atoms are very far apart (r large), the energy is just the sum of the energies of the two individual atoms. As the atoms approach one another there is for stable states a slight attraction which increases with decreasing r, as is shown by the curvature of U in Figure 34–2. For stable molecules, U must have a minimum value at the equilibrium separation $r = r_e$. For smaller values of r, U rises rapidly, corresponding to the high repulsion of atoms "in contact."

For most molecules in their lower states of vibration it will be found that the wave function has an appreciable value only in a rather narrow region near the equilibrium position, this having the significance that the amplitude of vibration of most molecules is small compared to the equilibrium separation. This is important because it means that for these lower levels the nature of the potential function near the minimum is more important than its behavior in other regions.

However, for higher vibrational levels, that is, for larger amplitudes of vibration, the complete potential function is of importance. The behavior of U in approaching a constant value for larger values of r is of particular significance for these higher levels and is responsible for the fact that if sufficient energy is transferred to the molecule it will dissociate into two atoms.

In the following sections two approximations for $U(r)$ will be introduced, the first of which is very simple and the second somewhat more complicated but also more accurate.

35c. A Simple Potential Function for Diatomic Molecules.— The simplest assumption which can be made concerning the force between the atoms of a diatomic molecule is that it is proportional to the displacement of the internuclear distance from its equilibrium value r_e. This corresponds to the potential function

$$U(r) = \tfrac{1}{2}k(r - r_e)^2, \qquad (35\text{–}14)$$

which is plotted in Figure 35–1. k is the force constant for the molecule, the value of which can be determined empirically from the observed energy levels. A potential-energy function of this type is called a *Hooke's-law potential energy function.*

It is obvious from a comparison of Figures 34–2 and 35–1 that this simple function is not at all correct for large internuclear distances. Nevertheless, by a proper choice of k a fair approximation to the true $U(r)$ can be achieved in the neighborhood of $r = r_e$. This approximation corresponds to expanding the true $U(r)$ in a Taylor series in powers of $(r - r_e)$ and neglecting all powers above the second, a procedure which is justified only for small values of $r - r_e$. The coefficient of $(r - r_e)^0$ (that is, the constant term) in this expansion can be conveniently set equal to zero without loss of generality so far as the solution of the wave equation is concerned. The linear term in the

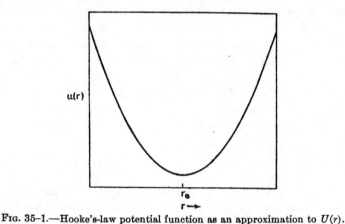

Fig. 35-1.—Hooke's-law potential function as an approximation to $U(r)$.

expansion vanishes, inasmuch as $U(r)$ has a minimum at $r = r_e$, and so the series begins with the term $\frac{1}{2}\left(\dfrac{d^2U}{dr^2}\right)_{r=r_e} (r - r_e)^2$. Comparison with Equation 35-14 shows that the force constant k is equal to $\left(\dfrac{d^2U}{dr^2}\right)_{r=r_e}$.

Insertion of this form for $U(r)$ into the radial equation 35–13 yields the equation

$$\frac{d^2S}{dr^2} + \left[-\frac{K(K+1)}{r^2} + \frac{8\pi^2\mu}{h^2}\left\{ W - \frac{1}{2}k(r - r_e)^2\right\}\right]S = 0,$$

$$(35\text{–}15)$$

which may be transformed by the introduction of the new independent variable $\rho = r - r_e$ (the displacement from the equilibrium separation) into the equation

$$\frac{d^2S}{d\rho^2} + \frac{8\pi^2\mu}{h^2}\left\{ W - \frac{1}{2}k\rho^2 - \frac{h^2}{8\pi^2\mu}\frac{K(K+1)}{(r_e + \rho)^2}\right\}S = 0.$$

Since the approximation which we have used for $U(r)$ is good only for ρ small compared to r_e, it is legitimate to introduce the expansion

$$\frac{1}{(r_e + \rho)^2} = \frac{1}{r_e^2}\left(1 - 2\frac{\rho}{r_e} + 3\frac{\rho^2}{r_e^2} - \cdots \right),$$

a step which leads to the result

$$\frac{d^2S}{d\rho^2} + \frac{8\pi^2\mu}{h^2}\left\{ W - K(K + 1)\sigma + 2K(K + 1)\frac{\sigma}{r_e}\rho - \right.$$
$$\left. 3K(K + 1)\frac{\sigma}{r_e^2}\rho^2 - \frac{1}{2}k\rho^2 \right\}S = 0, \quad (35\text{-}16)$$

in which powers of ρ/r_e greater than the second have been neglected, and the symbol σ has been introduced, with

$$\sigma = \frac{h^2}{8\pi^2\mu r_e^2} = \frac{h^2}{8\pi^2 I_e}, \quad (35\text{-}17)$$

and $I_e = \mu r_e^2$. I_e is called the *equilibrium moment of inertia* of the molecule.

By making a suitable transformation $\rho = \zeta + a$, we can eliminate the term containing the first power in the independent variable, obtaining thereby an equation of the same form as Equation 11-1, the wave equation for the harmonic oscillator, which we have previously solved. It is easily verified that the proper value for a is

$$a = \frac{K(K + 1)\sigma r_e}{3K(K + 1)\sigma + \frac{1}{2}k r_e^2},$$

and that the introduction of this transformation into Equation 35-16 yields the equation

$$\frac{d^2S}{d\zeta^2} + \frac{8\pi^2\mu}{h^2}\left\{\left[W - K(K + 1)\sigma + \frac{\{K(K + 1)\sigma\}^2}{3K(K + 1)\sigma + \frac{1}{2}k r^2} \right]\right.$$
$$\left. - \left[\frac{1}{2}k + 3K(K + 1)\frac{\sigma}{r_e^2}\right]\zeta^2 \right\}S = 0. \quad (35\text{-}18)$$

We seek the solutions of this equation which make $\psi(r, \vartheta, \varphi)$ of Equation 35-7 a satisfactory wave function. This requires that S vanish for $r = 0$ and $r = \infty$, the former condition entering because of the relation $R = \frac{1}{r}S$. We know the solutions of the equation which vanish for $\zeta = -\infty$ and $\zeta = +\infty$, since for these boundary conditions the problem is analogous to that of the linear harmonic oscillator (Sec. 11). Because of the rapid decrease in the harmonic oscillator functions outside of the classically permitted region (see Fig. 11–3), it does not introduce a serious error to consider that the two sets of boundary conditions are practically equivalent, so that as an approximation we may use the harmonic oscillator wave functions for the functions S.

The energy levels are, therefore, using the results of Section 11a,

$$W_{v,K} = K(K+1)\sigma - \frac{\{K(K+1)\sigma\}^2}{3K(K+1)\sigma + \frac{1}{2}kr_e^2} + \left(v + \frac{1}{2}\right)h\nu_e', \tag{35–19}$$

in which

$$\nu_e' = \frac{1}{2\pi}\left\{\frac{kr_e^2 + 6K(K+1)\sigma}{\mu r_e^2}\right\}^{\frac{1}{2}} \tag{35–20}$$

and v is the vibrational quantum number (corresponding to the quantum number n for the harmonic oscillator), which can take on the values 0, 1, 2, $\cdots$. The functions $S(\zeta)$ are (Sec. 11)

$$S_v(\zeta) = \left\{\left(\frac{\alpha}{\pi}\right)^{\frac{1}{2}} \frac{1}{2^v v!}\right\}^{\frac{1}{2}} e^{-\frac{\alpha}{2}\zeta^2} H_v(\sqrt{\alpha}\zeta), \tag{35–21}$$

in which $\alpha = 4\pi^2 \mu \nu_e'/h$ and $\zeta = \rho - a = r - r_e - a$, and H_v is the vth Hermite polynomial.

The values of k, r_e, and σ for actual molecules are of such magnitudes that the expression for W can be considerably simplified without loss of accuracy by the use of the expansions

$$\frac{1}{3K(K+1)\sigma + \frac{1}{2}kr_e^2} = \frac{1}{\frac{1}{2}kr_e^2}\left\{1 - \frac{6K(K+1)\sigma}{kr_e^2} + \cdots\right\},$$

$$\nu_e' = \frac{1}{2\pi}\left\{\frac{kr_e^2 + 6K(K+1)\sigma}{\mu r_e^2}\right\}^{\frac{1}{2}} =$$

$$\frac{1}{2\pi}\left(\frac{k}{\mu}\right)^{\frac{1}{2}}\left\{1 + \frac{3K(K+1)\sigma}{kr_e^2} + \cdots\right\}.$$

Introducing these into Equation 35–19, we obtain for W the expression

$$W_{v,K} = \left(v + \frac{1}{2}\right)h\nu_e + K(K+1)\sigma - \frac{\{K(K+1)\sigma\}^2}{\frac{1}{2}kr_e^2},$$

in which only the first terms of the expansions have been used and the symbol ν_e is given by

$$\nu_e = \frac{1}{2\pi}\sqrt{\frac{k}{\mu}}. \tag{35-22}$$

Replacing k by its expression in terms of ν_e and introducing the value of Equation 35–17 for σ, we finally obtain for W the expression

$$W_{v,K} = \left(v + \frac{1}{2}\right)h\nu_e + K(K+1)\frac{h^2}{8\pi^2 I_e} - \frac{K^2(K+1)^2h^4}{128\pi^6\nu_e^2 I_e^3}. \tag{35-23}$$

The first term is evidently the vibrational energy of the molecule, considered as a harmonic oscillator. The second term is the energy of rotation, assuming that the molecule is a rigid body,[1] while the third term is the correction which takes account of the stretching of the actual, non-rigid molecule due to the rotation. The terms of higher order are unreliable because of the inaccuracy of the assumed potential function.

The experimental data for most molecules fit Equation 35–23 fairly well. For more refined work additional correction terms are needed, one of which will be obtained in the next section.

35d. A More Accurate Treatment. The Morse Function.— The simple treatment which we have just given fails to agree with experiment in that it yields equally spaced levels, whereas the observed vibrational levels show a convergence for increasing values of v. In order to obtain this feature a potential function $U(r)$ is required which is closer to the true $U(r)$ described in

---

[1] This is seen by allowing k to become infinite, causing the third term to vanish (because $\nu_e \to \infty$). A rigid molecule would have no vibrational energy, so the first term would become an additive constant. The *rigid rotator* is often discussed as a separate problem, with the wave equation

$$\frac{1}{\sin\vartheta}\frac{\partial}{\partial\vartheta}\left(\sin\vartheta\frac{\partial\psi}{\partial\vartheta}\right) + \frac{1}{\sin^2\vartheta}\frac{\partial^2\psi}{\partial\varphi^2} + \frac{8\pi^2 I}{h^2}W\psi = 0, \tag{35-24}$$

the solutions of which are $\psi = \Phi_M(\varphi)\Theta_{KM}(\vartheta)$, in which Φ and Θ are given by Equations 35–5 and 35–6. The energy levels are $W_K = K(K+1)\frac{h^2}{8\pi^2 I}.$
The rigid rotator is of course an idealization which does not occur in nature.

Another idealized problem is the *rigid rotator in a plane*, for which the wave equation is

$$\frac{d^2\psi}{d\varphi^2} + \frac{8\pi^2 I}{h^2}W\psi = 0. \tag{35-25}$$

The solutions are $\psi = \sin M\varphi$ and $\psi = \cos M\varphi$, $M = 0, 1, 2, \cdots$, and the energy levels are $W_M = M^2h^2/8\pi^2 I$ (Sec. 25a).

Section 35b, especially with regard to its behavior for large values of r.

Morse[1] proposed a function of the form

$$U(r) = D\{1 - e^{-a(r-r_e)}\}^2, \tag{35-26}$$

which is plotted in Figure 34–2. It has a minimum value of zero at $r = r_e$ and approaches a finite value D for r large. It therefore agrees with the qualitative considerations of Section 35b except for its behavior at $r = 0$. At this point the true $U(r)$ is infinite, whereas the Morse function is finite. However, the Morse function is very large at this point, and this deficiency is not a serious one.

With the introduction of this function, the radial equation 35–13 becomes

$$\frac{d^2S}{dr^2} + \left\{-\frac{K(K+1)}{r^2} + \frac{8\pi^2\mu}{h^2}(W - D - De^{-2a(r-r_e)} + \right.$$

$$\left. 2De^{-a(r-r)}\right)\right\}S = 0. \tag{35-27}$$

If we make the substitutions

$$y = e^{-a(r-r_e)} \quad \text{and} \quad A = K(K+1)\frac{h^2}{8\pi^2\mu r_e^2}, \tag{35-28}$$

the radial equation becomes

$$\frac{d^2S}{dy^2} + \frac{1}{y}\frac{dS}{dy} + \frac{8\pi^2\mu}{a^2h^2}\left(\frac{W-D}{y^2} + \frac{2D}{y} - D - \frac{Ar_e^2}{y^2r^2}\right)S = 0. \tag{35-29}$$

The quantity r_e^2/r^2 may be expanded in terms of y in the following way:[2]

$$\frac{r_e^2}{r^2} = \frac{1}{\left(1 - \dfrac{\ln y}{ar_e}\right)^2} = 1 + \frac{2}{ar_e}(y-1) + \left(-\frac{1}{ar_e} + \frac{3}{a^2r_e^2}\right)(y-1)^2$$

$$+ \cdots, \tag{35-30}$$

the series being the Taylor expansion of the second expression in powers of $(y-1)$. Using the first three terms of this expansion in Equation 35–29 we obtain the result

---

[1] P. M. MORSE, *Phys. Rev.* **34**, 57 (1929).

[2] This treatment is due to C. L. Pekeris, *Phys. Rev.* **45**, 98 (1934). Morse solved the equation for the case $K = 0$ only.

$$\frac{d^2S}{dy^2} + \frac{1}{y}\frac{dS}{dy} + \frac{8\pi^2\mu}{a^2h^2}\left(\frac{W - D - c_0}{y^2} + \frac{2D - c_1}{y} - D - c_2\right)S = 0,$$

$$(35\text{-}31)$$

in which

$$c_0 = A\left(1 - \frac{3}{ar_e} + \frac{3}{a^2r_e^2}\right),$$

$$c_1 = A\left(\frac{4}{ar_e} - \frac{6}{a^2r_e^2}\right),$$

$$c_2 = A\left(-\frac{1}{ar_e} + \frac{3}{a^2r_e^2}\right).$$

$$(35\text{-}32)$$

The substitutions

$$S(y) = e^{-\frac{z}{2}}z^{\frac{b}{2}}F(z),$$

$$z = 2dy,$$

$$d^2 = \frac{8\pi^2\mu}{a^2h^2}(D + c_2),$$

$$b^2 = -\frac{32\pi^2\mu}{a^2h^2}(W - D - c_0)$$

$$(35\text{-}33)$$

simplify Equation 35-31 considerably, yielding the equation

$$\frac{d^2F}{dz^2} + \left(\frac{b + 1}{z} - 1\right)\frac{dF}{dz} + \frac{v}{z}F = 0,$$

$$(35\text{-}34)$$

in which

$$v = \frac{4\pi^2\mu}{a^2h^2d}(2D - c_1) - \frac{1}{2}(b + 1).$$

$$(35\text{-}35)$$

Equation 35-34 is closely related to the radial equation 18-37 of the hydrogen atom and may be solved in exactly the same manner. If this is done, it is found that it is necessary to restrict v to the values 0, 1, 2, $\cdots$ in order to obtain a polynomial solution.[1] If we solve for W by means of Equations 35-35 and the definitions of Equations 35-33, 35-32, and 35-28, we obtain the equation

$$W_{K,v} = D + c_0 - \frac{(D - \frac{1}{2}c_1)^2}{(D + c_2)} + \frac{ah(D - \frac{1}{2}c_1)}{\pi\sqrt{2\mu}\sqrt{D + c_2}}\left(v + \frac{1}{2}\right) - \frac{a^2h^2}{8\pi^2\mu}\left(v + \frac{1}{2}\right)^2.$$

[1] The solutions for v integral satisfy the boundary conditions $F \to 0$ as $r \to -\infty$ instead of as $r \to 0$ (Sec. **35c**).

By expanding in terms of powers of c_1/D and c_2/D, this relation may be brought into the form usually employed in the study of observed spectra; namely,

$$\frac{W_{K,v}}{hc} = \bar{\nu}_e\left(v + \frac{1}{2}\right) - x_e\bar{\nu}_e\left(v + \frac{1}{2}\right)^2 + K(K+1)B_e +$$
$$D_e K^2(K+1)^2 - \alpha_e(v + \tfrac{1}{2})K(K+1), \quad (35\text{-}36)$$

in which c is the velocity of light, and[1]

$$\left.\begin{aligned}
\bar{\nu}_e &= \frac{a}{2\pi c}\sqrt{\frac{2\bar{D}}{\mu}}, \\
x_e &= \frac{h\bar{\nu}_e c}{4D}, \\
B_e &= \frac{h}{8\pi^2 I_e c}, \\
D_e &= -\frac{h^3}{128\pi^6\mu^3\bar{\nu}_e^2 c^3 r_e^6}, \\
\alpha_e &= \frac{3h^2\bar{\nu}_e}{16\pi^2\mu r_e^2 D}\left(\frac{1}{ar_e} - \frac{1}{a^2 r_e^2}\right).
\end{aligned}\right\} \qquad (35\text{-}37)$$

For nearly all molecules this relation gives very accurate values for the energy levels; for a few molecules only is it necessary to consider further refinements.

We shall not discuss the wave functions for this problem. They are given in the two references quoted.

Problem 35–1. Another approximate potential function which has been used for diatomic molecules[2] is

$$U(r) = \frac{B}{r^2} - \frac{Ze^2}{r}.$$

Obtain the energy levels for a diatomic molecule with such a potential function, using the polynomial method. (*Hint:* Follow the procedure of Sec. 18 closely.) Expand the expression for the energy so obtained in powers of $(K+1)^2\dfrac{8\pi^2\mu}{h^2 B}$ and compare with Equation 35–23. Also obtain the position of the minimum of $U(r)$ and the curvature of $U(r)$ at the minimum.

Problem 35–2. Solve Equation 35–35 for the energy levels.

---

[1] The symbol ω_e is often used in place of $\bar{\nu}_e$.
[2] E. Fues, *Ann. d. Phys.* **80**, 367 (1926).

36. THE ROTATION OF POLYATOMIC MOLECULES

The straightforward way to treat the rotational and vibrational motion of a polyatomic molecule would be to set up the wave equation for $\psi_{n,\nu}(\xi)$ (Eq. 34–4), introducing for $U_n(\xi)$ an expression obtained either by solution of the electronic wave equation 34–3 or by some empirical method, and then to solve this nuclear wave equation, using some approximation method if necessary. This treatment, however, has proved to be so difficult that it is customary to begin by making the approximation of neglecting all interaction between the rotational motion and the vibrational motion of the molecule.[1] The nuclear wave equation can then be separated into two equations, one, called the *rotational wave equation*, representing the rotational motion of a rigid body. In the following paragraphs we shall discuss this equation, first for the special case of the so-called *symmetrical-top molecules*, for which two of the principal moments of inertia are equal (Sec. 36a), and then for the *unsymmetrical-top molecules*, for which the three principal moments of inertia are unequal (Sec. 36b). The second of the two equations into which the nuclear wave equation is separated is the *vibrational wave equation*, representing the vibrational motion of the non-rotating molecule. This equation will be treated in Section 37, with the usual simplifying assumption of Hooke's-law forces, the potential energy being expressed as a quadratic function of the nuclear coordinates.

36a. The Rotation of Symmetrical-top Molecules.—A rigid body in which two of the three principal moments of inertia[2]

---

[1] See, however, C. ECKART, *Phys. Rev.* **47**, 552 (1935); J. H. VAN VLECK, *ibid.* **47**, 487 (1935); D. M. DENNISON and M. JOHNSON, *ibid.* **47**, 93 (1935).

[2] Every body has three axes the use of which permits the kinetic energy to be expressed in a particularly simple form. These are called the *principal axes of inertia*. The *moment of inertia* about a principal axis is defined by the expression $\int \rho r^2 d\tau$, in which ρ is the density of matter in a given volume element $d\tau$, r is the perpendicular distance of this element from the axis in question, and the integration is over the entire volume of the solid. For a discussion of this question see J. C. Slater and N. H. Frank, " Introduction to Theoretical Physics," p. 94, McGraw-Hill Book Company, Inc., New York, 1933.

In case that a molecule possesses an n-fold symmetry axis with n greater than 2 (such as ammonia, with a three-fold axis), then two principal moments of inertia about axes perpendicular to this symmetry axis are equal, and the are equal is called a *symmetrical top*.   Its position in space is best described by the use of the three Eulerian angles ϑ, φ, and χ shown in Figure 36–1. ϑ and φ are the ordinary polar-coordinate angles of the axis of the top while χ (usually called ψ) is the angle measuring the rotation about this axis.

Since we have considered only assemblages of point particles heretofore, we have not given the rules for setting up the wave equation for a rigid body.   We shall not discuss these rules here[1] but shall take the wave equation for the symmetrical top

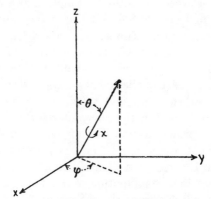

Fig. 36–1.—Diagram showing Eulerian angles.

from the work of others.[2]   Using C to represent the moment of inertia about the symmetry axis and A the two other equal moments of inertia, this wave equation is

---

molecule is a symmetrical top.   A two-fold axis does not produce a symmetrical-top molecule (example, water).   If the molecule possesses two or more symmetry axes with n greater than 2, it is called a *spherical-top molecule*, all three moments of inertia being equal.

[1] Since the dynamics of rigid bodies is based on the dynamics of particles, these rules must be related to the rules given in Chapter IV.   For a discussion of a method of finding the wave equation for a system whose Hamiltonian is not expressed in Cartesian coordinates, see B. Podolsky, *Phys. Rev.* **32**, 812 (1928), and for the specific application to the symmetrical top see the references below.

[2] F. Reiche and H. Rademacher, *Z. f. Phys.* **39**, 444 (1926); **41**, 453 (1927); R. de L. Kronig and I. I. Rabi, *Phys. Rev.* **29**, 262 (1927).   D. M. Dennison, *Phys. Rev.* **28**, 318 (1926), was the first to obtain the energy levels for this system, using matrix mechanics rather than wave mechanics.

$$\frac{1}{\sin\vartheta}\frac{\partial}{\partial\vartheta}\left(\sin\vartheta\frac{\partial\psi}{\partial\vartheta}\right) + \frac{1}{\sin^2\vartheta}\frac{\partial^2\psi}{\partial\varphi^2} + \left(\frac{\cos^2\vartheta}{\sin^2\vartheta} + \frac{A}{C}\right)\frac{\partial^2\psi}{\partial\chi^2}$$
$$- \frac{2\cos\vartheta}{\sin^2\vartheta}\frac{\partial^2\psi}{\partial\chi\partial\varphi} + \frac{8\pi^2AW}{h^2}\psi = 0. \quad (36\text{-}1)$$

The angles χ and φ do not occur in this equation, although derivatives with respect to them do. They are therefore cyclic coordinates (Sec. 17), and we know that they enter the wave function in the following manner:

$$\psi = \Theta(\vartheta)e^{iM\varphi}e^{iK\chi}, \quad (36\text{-}2)$$

in which M and K have the integral values $0, \pm1, \pm2, \cdot\cdot\cdot$. Substitution of this expression in the wave equation confirms this, yielding as the equation in ϑ

$$\frac{1}{\sin\vartheta}\frac{d}{d\vartheta}\left(\sin\vartheta\frac{d\Theta}{d\vartheta}\right) - \left\{\frac{M^2}{\sin^2\vartheta} + \left(\frac{\cos^2\vartheta}{\sin^2\vartheta} + \frac{A}{C}\right)K^2\right.$$
$$\left. - 2\frac{\cos\vartheta}{\sin^2\vartheta}KM - \frac{8\pi^2A}{h^2}W\right\}\Theta = 0. \quad (36\text{-}3)$$

We see that $\vartheta = 0$ and $\vartheta = \pi$ are singular points for this equation (Sec. 17). It is convenient to eliminate the trigonometric functions by the change of variables

$$\begin{aligned} x &= \tfrac{1}{2}(1 - \cos\vartheta), \\ \Theta(\vartheta) &= T(x), \end{aligned} \Bigg\} \quad (36\text{-}4)$$

at the same time introducing the abbreviation

$$\lambda = \frac{8\pi^2AW}{h^2} - \frac{A}{C}K^2, \quad (36\text{-}5)$$

the result being

$$\frac{d}{dx}\left\{x(1-x)\frac{dT}{dx}\right\} + \left[\lambda - \frac{\{M + K(2x-1)\}^2}{4x(1-x)}\right]T = 0. \quad (36\text{-}6)$$

The singular points, which are regular points, have now been shifted to the points 0 and 1 of x, so that the indicial equation must be obtained at each of these points. Making the substitution $T(x) = x^sG(x)$, we find by the procedure of Section 17 that s equals $\tfrac{1}{2}|K - M|$, while the substitution

$$T(x) = (1 - x)^{s'}H(1 - x)$$

yields a value of $\frac{1}{2}|K + M|$ for s'. Following the method of Section 18c we therefore make the substitution

$$\Theta(\vartheta) = T(x) = x^{\frac{1}{2}|K-M|}(1 - x)^{\frac{1}{2}|K+M|}F(x), \qquad (36\text{-}7)$$

which leads to the equation[1] for F

$$x(1 - x)\frac{d^2F}{dx^2} + (\alpha - \beta x)\frac{dF}{dx} + \gamma F = 0, \qquad (36\text{-}8)$$

in which

$\alpha = |K - M| + 1$,
$\beta = |K + M| + |K - M| + 2$,
and
$\gamma = \lambda + K^2 - (\frac{1}{2}|K + M| + \frac{1}{2}|K - M|)(\frac{1}{2}|K + M| + \frac{1}{2}|K - M| + 1)$.

We can now apply the polynomial method to this equation by substituting the series expression

$$F(x) = \sum_{\nu=0}^{\infty} a_\nu x^\nu$$

in Equation 36–8. The recursion formula which results is

$$a_{j+1} = \frac{j(j - 1) + \beta j - \gamma}{(j + 1)(j + \alpha)}a_j. \qquad 36\text{-}9)$$

For this to break off after the jth term (the series is not an acceptable wave function unless it terminates), it is necessary for the numerator of Equation 36–9 to vanish, a condition which leads to the equation for the energy levels

$$W_{J,K} = \frac{h^2}{8\pi^2}\left\{\frac{J(J + 1)}{A} + K^2\left(\frac{1}{C} - \frac{1}{A}\right)\right\}, \qquad (36\text{-}10)$$

in which

$$J = j + \frac{1}{2}|K + M| + \frac{1}{2}|K - M|, \qquad (36\text{-}11)$$

that is, J is equal to, or larger than, the larger of the two quantities $|K|$ and $|M|$. The quantum number J is therefore zero or a positive integer, so that we have as the allowed values of the three quantum numbers

$$\left.\begin{aligned} J &= 0, 1, 2, \cdots, \\ K &= 0, \pm1, \pm2, \cdots, \pm J, \\ M &= 0, \pm1, \pm2, \cdots, \pm J. \end{aligned}\right\} \qquad (36\text{-}12)$$

---

[1] This equation is well known to mathematicians as the *hypergeometric equation*.

It can be shown that $J(J+1)\dfrac{h^2}{4\pi^2}$ is the square of the total angular momentum, while $Kh/2\pi$ is the component of angular momentum along the symmetry axis of the top and $Mh/2\pi$ the component along an arbitrary axis fixed in space.

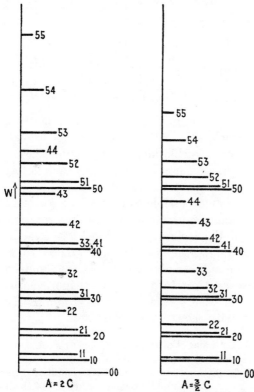

Fig. 36–2.—Energy-level diagram for symmetrical-top molecule, with $A = 2C$ and with $A = \tfrac{3}{2}C$. Values of the quantum numbers J and K are given for each level.

When K is zero, the expression for W reduces to that for the simple rotator in space, given in a footnote in Section 35c. The energy does not depend on M or on the sign of K, and hence the degeneracy of a level with given J is $2J + 1$ or $4J + 2$, depending on whether K is equal to zero or not. The appearance[1] of the set of energy levels depends on the relative magnitudes of A and C, as shown in Figure 36–2.

[1] For a discussion of the nature of these energy levels and of the spectral lines arising from them, see D. M. Dennison, *Rev. Mod. Phys.* **3**, 280 (1931).

The wave functions can be constructed by the use of the recursion formula 36–9. In terms of the hypergeometric functions[1] $F(a, b; c; x)$, the wave function is

$$\psi_{JKM}(\vartheta, \varphi, \chi) = N_{JKM}x^{\frac{1}{2}|K-M|}(1 - x)^{\frac{1}{2}|K+M|}e^{i(M\varphi+K\chi)}$$
$$F(-J + \tfrac{1}{2}\beta - 1, J + \tfrac{1}{2}\beta; 1 + |K - M|; x), \quad (36\text{–}13)$$

in which $x = \frac{1}{2}(1 - \cos \vartheta)$ and

$$N_{JKM} = \left\{ \frac{(2J+1)(J+\frac{1}{2}|K+M|+\frac{1}{2}|K-M|)!}{8\pi^2(J-\frac{1}{2}|K+M|-\frac{1}{2}|K-M|)!(|K-M|!)^2} \right.$$
$$\left. \frac{(J-\frac{1}{2}|K+M|+\frac{1}{2}|K-M|)!}{(J+\frac{1}{2}|K+M|-\frac{1}{2}|K-M|)!} \right\}^{\frac{1}{2}}. \quad (36\text{–}14)$$

In case that all three principal moments of inertia of a molecule are equal, the molecule is called a *spherical-top molecule* (examples: methane, carbon tetrachloride, sulfur hexafluoride). The energy levels in this case assume a particularly simple form (Problem 36–2).

It has been found possible to discuss the rotational motion of molecules containing parts capable of free rotation relative to other parts of the molecule. Nielsen[2] has treated the ethane molecule, assuming the two methyl groups to rotate freely relative to one another about the C-C axis, and La Coste[3] has similarly discussed the tetramethylmethane molecule, assuming free rotation of each of the four methyl groups about the axis connecting it with the central carbon atom.

Problem 36–1. Using Equation 36–9, construct the polynomial $F(x)$ for the first few sets of quantum numbers.

Problem 36–2. Set up the expression for the rotational energy levels for a spherical-top molecule, and discuss the degeneracy of the levels. Calculate the term values for the six lowest levels for the methane molecule, assuming the C-H distance to be 1.06 Å.

36b. The Rotation of Unsymmetrical-top Molecules.—The treatment of the rotational motion of a molecule with all three principal moments of inertia different (called an *unsymmetrical-top molecule*) is a much more difficult problem than that of the preceding section for the symmetrical top. We shall outline a

[1] The hypergeometric function is discussed in Whittaker and Watson, "Modern Analysis," Chap. XIV.

[2] H. H. NIELSEN, *Phys. Rev.* **40**, 445 (1932).

[3] L. J. B. LA COSTE, *Phys. Rev.* **46**, 718 (1934).

procedure which has been used with success in the interpretation of the spectra of molecules of this type.

Let us write the wave equation symbolically as

$$H\psi = W\psi. \tag{36-15}$$

Inasmuch as the known solutions of the wave equation for a symmetrical-top molecule form a complete set of orthogonal functions (discussed in the preceding section), we can expand the wave function ψ in terms of them, writing

$$\psi = \sum_{JKM} a_{JKM}\psi^0_{JKM}, \tag{36-16}$$

in which we use the symbol ψ^0_{JKM} to represent the symmetrical-top wave functions for a hypothetical molecule with moments of inertia A_0, $B_0(= A_0)$, and C_0. If we now set up the secular equation corresponding to the use of the series of Equation 36–16 as a solution of the unsymmetrical-top wave equation (Sec. 27a), we find that the only integrals which are not zero are those between functions with the same values of J and M, so that the secular equation immediately factors into equations corresponding to variation functions of the type[1]

$$\psi_{J\sigma M} = \sum_{K=-J}^{+J} a_{JKM}\psi^0_{JKM}. \tag{36-17}$$

On substituting this expression in the wave equation 36–15, we obtain the equation

$$\sum_K a_K H\psi^0_K = W\sum_K a_K\psi^0_K, \tag{36-18}$$

in which for simplicity we have omitted the subscripts J and M, the argument from now on being understood to refer to definite values of these two quantum numbers. On multiplication by $\psi^0_L{}^*$ and integration, this equation leads to the following set of simultaneous homogeneous linear equations in the coefficients a_K:

$$\sum_K a_K(H_{LK} - \delta_{LK}W) = 0, \qquad L = -J, -J+1, \cdots, +J, \tag{36-19}$$

in which δ_{LK} has the value 1 for $L = K$ and 0 otherwise, and H_{LK}

---

[1] The same result follows from the observation that J and M correspond to the total angular momentum of the system and its component along a fixed axis in space (see Sec. 52, Chap. XV).

represents the integral $\int \psi_L^0 {}^* H \psi_K^0 d\tau$. This set of equations has a solution only for values of W satisfying the secular equation

$$
\begin{vmatrix}
H_{-J,-J} - W & H_{-J,-J+1} & \cdots & H_{-J,J} \\
H_{-J+1,-J} & H_{-J+1,-J+1} - W & \cdots & H_{-J+1,J} \\
\cdots & \cdots & \cdots & \cdots \\
H_{J,-J} & H_{J,-J+1} & \cdots & H_{J,J} - W
\end{vmatrix} = 0.
$$

(36-20)

These values of W are then the allowed values for the rotational energy of the unsymmetrical-top molecule. Wang[1] has evaluated the integrals H_{LK} and shown that the secular equation can be further simplified. The application in the interpretation of the rotational fine structure of spectra has been carried out in several cases, including water,[2] hydrogen sulfide,[3] and formaldehyde.[4]

37. THE VIBRATION OF POLYATOMIC MOLECULES

The vibrational motion of polyatomic molecules is usually treated with an accuracy equivalent to that of the simple discussion of diatomic molecules given in Section 35c, that is, with the assumption of Hooke's-law forces between the atoms. When greater accuracy is needed, perturbation methods are employed.

Having made the assumption of Hooke's-law forces, we employ the method of *normal coordinates* to reduce the problem to soluble form. This method is applicable whether we use classical mechanics or quantum mechanics. Inasmuch as the former provides a simpler introduction to the method, we shall consider it first.

37a. Normal Coordinates in Classical Mechanics.—Let the positions of the n nuclei in the molecule be described by giving the Cartesian coordinates of each nucleus referred to the equilibrium position of that nucleus as origin, as shown in Figure 37-1. Let us call these coordinates q_1', q_2', $\cdots$, q_{3n}'. In terms of them we may write the kinetic energy of the molecule in the form

[1] S. C. WANG, *Phys. Rev.* **34**, 243 (1929). See also H. A. KRAMERS and G. P. ITTMANN, *Z. f. Phys.* **53**, 553 (1929); **58**, 217 (1929); **60**, 663 (1930); O. KLEIN, *Z. f. Phys.* **58**, 730 (1929); E. E. WITMER, *Proc. Nat. Acad. Sci.* **13**, 60 (1927); H. H. NIELSEN, *Phys. Rev.* **38**, 1432 (1931).

[2] R. MECKE, *Z. f. Phys.* **81**, 313 (1933).

[3] P. C. CROSS, *Phys. Rev.* **47**, 7 (1935).

[4] G. H. DIEKE and G. B. KISTIAKOWSKY, *Phys. Rev.* **45**, 4 (1934).

$$T = \tfrac{1}{2} \sum_{i=1}^{3n} M_i \dot{q}_i'^2, \tag{37-1}$$

in which M_i is the mass of the nucleus with coordinate q_i'. By changing the scale of the coordinates by means of the relation

$$q_i = \sqrt{M_i} q_i', \qquad i = 1, 2, \cdots, 3n, \tag{37-2}$$

we can eliminate the masses from the kinetic energy expression, obtaining

$$T = \tfrac{1}{2} \sum \dot{q}_i^2. \tag{37-3}$$

The potential energy V depends on the mutual positions of the nuclei and therefore upon the coordinates q_i. If we restrict

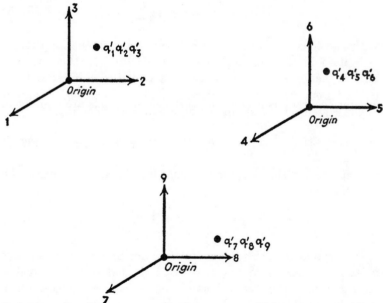

FIG. 37-1.—Coordinates $q_1' \cdots q_{3n}'$ of atoms measured relative to equilibrium positions.

ourselves to a discussion of small vibrations, we may expand V as a Taylor series in powers of the q's,

$$V(q_1 q_2 \cdots q_{3n}) = V_0 + \sum_i \left(\frac{\partial V}{\partial q_i}\right)_0 q_i + \frac{1}{2} \sum_{ij} b_{ij} q_i q_j + \cdots, \tag{37-4}$$

in which b_{ij} is given by

$$b_{ij} = \left(\frac{\partial^2 V}{\partial q_i \partial q_j}\right)_0,$$

and the subscript 0 means that the derivatives are evaluated at the point $q_1 = 0$, $q_2 = 0$, etc. If we choose our zero of energy so that V equals zero when q_1, q_2, etc. are zero, then V_0 is zero. Likewise the second term is zero, because by our choice of coordinate axes the equilibrium position is the configuration $q_1 = 0$, $q_2 = 0$, etc., and the condition for equilibrium is

$$\left(\frac{\partial V}{\partial q_i}\right)_0 = 0, \qquad i = 1, 2, \cdots, 3n. \qquad (37\text{--}5)$$

Neglecting higher terms, we therefore write

$$V(q_1 q_2 \cdots q_{3n}) = \tfrac{1}{2}\sum_{ij} b_{ij} q_i q_j. \qquad (37\text{--}6)$$

Using the coordinates q_i, we now set up the classical equations of motion in the Lagrangian form (Sec. 1c). In this case the kinetic energy T is a function of the velocities q_i only, and the potential energy V is a function of the coordinates q_i only, and in consequence the Lagrangian equations have the form

$$\frac{d}{dt}\left(\frac{\partial T}{\partial \dot{q}_k}\right) + \frac{\partial V}{\partial q_k} = 0, \qquad k = 1, 2, \cdots, 3n. \qquad (37\text{--}7)$$

On introducing the above expressions for T and V we obtain the equations of motion

$$\ddot{q}_k + \sum_i b_{ik} q_i = 0, \qquad k = 1, 2, \cdots, 3n. \qquad (37\text{--}8)$$

In case that the potential-energy function involves only squares q_i^2 and no cross-products $q_i q_j$ with $i \neq j$; that is, if b_{ij} vanishes for $i \neq j$, then these equations of motion can be solved at once. They have the form

$$\ddot{q}_k + b_{kk} q_k = 0, \qquad k = 1, 2, \cdots, 3n, \qquad (37\text{--}9)$$

the solutions of which are (Sec. 1a)

$$q_k = q_k^0 \sin(\sqrt{b_{kk}}\, t + \delta_k), \qquad k = 1, 2, \cdots, 3n, \qquad (37\text{--}10)$$

in which the q_k^0's are amplitude constants and the δ_k's phase constants of integration. In this special case, then, each of the coordinates q_k undergoes harmonic oscillation, the frequency being determined by the constant b_{kk}.

Now it is always possible by a simple transformation of variables to change the equations of motion from the form 37–8 to the form 37–9; that is, to eliminate the cross-products from the potential energy and at the same time retain the form 37–3 for the kinetic energy. Let us call these new coordinates $Q_l(l = 1, 2, \cdots, 3n)$. In terms of them the kinetic and the potential energy would be written

$$T = \tfrac{1}{2} \sum_l \dot{Q}_l^2 \qquad (37\text{--}11)$$

and

$$V = \tfrac{1}{2} \sum_l \lambda_l Q_l^2, \qquad (37\text{--}12)$$

and the solutions of the equations of motion would be

$$Q_l = Q_l^0 \sin \left(\sqrt{\lambda_l}\, t + \delta_l\right), \qquad l = 1, 2, \cdots, 3n. \quad (37\text{--}13)$$

Instead of finding the equations of transformation from the q's to the Q's by the consideration of the kinetic and potential energy functions, we shall make use of the equations of motion. In case that all of the amplitude constants Q_l^0 are zero except one, Q_1^0, say, then Q_1 will vary with the time in accordance with Equation 37–13, and, inasmuch as the q's are related to the Q's by the linear relation

$$q_k = \sum_{l=1}^{3n} B_{kl} Q_l, \qquad (37\text{--}14)$$

each of the q's will vary with the time in the same way, namely,

$$q_k = A_k \sin \left(\sqrt{\lambda}\, t + \delta_1\right), \qquad k = 1, 2, \cdots, 3n. \quad (37\text{--}15)$$

In these equations A_k represents the product $B_{k1} Q_1^0$, and λ the quantity λ_1, inasmuch as we selected Q_1 as the excited coordinate; the new symbols are introduced for generality. On substituting these expressions in the equations of motion 37–8, we obtain the set of equations

$$-\lambda A_k + \sum_{i=1}^{3n} b_{ik} A_i = 0, \qquad k = 1, 2, \cdots, 3n. \quad (37\text{--}16)$$

This is a set of $3n$ simultaneous linear homogeneous equations in the $3n$ unknown quantities A_k. As we know well by this time (after Secs. 24, 26d, etc.), this set of equations possesses a solution other than the trivial one $A_1 = A_2 = \cdots = 0$ only when the corresponding determinantal equation (the secular equation of perturbation and variation problems) is satisfied. This equation is

$$\begin{vmatrix} b_{11} - \lambda & b_{12} & \cdots & b_{1 3n} \\ b_{21} & b_{22} - \lambda & \cdots & b_{2 3n} \\ \cdots & \cdots & \cdots & \cdots \\ b_{3n1} & b_{3n\,2} & \cdots & b_{3n3n} - \lambda \end{vmatrix} = 0. \quad (37\text{-}17)$$

In other words, Equation 37-15 can represent a solution of the equations of motion only when λ has one of the $3n$ values which satisfy Equation 37-17. (Some of these roots may be equal.) Having found one of these roots, we can substitute it in Equation 37-16 and solve for the ratios[1] of the A's. If we put

$$A_{kl} = B_{kl}Q_l^0, \quad (37\text{-}18)$$

and introduce the extra condition

$$\sum_k B_{kl}^2 = 1, \quad (37\text{-}19)$$

in which the subscript l specifies which root λ_l of the secular equation has been used, then we can determine the values of the B_{kl}'s, Q_l^0 being left arbitrary.

By this procedure we have obtained $3n$ particular solutions of the equations of motion, one for each root of the secular equation. A general solution may be obtained by adding all of these together, a process which yields the equations

$$q_k = \sum_{l=1}^{3n} Q_l^0 B_{kl} \sin (\sqrt{\lambda_l} t + \delta_l). \quad (37\text{-}20)$$

This solution of the equations of motion contains $6n$ arbitrary constants, the *amplitudes* Q_l^0 and the *phases* δ_l, which in any particular case are determined from a knowledge of the initial positions and velocities of the n nuclei.

We have thus solved the classical problem of determining the positions of the nuclei as a function of the time, given any set of initial conditions. Let us now discuss the nature of the

---

[1] These equations are homogeneous, so that only the ratios of the A's can be determined. The extra condition 37-19 on the B_{kl}'s then allows them to be completely determined.

solution. As mentioned above, if we start the molecule vibrating in such a way that all the Q_l^0's except one, say Q_1^0, are zero, the solution is

$$q_k = Q_1^0 B_{k1} \sin (\sqrt{\lambda_1} t + \delta_1), \qquad k = 1, 2, \cdots, 3n, \quad (37\text{--}21)$$

which shows that each of the nuclei carries out a simple harmonic oscillation about its equilibrium position with the frequency

$$\nu_1 = \frac{\sqrt{\lambda_1}}{2\pi}. \qquad (37\text{--}22)$$

All of the nuclei move with the same frequency and the same phase; that is, they all pass through their equilibrium positions at the same time and reach their positions of maximum amplitude at the same time. These amplitudes, however, are not the same for the different nuclei but depend on the values of the B_{k1}'s and upon the initial amplitude, which is determined by Q_1^0. A vibration governed by Equation 37–21 and therefore having these properties is called a *normal mode of vibration* of the system (see Fig. 37–2).

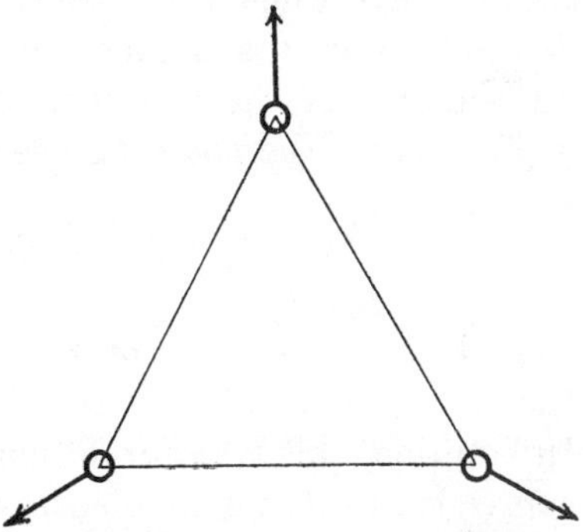

It is not required, however, that the nuclei have initial amplitudes and velocities such that the molecule undergo such a special motion. We can start the molecule off in any desired manner, with the general result that many of the constants Q_l^0 will be different from zero. In such a case the subsequent motion of the nuclei may be thought of as corresponding to a superposition of normal vibrations, each with its own frequency $\sqrt{\lambda_l}/2\pi$ and amplitude Q_l^0. The actual motion may be very complicated, although the normal modes of vibration themselves are frequently quite simple.

The *normal coordinates* of the system are the coordinates Q_l, which we introduced in Equation 37–14. These coordinates specify the configuration of the system just as definitely as the original coordinates q_i.

Fig. 37–2.—One of the normal modes of vibration of a symmetrical triatomic molecule. Each of the atoms moves in and out along a radial direction as shown by the arrows. All the atoms move with the same frequency and phase, and in this special case with the same amplitude.

The expansion of V given in Equation 37–4 is not valid except when the nuclei stay near their equilibrium positions. That is, we have assumed that the molecule is not undergoing translational or rotational motion as a whole. Closely related to this is the fact, which we shall not prove, that zero occurs six[1] times among the roots λ_l of the secular equation. The six normal modes of motion corresponding to these roots, which are not modes of vibration because they have zero frequency, are the three motions of translation in the x, y, and z directions and the three motions of rotation about the x, y, and z axes.

37b. Normal Coordinates in Quantum Mechanics.—It can be shown[2] that when the coefficients B_{kl} of Equation 37–14 are determined in the manner described in the last section, the introduction of the transformation 37–14 for the q_k's into the expression for the potential energy yields the result

$$V = \tfrac{1}{2}\sum_{ij}b_{ij}q_iq_j = \tfrac{1}{2}\sum_l\lambda_lQ_l^2; \qquad (37\text{–}23)$$

that is, the transformation to normal coordinates has eliminated the cross-products from the expression for the potential energy. In addition, this transformation has the property of leaving the expression for the kinetic energy unchanged in form;[3] i.e.,

$$T = \tfrac{1}{2}\sum_i\dot{q}_i^2 = \tfrac{1}{2}\sum_l\dot{Q}_l^2. \qquad (37\text{–}24)$$

These properties of the normal coordinates enable us to treat the problem of the vibrations of polyatomic molecules by the methods of quantum mechanics.

The wave equation for the nuclear motion of a molecule is

$$\sum_{j=1}^{n}\frac{1}{M_j}\nabla_j^2\psi + \frac{8\pi^2}{h^2}(W - V)\psi = 0, \qquad (37\text{–}25)$$

in which ψ represents the nuclear wave function $\psi_{n,\nu}(\xi)$ of Equa-

---

[1] This becomes five for linear molecules, which have only two degrees of rotational freedom.

[2] For a proof of this see E. T. Whittaker, "Analytical Dynamics," Sec. 77, Cambridge University Press, 1927.

[3] A transformation which leaves a simple sum of squares unaltered is called an *orthogonal transformation*.

tion 34–4. In terms of the Cartesian coordinates q_i' previously described (Fig. 37–1), we write

$$\sum_{j=1}^{n} \frac{1}{M_j} \nabla_j^2 \psi = \sum_{i=1}^{3n} \frac{1}{M_i} \frac{\partial^2 \psi}{\partial q_i'^2}. \tag{37-26}$$

By changing the scale of the coordinates as indicated by Equation 37–2 we eliminate the M's, obtaining for the wave equation the expression

$$\sum_{i=1}^{3n} \frac{\partial^2 \psi}{\partial q_i^2} + \frac{8\pi^2}{h^2}(W - V)\psi = 0. \tag{37-27}$$

We now introduce the normal coordinates Q_l. The reader can easily convince himself that an orthogonal transformation will leave the form of the first sum in the wave equation unaltered, so that, using also Equation 37–23, we obtain the wave equation in the form

$$\sum_{l=1}^{3n} \frac{\partial^2 \psi}{\partial Q_l^2} + \frac{8\pi^2}{h^2}\left(W - \frac{1}{2}\sum_{l=1}^{3n}\lambda_l Q_l^2\right)\psi = 0. \tag{37-28}$$

This equation, however, is immediately separable into $3n$ one-dimensional equations. We put

$$\psi = \psi_1(Q_1)\psi_2(Q_2) \cdots \psi_{3n}(Q_{3n}), \tag{37-29}$$

and obtain the equations

$$\frac{d^2\psi_k}{dQ_k^2} + \frac{8\pi^2}{h^2}\left(W_k - \frac{1}{2}\lambda_k Q_k^2\right)\psi_k = 0, \tag{37-30}$$

each of which is identical with the equation for the one-dimensional harmonic oscillator (Sec. 11a). The total energy W is the sum of the energies W_k associated with each normal coordinate; that is,

$$W = \sum_{k=1}^{3n} W_k. \tag{37-31}$$

The energy levels of the harmonic oscillator were found in Section 11a to have the values $(v + \frac{1}{2})h\nu_0$, where v is the quantum number and ν_0 the classical frequency of the oscillator. Applying this to the problem of the polyatomic molecule, we see that

$$W = \sum_k W_k = \sum_k (v_k + \tfrac{1}{2})h\nu_k, \qquad (37\text{--}32)$$

in which v_k is the quantum number ($v_k = 0, 1, 2, \cdots$) and ν_k is the classical frequency of the kth normal mode of vibration We have already seen (from Eq. 37–22) that

$$\nu_k = \frac{\sqrt{\lambda_k}}{2\pi}. \qquad (37\text{--}33)$$

The energy-level diagram of a polyatomic molecule is therefore quite complex. If, however, we consider only the *fundamental frequencies* emitted or absorbed by such a molecule; that is, the frequencies due to a change of only one quantum number v_k by one unit, we see that these frequencies are $\nu_1, \nu_2, \cdots, \nu_{3n}$; that is, they are the classical frequencies of motion of the molecule.

This type of treatment has been very useful as a basis for the interpretation of the vibrational spectra of polyatomic molecules. Symmetry considerations have been widely employed to simplify the solution of the secular equation and in that connection the branch of mathematics known as group theory has been very helpful.[1]

38. THE ROTATION OF MOLECULES IN CRYSTALS

In the previous sections we have discussed the rotation and vibration of free molecules, that is, of molecules in the gas phase. There is strong evidence[2] that molecules and parts of molecules in many crystals can rotate if the temperature is sufficiently high. The application[2,3] of quantum mechanics to this problem has led to a clarification of the nature of the motion of a molecule within a crystal which is of some interest. The problem is closely related to that dealing with the rotation of one part of a molecule relative to the other parts, such as the rotation of methyl groups in hydrocarbon molecules.[4]

---

[1] C. J. BRESTER, *Z. f. Phys.* **24**, 324 (1924); E. WIGNER, *Göttinger Nachr.* 133 (1930); G. PLACZEK, *Z. f. Phys.* **70**, 84 (1931); E. B. WILSON, JR., *Phys. Rev.* **45**, 706 (1934); *J. Chem. Phys.* **2**, 432 (1934); and others.

[2] L. PAULING, *Phys. Rev.* **36**, 430 (1930). This paper discusses the mathematics of the plane rotator in a crystal as well as the empirical evidence for rotation.

[3] T. E. STERN, *Proc. Roy. Soc.* **A 130**, 551 (1931).

[4] E. TELLER and K. WEIGERT, *Göttinger Nachr.* 218 (1933); J. E. LENNARD-JONES and H. H. M. PIKE, *Trans. Faraday Soc.* **30**, 830 (1934).

The wave equation for a diatomic molecule in a crystal, considered as a rigid rotator, obtained by introducing V into the equation for the free rotator given in a footnote of Section 35c, is

$$\frac{1}{\sin\vartheta}\frac{\partial}{\partial\vartheta}\left(\sin\vartheta\,\frac{\partial\psi}{\partial\vartheta}\right) + \frac{1}{\sin^2\vartheta}\frac{\partial^2\psi}{\partial\varphi^2} + \frac{8\pi^2 I}{h^2}(W - V)\psi = 0, \qquad (38\text{-}1)$$

in which ϑ and φ are the polar coordinates of the axis and I is the moment of inertia of the molecule. The potential function V is introduced as an approximate description of the effects of the other molecules of the crystal upon the molecule in question.

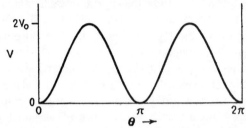

FIG. 38-1.—Idealized potential function for a symmetrical diatomic molecule in a crystal.

If the molecule being studied is made up of like atoms, such as is O_2 or H_2, then a reasonable form to assume for V is

$$V = V_0(1 - \cos 2\vartheta), \qquad (38\text{-}2)$$

which is shown in Figure 38-1. Turning a symmetrical molecule end for end does not change V, as is shown in the figure by the periodicity of V with period π.

The wave equation 38-1 with the above form for V has been studied by Stern,[1] who used the mathematical treatment given by A. H. Wilson.[2] We shall not reproduce their work, although the method of solution is of some interest. The first steps are exactly the same as in the solution of the equation discussed in Section 18c except that a three-term recursion formula is obtained. The method of obtaining the energy levels from this three-term formula is then similar to the one which is discussed in Section 42c, where a similar situation is encountered.

We have referred to the case of free rotation of methyl groups at the end of Section 36a.

[1] T. E. STERN, *loc. cit.*

[2] A. H. WILSON, *Proc. Roy. Soc.* **A 118,** 628 (1928).

The results obtained may best be described by starting with the two limiting cases. When the energy of the molecule is small compared with V_0 (i.e., at low temperatures), then the potential function can be regarded as parabolic in the neighborhood of the minima and we expect, as is actually found, that the energy levels will be those of a two-dimensional harmonic oscillator and that the wave functions will show that the molecule oscillates about either one of the two positions of equilibrium, with little tendency to turn end over end. When the molecule is in a state with energy large compared with V_0 (i.e., at high temperatures), the wave functions and energy levels approximate those of the free rotator (Sec. 35c, footnote), the end-over-end motion being only slightly influenced by the potential energy. In the intermediate region, the quantum-mechanical treatment shows that there is a fairly sharp but nevertheless continuous transition between oscillation and rotation. In other words, for a given energy there is a definite probability of turning end over end, in sharp contrast with the results of classical mechanics, which are that the molecule either has enough energy to rotate or only enough to oscillate.

The transition between rotation and oscillation takes place roughly at the temperature $T = 2V_0/k$, where k is Boltzmann's constant. This temperature lies below the melting point for a number of crystals, such as hydrogen chloride, methane, and the ammonium halides, and is recognizable experimentally as a transition point in the heat-capacity curve. For solid hydrogen even the lowest energy level is in the rotational region, a fact which is of considerable significance in the application of the third law of thermodynamics.

Problem 38-1. Considering the above system as a perturbed rigid rotator, study the splitting of the rotator levels by the field, indicating by an energy-level diagram the way in which the components of the rotator levels begin to change as the perturbation is increased.

General References

A. E. Ruark and H. C. Urey: "Atoms, Molecules and Quanta," Chap. XII, McGraw-Hill Book Company, Inc., New York, 1930. A general discussion of the types of molecular spectra found experimentally, with the theoretical treatment of some of them.

W. Weizel: "Bandenspektren," Handbuch der Experimentalphysik (Wien-Harms), Ergänzungsband I. A complete discussion of the theory and results for diatomic molecules, with some reference to polyatomic molecules.

D. M. DENNISON: *Rev. Mod. Phys.* **3,** 280 (1931). A discussion of the methods of treating the rotational and vibrational spectra of polyatomic molecules.

R. DE L. KRONIG: "Band Spectra and Molecular Structure," Cambridge University Press, 1930. The theory of diatomic spectra, including consideration of electron spins.

A. SCHAEFER and F. MATOSSI: "Das Ultrarote Spektrum," Juliuf Springer, Berlin, 1930. Methods, theory, and results of infrared spectroscopy.

K. W. F. KOHLRAUSCH: "Der Smekal-Raman-Effekt," Julius Springer, Berlin, 1931.

G. PLACZEK: "Rayleigh-Streuung und Ramaneffekt," Handbuch der Radiologie, Vol. VI, Akademische Verlagsgesellschaft, Leipzig, 1934.

E. TELLER: "Theorie der langwelligen Molekülspektren," Hand- und Jahrbuch der chemischen Physik, Vol. IX, Akademische Verlagsgesellschaft, Leipzig, 1934.

CHAPTER XI

PERTURBATION THEORY INVOLVING THE TIME, THE EMISSION AND ABSORPTION OF RADIATION, AND THE RESONANCE PHENOMENON

39. THE TREATMENT OF A TIME-DEPENDENT PERTURBATION BY THE METHOD OF VARIATION OF CONSTANTS

There have been developed two essentially different wave-mechanical perturbation theories. The first of these, due to Schrödinger, provides an approximate method of calculating energy values and wave functions for the stationary states of a system under the influence of a constant (time-independent) perturbation. We have discussed this theory in Chapter VI. The second perturbation theory, which we shall treat in the following paragraphs, deals with the time behavior of a system under the influence of a perturbation; it permits us to discuss such questions as the probability of transition of the system from one unperturbed stationary state to another as the result of the perturbation. (In Section 40 we shall apply the theory to the problem of the emission and absorption of radiation.) The theory was developed by Dirac.[1] It is often called *the theory of the variation of constants;* the reason for this name will be evident from the following discussion.

Let us consider an unperturbed system with wave equation including the time

$$H^0\Psi^0 = -\frac{h}{2\pi i}\frac{\partial\Psi^0}{\partial t}, \tag{39-1}$$

the normalized general solution of which is

$$\Psi^0 = \sum_{n=0}^{\infty}a_n\Psi_n^0, \tag{39-2}$$

---

[1] P. A. M. DIRAC, *Proc. Roy. Soc.* A **112**, 661 (1926); A **114**, 243 (1927). Less general discussions were also given by Schrödinger in his fourth 1926 paper and by J. C. Slater, *Proc. Nat. Acad. Sci.* **13**, 7 (1927).

in which the a_n's are constants, with $\sum_n a_n^* a_n = 1$, and the Ψ_n^0's are the time-dependent wave functions for the stationary states, the corresponding energy values being W_0^0, W_1^0, $\cdots$, W_n^0, $\cdots$. Now let us assume that the Hamiltonian for the actual system contains in addition to H^0 (which is independent of t) a perturbing term H', which may be a function of the time as well as of the coordinates of the system.[1] (For example, H' might be zero except during the period $t_1 < t < t_2$, the perturbation then being effective only during this period.) Since we desire to express our results in terms of the unperturbed wave functions including the time, we must consider the Schrödinger time equation for the system. This equation is

$$(H^0 + H')\Psi = -\frac{h}{2\pi i}\frac{\partial \Psi}{\partial t}. \tag{39-3}$$

A wave function satisfying this equation is a function of the time and of the coordinates of the system. For a given value of t, say t', $\Psi(t')$ is a function of the coordinates alone. By the general expansion theorem of Section 22 it can be represented as a series involving the complete set of orthogonal wave functions for the unperturbed system,

$$\Psi(x_1, \cdots, z_N, t') = \sum_n a_n \Psi_n^0(x_1, \cdots, z_N, t'), \tag{39-4}$$

the symbol $\Psi_n^0(x_1, \cdots, z_N, t')$ indicating that t' is introduced in place of t in the exponential time factors. The quantities a_n are constants. For any other value of t a similar expansion can be made, involving different values of the constants a_n. A general solution of the wave equation 39-3 can accordingly be written as

$$\Psi(x_1, \cdots, z_N, t) = \sum_n a_n(t) \Psi_n^0(x_1, \cdots, z_N, t), \tag{39-5}$$

the quantities $a_n(t)$ being functions of t alone, such as to cause Ψ to satisfy the wave equation 39-3.

The nature of these functions is found by substituting the expression 39-5 in the wave equation 39-3, which gives

---

[1] H' might also be a function of the momenta $p_{x_1}, \cdots$, which would then be replaced by $\dfrac{h}{2\pi i}\dfrac{\partial}{\partial x_1}$, $\cdots$.

$$\sum_n a_n(t)H^0\Psi_n^0 + \sum_n a_n(t)H'\Psi_n^0 =$$

$$-\frac{h}{2\pi i}\sum_n \dot{a}_n(t)\Psi_n^0 - \frac{h}{2\pi i}\sum_n a_n(t)\frac{\partial\Psi_n^0}{\partial t}.$$

The first and last terms in this equation cancel (by Eqs. 39-1 and 39-2), leaving

$$-\frac{h}{2\pi i}\sum_n \dot{a}_n(t)\Psi_n^0 = \sum_n a_n(t)H'\Psi_n^0.$$

If we now multiply by Ψ_m^{0*} and integrate over configuration space, noting that all terms on the left vanish except that for $n = m$ because of the orthogonality properties of the wave functions, we obtain

$$\dot{a}_m(t) = -\frac{2\pi i}{h}\sum_{n=0}^{\infty} a_n(t)\int \Psi_m^{0*}H'\Psi_n^0 d\tau, \qquad m = 0, 1, 2, \cdots .$$

$$(39\text{-}6)$$

This is a set of simultaneous differential equations in the functions $a_m(t)$, by means of which these functions can be evaluated in particular cases.

39a. A Simple Example.—As an illustration of the use of the set of equations 39-6, let us consider that at the time $t = 0$ we know that a system in which we are interested is in a particular stationary state, our knowledge perhaps having been obtained by a measurement of the energy of the system. The wave function representing the system is then Ψ_l^0, in which l has a particular value. If a small perturbation H' acts on the system for a short time t', H' being independent of t during this period, we may solve the equations 39-6 by neglecting all terms on the right side except that with $n = l$; that is, by assuming that only the term in $a_l(t)$ need be retained on the right side of these equations. It is first necessary for us to discuss the equation for a_l itself. This equation (Eq. 39-6 with $m = l$ and $a_n = 0$ for $n \neq l$) is

$$\frac{da_l(t)}{dt} = -\frac{2\pi i}{h}a_l(t)H'_{ll},$$

in which $H'_{ll} = \int \psi_l^{0*}H'\psi_l^0 d\tau$, which can be integrated at once to give

$$a_l(t) = e^{-2\pi i H'_{ll}t/h}, \qquad 0 \leqslant t \leqslant t', \qquad (39\text{-}7)$$

the assumption being made that $a_l = 1$ at the time $t = 0$.

This expression shows the way that the coefficient a_l changes during the time that the perturbation is acting. During this time the wave function, neglecting the terms with $m \neq l$, is

$$a_l(t)\Psi_l^0 = \psi_l^0 e^{-\frac{2\pi i}{h}(W^0 + H_{ll}')t}.$$

It will be observed that the time-dependent factor contains the first-order energy $W^0 + H_{ll}'$, as given by the Schrödinger perturbation theory; this illustrates the intimate relation of the two perturbation theories.

Now let us consider the remaining equations of the set 39–6, determining the behavior of the coefficients $a_m(t)$ with $m \neq l$. Replacing a_l on the right side of 39–6 by its initial value[1] $a_l(0) = 1$, and neglecting all other a_n's, we obtain the set of approximate equations

$$\frac{da_m(t)}{dt} = -\frac{2\pi i}{h} \int \Psi_m^{0*} H' \Psi_l^0 d\tau$$

This can be written as

$$da_m(t) = -\frac{2\pi i}{h} H_{ml}' e^{-\frac{2\pi i (W_l - W_m)t}{h}} dt, \qquad 0 \leqslant t \leqslant t', \ m \neq l,$$

in which

$$H_{ml}' = \int \psi_m^{0*} H' \psi_l^0 d\tau, \tag{39-8}$$

and H_{ml}' is independent of t, since we have considered H' to be independent of t during the period $0 \leqslant t \leqslant t'$, and have replaced the time-containing wave functions Ψ_m^{0*} and Ψ_l^0 by the amplitude functions ψ_m^{0*} and ψ_l^0 and the corresponding time factors, the latter being now represented explicitly by the exponential functions. These equations can be integrated at once; on introducing the limits, and noting that $a_m(0) = 0$ for $m \neq l$, we obtain

$$a_m(t') = H_{ml}' \frac{1 - e^{\frac{2\pi i (W_m - W_l)t'}{h}}}{W_m - W_l}, \qquad m \neq l, \tag{39-9}$$

in which, it is remembered, the subscript l refers to the state initially occupied and m to other states. In case that the time t' is small compared with the time $h/(W_m - W_l)$, the expression can be expanded, giving

---

[1] The expression for $a_l(t)$ given by Equation 39–7 could be introduced in place of $a_l(0) = 1$, with, however, no essential improvement in the result.

$$a_m(t') = -\frac{2\pi i}{h}H'_{ml}t', \qquad m \neq l. \tag{39-10}$$

At the time t' the wave function for the system (which was Ψ_l^0 at time $t = 0$) is approximately

$$\Psi(t') = a_l(t')\Psi_l^0 + \sum_m{}' a_m(t')\Psi_m^0 \tag{39-11}$$

(the prime on the summation sign indicating that the term $m = l$ is omitted), in which a_l is nearly equal to 1 and the a_m's are very small. This wave function continues to represent the system at times later than t', so long as it remains isolated. We could now carry out a measurement (of the energy, say) to determine the stationary state of the system. The probability that the system would be found in the mth stationary state is $a_m^* a_m$.

This statement requires the extension of the postulates regarding the physical interpretation of the wave equation given in Sections 10c and 12d. It was shown there that an average value could be predicted for a dynamical function for a system at time t from a knowledge of the wave function representing the system. The average value predicted for the energy of a system with wave function $\Psi = \sum_n a_n\Psi_n^0$ is $\overline{W} = \sum_n a_n^* a_n W_n^0$. However, an actual individual measurement of the energy must give one of the values W_0^0, W_1^0, W_2^0, etc., inasmuch as it is only for wave functions corresponding to stationary states that the energy has a definite value (Sec. 10c). Hence when a measurement of the energy has been made, the wave function representing the system is no longer $\Psi = \sum_n a_n\Psi_n^0$, but is one of the functions Ψ_0^0, Ψ_1^0, Ψ_2^0, etc.

This shows how a wave function does not really represent the system but rather our knowledge of the system. At time $t = 0$ we knew the energy of the system to be W_l^0, and hence we write Ψ_l^0 for the wave function. (We do not know everything about the system, however; thus we do not know the configuration of the system but only the probability distribution function $\Psi_l^{0*}\Psi_l^0$.) At time t' we know that at time $t = 0$ the wave function was Ψ_l^0, and that the perturbation H' was acting between times $t = 0$ and t'. From this information we obtain the wave function of Equations 39–11, 39–10, and 39–8 as representing our knowledge of the system. With it we predict that the probability that the system is in the mth stationary state is $a_m^* a_m$. So long as we leave the system isolated, this wave function represents the system. If we allow the system to be affected by a known perturbation, we can find a new wave function by the foregoing methods. If we now further perturb the system by an unknown amount in the process of making a measurement of the energy, we can no longer apply these methods;

instead, we assign to the system a new wave function compatible with our new knowledge of the result of the experiment. A more detailed discussion of these points will be given in Chapter XV.

Equation 39–10 shows that in case t' is small the probability of finding the system in the stationary state m as a result of transition from the original state l is

$$a_m^* a_m = \frac{4\pi^2}{h^2} H_{ml}'^* H_{ml}' t'^2, \qquad (39\text{–}12)$$

being thus proportional to the square of the time t' rather than to the first power as might have been expected. In most cases the nature of the system is such that experiments can be designed to measure not the probability of transition to a single state but rather the integrated probability of transition to a group of adjacent states; it is found on carrying out the solution of the fundamental equations 39–6 and subsequent integration that for small values of t' the integrated probability of transition is proportional to the first power of the time t'. An example of a calculation of a related type will be given in Section 40b.

40. THE EMISSION AND ABSORPTION OF RADIATION

Inasmuch as a thoroughly satisfactory quantum-mechanical theory of systems containing radiation as well as matter has not yet been developed, we must base our discussion of the emission and absorption of radiation by atoms and molecules on an approximate method of treatment, drawing upon classical electromagnetic theory for aid. The most satisfactory treatment of this type is that of Dirac,[1] which leads directly to the formulas for spontaneous emission as well as absorption and induced emission of radiation. Because of the complexity of this theory, however, we shall give a simpler one, in which only absorption and induced emission are treated, prefacing this by a general discussion of the Einstein coefficients of emission and absorption of radiation in order to show the relation that spontaneous emission bears to the other two phenomena.

40a. The Einstein Transition Probabilities.—According to classical electromagnetic theory, a system of accelerated electrically charged particles emits radiant energy. In a bath of

[1] P. A. M. Dirac, *Proc. Roy. Soc.* **A112**, 661 (1926); **A114**, 243 (1927); J. C. Slater, *Proc. Nat. Acad. Sci.* **13**, 7 (1927).

radiation at temperature T it also absorbs radiant energy, the rates of absorption and of emission being given by the classical laws. These opposing processes might be expected to lead to a state of equilibrium. The following treatment of the corresponding problem for quantized systems (atoms or molecules) was given by Einstein[1] in 1916.

Let us consider two non-degenerate stationary states m and n of a system, with energy values W_m and W_n such that W_m is greater than W_n. According to the Bohr frequency rule, transition from one state to another will be accompanied by the emission or absorption of radiation of frequency

$$\nu_{mn} = \frac{W_m - W_n}{h}.$$

We assume that the system is in the lower state n in a bath of radiation of density $\rho(\nu_{mn})$ in this frequency region (the energy of radiation between frequencies ν and $\nu + d\nu$ in unit volume being $\rho(\nu)d\nu$). The probability that it will absorb a quantum of energy of radiation and undergo transition to the upper state in unit time is

$$B_{n \to m}\rho(\nu_{mn}).$$

$B_{n \to m}$ is called *Einstein's coefficient of absorption.* The probability of absorption of radiation is thus assumed to be proportional to the density of radiation. On the other hand, it is necessary in order to carry through the following argument to postulate[2] that the probability of emission is the sum of two parts, one of which is independent of the radiation density and the other proportional to it. We therefore assume that the probability that the system in the upper state m will undergo transition to the lower state with the emission of radiant energy is

$$A_{m \to n} + B_{m \to n}\rho(\nu_{mn}).$$

$A_{m \to n}$ is *Einstein's coefficient of spontaneous emission* and $B_{m \to n}$ is *Einstein's coefficient of induced emission.*

---

[1] A. EINSTEIN, *Verh. d. Deutsch. Phys. Ges.* **18**, 318 (1916); *Phys. Z.* **18**, 121 (1917).

[2] This postulate is of course closely analogous to the classical theory, according to which an oscillator interacting with an electromagnetic wave could either absorb energy from the field or lose energy to it, depending on the relative phases of oscillator and wave.

We now consider a large number of identical systems of this type in equilibrium with radiation at temperature T. The density of radiant energy is known to be given by Planck's radiation law as

$$\rho(\nu) = \frac{8\pi h \nu^3}{c^3} \frac{1}{e^{\frac{h\nu}{kT}} - 1}, \tag{40-1}$$

in which k is the Boltzmann constant. Let the number of systems in state m be N_m, and that in state n be N_n. The number of systems undergoing transition in unit time from state n to state m is then

$$N_n B_{n \to m} \rho(\nu_{mn}),$$

and the number undergoing the reverse transition is

$$N_m \{A_{m \to n} + B_{m \to n} \rho(\nu_{mn})\}.$$

At equilibrium these two numbers are equal, giving

$$\frac{N_n}{N_m} = \frac{A_{m \to n} + B_{m \to n} \rho(\nu_{mn})}{B_{n \to m} \rho(\nu_{mn})}. \tag{40-2}$$

The equations of quantum statistical mechanics (Sec. 49) require that the ratio N_n/N_m be given by the equation

$$\frac{N_n}{N_m} = e^{-\frac{(W_n - W_m)}{kT}} = e^{h\nu_{mn}/kT}. \tag{40-3}$$

From Equations 40-2 and 40-3 we find for $\rho(\nu_{mn})$ the expression

$$\rho(\nu_{mn}) = \frac{A_{m \to n}}{B_{n \to m} e^{h\nu_{mn}/kT} - B_{m \to n}}. \tag{40-4}$$

In order for this to be identical with Equation 40-1, we must assume that the three Einstein coefficients are related by the equations

$$B_{n \to m} = B_{m \to n} \tag{40-5a}$$

and

$$A_{m \to n} = \frac{8\pi h \nu_{mn}^3}{c^3} B_{m \to n}; \tag{40-5b}$$

that is, the coefficients of absorption and induced emission are equal and the coefficient of spontaneous emission[1] differs from them by the factor $8\pi h \nu_{mn}^3/c^3$.

---

[1] It is interesting to note that at the temperature $T = \dfrac{h\nu_{nm}}{k \log 2}$ the probabilities of spontaneous emission and induced emission are equal.

40b. The Calculation of the Einstein Transition Probabilities by Perturbation Theory.—According to classical electromagnetic theory, the density of energy of radiation of frequency ν in space, with unit dielectric constant and magnetic permeability, is given by the expression

$$\rho(\nu) = \frac{1}{4\pi}\overline{E^2(\nu)}, \qquad (40\text{--}6)$$

in which $\overline{E^2(\nu)}$ represents the average value of the square of the electric field strength corresponding to this radiation. The distribution of radiation being isotropic, we can write

$$\tfrac{1}{3}\overline{E^2(\nu)} = \overline{E_x^2(\nu)} = \overline{E_y^2(\nu)} = \overline{E_z^2(\nu)}, \qquad (40\text{--}7)$$

$E_x(\nu)$ representing the component of the electric field in the x direction, etc. We may conveniently introduce the time variation of the radiation by writing

$$E_x(\nu) = 2E_x^0(\nu)\cos 2\pi\nu t = E_x^0(\nu)(e^{2\pi i\nu t} + e^{-2\pi i\nu t}), \quad (40\text{--}8)$$

the complex exponential form being particularly convenient for calculation. Since the average value of $\cos^2 2\pi\nu t$ is $\tfrac{1}{2}$, we see that

$$\rho(\nu) = \frac{1}{4\pi}\overline{E^2(\nu)} = \frac{3}{4\pi}\overline{E_x^2(\nu)} = \frac{6}{4\pi}E_x^0{}^2(\nu). \qquad (40\text{--}9)$$

Let us now consider two stationary states m and n of an unperturbed system, represented by the wave functions Ψ_m^0 and Ψ_n^0, and such that W_m is greater than W_n. Let us assume that at the time $t = 0$ the system is in the state n, and that at this time the system comes under the perturbing influence of radiation of a range of frequencies in the neighborhood of ν_{mn}, the electric field strength for each frequency being given by Equation 40–8. We shall calculate the probability of transition to the state m as a result of this perturbation, using the method of Section 39. The perturbation energy for a system of electrically charged particles in an electric field E_x parallel to the x axis is

$$H' = E_x\sum_j e_j x_j, \qquad (40\text{--}10)$$

in which e_j represents the charge and x_j the x coordinate of the jth particle of the system. The expression $\sum_j e_j x_j$ (the sum being taken over all particles in the system) is called the *component of electric dipole moment* of the system along the x axis and is often represented by the symbol μ_x. We now make the approximation that the dimensions of the entire system (a molecule, say) are small compared with the wave length of the radiation, so that the electric field of the radiation may be considered constant over the system. In the case under consideration the field strength E_x is given by the expression

$$E_x = \int E_x^0(\nu)(e^{2\pi i\nu t} + e^{-2\pi i\nu t})d\nu.$$

Let us temporarily consider the perturbation as due to a single frequency ν. Introducing $a_m(0) = 0$ and $a_n(0) = 1$ in the right side of Equation 39–6 (a_n being the coefficient of a particular state and all the other coefficients in the sum being zero), this equation becomes

$$\dot{a}_m(t) = -\frac{2\pi i}{h}\int \Psi_m^0{}^* H' \Psi_n^0 d\tau = -\frac{2\pi i}{h}\int \psi_m^0{}^* e^{\frac{2\pi i}{h}W_m t} E_x^0(\nu)(e^{2\pi i\nu t} +$$

$$e^{-2\pi i\nu t})\sum_j e_j x_j \psi_n^0 e^{-\frac{2\pi i}{h}W_n t} d\tau.$$

If we now introduce the symbol $\mu_{x_{mn}}$ to represent the integral

$$\mu_{x_{mn}} = \int \psi_m^0{}^* \sum_j e_j x_j \psi_n^0 d\tau = \int \psi_m^0{}^* \mu_x \psi_n^0 d\tau, \qquad (40\text{--}11)$$

we obtain the equation

$$\frac{da_m(t)}{dt} = -\frac{2\pi i}{h}\mu_{x_{mn}}E_x^0(\nu)\left\{ e^{\frac{2\pi i}{h}(W_m - W_n + h\nu)t} + e^{\frac{2\pi i}{h}(W_m - W_n - h\nu)t}\right\},$$

which gives, on integration,

$$a_m(t) = \mu_{x_{mn}}E_x^0(\nu)\left\{ \frac{1 - e^{\frac{2\pi i}{h}(W_m - W_n + h\nu)t}}{W_m - W_n + h\nu} + \right.$$

$$\left. \frac{1 - e^{\frac{2\pi i}{h}(W_m - W_n - h\nu)t}}{W_m - W_n - h\nu}\right\}. \qquad (40\text{--}12)$$

Of the two terms of Equation 40–12, only one is important, and that one only if the frequency ν happens to lie close to $\nu_{mn} = (W_m - W_n)/h$. The numerator in each fraction can vary in absolute magnitude only between 0 and 2, and, inasmuch as for a single frequency the term $\mu_{x_{mn}}E_x^0(\nu)$ is always small, the

expression will be small unless the denominator is also very
small; that is, unless $h\nu$ is approximately equal to $W_m - W_n$.
In other words, the presence of the so-called *resonance denominator* $W_m - W_n - h\nu$ causes the influence of the perturbation
in changing the system from the state n to the higher state m
to be large only when the frequency of the light is close to that
given by the Bohr frequency rule. In this case of absorption, it
is the second term which is important; for induced emission of
radiation (with $W_m - W_n$ negative), the first term would play
the same role.

Neglecting the first term, we obtain for $a_m^*(t)a_m(t)$, after slight
rearrangement, the expression

$$a_m^*(t)a_m(t) = 4(\mu_{x_{mn}})^2 E_x^{02}(\nu)\frac{\sin^2\left\{\frac{\pi}{h}(W_m - W_n - h\nu)t\right\}}{(W_m - W_n - h\nu)^2}.$$

(If $\mu_{x_{mn}}$ is complex, the square of its absolute value is to be used
in this equation.) This expression, however, includes only the
terms due to a single frequency. In practice we deal always
with a range of frequencies. It is found, on carrying through
the treatment, that the effects of light of different frequencies are
additive, so that we now need only to integrate the above
expression over the range of frequencies concerned. The
integrand is seen to make a significant contribution only over the
region of ν near ν_{mn}, so that we are justified in replacing $E_x^0(\nu)$
by the constant $E_x^0(\nu_{mn})$, obtaining

$$a_m^*(t)a_m(t) = 4(\mu_{x_{mn}})^2 E_x^{02}(\nu_{mn})\int\frac{\sin^2\left\{\frac{\pi}{h}(W_m - W_n - h\nu)t\right\}}{(W_m - W_n - h\nu)^2}d\nu.$$

This integral can be taken from $-\infty$ to $+\infty$, inasmuch as the
value of the integrand is very small except in one region; and
making use of the relation $\int_{-\infty}^{+\infty}\frac{\sin^2 x}{x^2}dx = \pi$, we can obtain the
equation

$$a_m^*(t)a_m(t) = \frac{4\pi^2}{h^2}(\mu_{x_{mn}})^2 E_x^{02}(\nu_{mn})t. \tag{40-13}$$

It is seen that, as the result of the integration over a range of
values of ν, the probability of transition to the state m in time t
is proportional to t, the coefficient being the transition probability

as usually defined. With the use of Equation 40–9 we may now introduce the energy density $\rho(\nu_{mn})$, obtaining as the probability of transition in unit time from state n to state m under the influence of radiation polarized in the x direction the expression

$$\frac{8\pi^3}{3h^2}(\mu_{x_{mn}})^2\rho(\nu_{mn}).$$

The expressions for the y and z directions are similar. Thus we obtain for the Einstein coefficient of absorption $B_{n\to m}$ the equation

$$B_{n\to m} = \frac{8\pi^3}{3h^2}\{(\mu_{x_{mn}})^2 + (\mu_{y_{mn}})^2 + (\mu_{z_{mn}})^2\}. \quad (40\text{–}14a)$$

By a completely analogous treatment in which the values $a_n(0) = 0$, $a_m(0) = 1$ are used, the Einstein coefficient of induced emission $B_{m\to n}$ is found to be given by the equation

$$B_{m\to n} = \frac{8\pi^3}{3h^2}\{(\mu_{x_{mn}})^2 + (\mu_{y_{mn}})^2 + (\mu_{z_{mn}})^2\}, \quad (40\text{–}14b)$$

as, indeed, is required by Equation 40–5a.

Our treatment does not include the phenomenon of spontaneous emission of radiation. Its extension to include this is not easy; Dirac's treatment is reasonably satisfactory, and we may hope that the efforts of theoretical physicists will soon provide us with a thoroughly satisfactory discussion of radiation. For the present we content ourselves with using Equation 40–5b in combination with the above equations to obtain

$$A_{m\to n} = \frac{64\pi^4\nu_{mn}^3}{3hc^3}\{(\mu_{x_{mn}})^2 + (\mu_{y_{mn}})^2 + (\mu_{z_{mn}})^2\} \quad (40\text{–}15)$$

as the equation for the Einstein coefficient of spontaneous emission.

As a result of the foregoing considerations, the wave-mechanical calculation of the intensities of spectral lines and the determination of selection rules are reduced to the consideration of the electric-moment integrals defined in Equation 40–11. We shall discuss the results for special problems in the following sections.

It is interesting to compare Equation 40–15 with the classical expression given by Equation 3–4 of Chapter I. Recalling that the energy change associated with each transition is $h\nu_{mn}$, we see that the wave-mechanical expression is to be correlated with the classical expression for the special case of the harmonic oscillator by interpreting $\mu_{x_{mn}}$ as one-half the maximum value ex_0 of the electric moment of the classical oscillator.

40c. Selection Rules and Intensities for the Harmonic Oscillator.—The electric dipole moment for a particle with electric charge e carrying out harmonic oscillational motion along the x axis (a neutralizing charge $-e$ being at the origin) has the components ex along this axis and zero along the y and z axes. The only non-vanishing dipole moment integrals $\mu_{x_{mn}} = ex_{mn}$ have been shown in Section 11c to be those with $m = n + 1$ or $m = n - 1$. Hence the only transitions which this system can undergo with the emission or absorption of radiation are those from a given stationary state to the two adjacent states.[1] The selection rule for the harmonic oscillator is therefore $\Delta n = \pm 1$, and the only frequency of light emitted or absorbed is ν_0. The expression for $x_{n,n-1}$ in Equation 11–25a corresponds to the value

$$A_{n \to n-1} = \frac{64\pi^4 \nu_0^3 e^2}{3hc^3} \frac{n}{2\alpha}, \qquad (40\text{–}16)$$

with $\alpha = 4\pi^2 m \nu_0 / h$, for the coefficient of spontaneous emission, with similar expressions for the other coefficients. An application of this formula will be given in Section 40e.

Problem 40–1. Show that for large values of n Equation 40–16 reduces to the classical expression for the same energy.

Problem 40–2. Discuss the selection rules and intensities for the three dimensional harmonic oscillator with characteristic frequencies ν_x, ν_y, and ν_z.

Problem 40–3. Using first-order perturbation theory, find perturbed wave functions for the anharmonic oscillator with $V = 2\pi^2 m \nu_0^2 x^2 + ax^3$, and with them discuss selection rules and transition probabilities.

40d. Selection Rules and Intensities for Surface-harmonic Wave Functions.—In Section 18 we showed that the wave functions for a system of two particles interacting with one another in the way corresponding to the potential function

---

[1] This statement is true only to within the degree of approximation of our treatment. A more complete discussion shows that transitions may also occur as a result of interactions corresponding to quadrupole terms and still higher terms, as mentioned in Section 3, and as a result of magnetic interactions.

$V(r)$, in which r is the distance between the two particles, are of the form

$$R_{nl}(r)\Theta_{lm}(\vartheta)\Phi_m(\varphi),$$

in which the ϑ,φ functions $\Theta_{lm}(\vartheta)\Phi_m(\varphi)$ are surface harmonics, independent of $V(r)$. We can hence discuss selection rules and intensities in their dependence on l and m for all systems of this type at one time.

The components of electric dipole moment along the x, y, and z axes are

$$\mu_x = \mu(r) \sin \vartheta \cos \varphi,$$
$$\mu_y = \mu(r) \sin \vartheta \sin \varphi,$$

and

$$\mu_z = \mu(r) \cos \vartheta,$$

in which $\mu(r)$ is a function of r alone, being equal to er for two particles with charges $+e$ and $-e$. Each of the dipole moment integrals, such as

$$\mu_{x_{nlmn'l'm'}} = \iiint R_{nl}^*(r)\Theta_{lm}^*(\vartheta)\Phi_m^*(\varphi)\mu(r) \sin \vartheta \cos \varphi$$
$$R_{n'l'}(r)\Theta_{l'm'}(\vartheta)\Phi_n(\varphi)r^2 \sin \vartheta d\varphi d\vartheta dr,$$

can accordingly be written as the product of three factors, one involving the integral in r, one the integral in ϑ, and one the integral in φ:

$$\left.\begin{aligned}
\mu_{x_{nlmn'l'm'}} &= \mu_{nln'l'}f_{x_{lml'm'}}g_{x_{mm'}},\\
\mu_{y_{nlmn'l'm'}} &= \mu_{nln'l'}f_{y_{lml'm'}}g_{y_{mm'}},\\
\mu_{z_{nlmn'l'm'}} &= \mu_{nln'l'}f_{z_{lml'm'}}g_{z_{mm'}},
\end{aligned}\right\} \qquad (40\text{-}17)$$

in which

$$\mu_{nln'l'} = \int_0^\infty R_{nl}^*(r)\mu(r)R_{n'l'}(r)r^2dr, \qquad (40\text{-}18)$$

$$\left.\begin{aligned}
f_{x_{lml'm'}}\\
f_{y_{lml'm'}}\\
f_{z_{lml'm'}}
\end{aligned}\right\} = \int_0^\pi \Theta_{lm}(\vartheta)\begin{Bmatrix}\sin \vartheta\\ \sin \vartheta\\ \cos \vartheta\end{Bmatrix}\Theta_{l'm'}(\vartheta) \sin \vartheta d\vartheta, \qquad (40\text{-}19)$$

and

$$\left.\begin{aligned}
g_{x_{mm'}}\\
g_{y_{mm'}}\\
g_{z_{mm'}}
\end{aligned}\right\} = \int_0^{2\pi} \Phi_m^*(\varphi)\begin{Bmatrix}\cos \varphi\\ \sin \varphi\\ 1\end{Bmatrix}\Phi_{m'}(\varphi)d\varphi. \qquad (40\text{-}20)$$

(In Equations 40–19 and 40–20 the subscripts x, y, and z are respectively associated with the three factors in braces.)

Let us first discuss the light polarized along the z axis, corresponding to the dipole moment μ_z. From the orthogonality and normalization integrals for $\Phi(\varphi)$ we see that

$$g_{z_{mm'}} = 0 \text{ for } m' \neq m$$

and

$$g_{z_{mm}} = 1.$$

In discussing $f_{z_{lml'm'}}$ we consequently need consider only the integrals with $m' = m$. It is found with the use of the recursion formula (Prob. 19–2)

$$\cos \vartheta P_l^{|m|}(\cos \vartheta) = \frac{(l + |m|)}{(2l + 1)} P_{l-1}^{|m|}(\cos \vartheta) + \frac{(l - |m| + 1)}{(2l + 1)} P_{l+1}^{|m|}(\cos \vartheta) \quad (40\text{--}21)$$

that $f_{z_{lml'm'}}$ vanishes except when l' is equal to $l + 1$ or $l - 1$.

A similar treatment of the integrals for x and y shows that light polarized along these axes is emitted only when m changes by $+1$ or -1, and l changes by $+1$ or -1.

We have thus obtained the selection rules $\Delta m = 0, +1,$ or -1 and $\Delta l = +1$ or -1. The selection rule for l is discussed in the following sections. That for m can be verified experimentally only by removing the degeneracy, as by the application of a magnetic field; it is found, in agreement with the theory, that in the Zeeman effect the light corresponding to $\Delta m = 0$ is polarized along the z axis (the axis of the magnetic field), and that corresponding to $\Delta m = \pm 1$ is polarized in the xy plane.

The values of the products of the factors f and g are

$$
\left.
\begin{aligned}
(fg)_{x_{l,|m|,l-1,|m|-1}} &= i(fg)_y = \frac{1}{2}\left\{\frac{(l + |m|)(l + |m| - 1)}{(2l + 1)(2l - 1)}\right\}^{1/2}, \\
(fg)_{x_{l,|m|-1,l-1,|m|}} &= i(fg)_y = \frac{1}{2}\left\{\frac{(l - |m|)(l - |m| + 1)}{(2l + 1)(2l - 1)}\right\}^{1/2}, \\
(fg)_{z_{l,|m|,l-1,|m|}} &= \left\{\frac{(l + |m|)(l - |m|)}{(2l + 1)(2l - 1)}\right\}^{1/2},
\end{aligned}
\right\} \quad (40\text{--}22)
$$

with similar expressions for the transitions l to $l + 1$, etc.

Problem 40–4. Using Equation 40–21, obtain selection rules and intensities for μ_z.

Problem 40–5. Similarly derive the other formulas of 40–22.

Problem 40–6. Calculate the total probability of transition from one level with given value of l to another, by summing over m. By separate summation for μ_x, μ_y, and μ_z show that the intensity of light polarized along these axes is the same.

40e. Selection Rules and Intensities for the Diatomic Molecule. The Franck-Condon Principle.—A very simple treatment of the emission and absorption of radiation for the diatomic molecule can be given, based on the approximate wave functions of Section 35c. For the complex system of two nuclei and several electrons the electric dipole moment $\mu(r)$ can be expanded as a series in $r - r_0$,

$$\mu(r) = \mu_0 + \epsilon(r - r_0) + \cdots, \tag{40-23}$$

in which ϵ is a constant. The permanent dipole moment μ_0 is the quantity which enters in the theory of the dielectric constant of dipole molecules; its value is known from dielectric constant measurements for many substances.

Introducing this expansion in Equation 40–18, we find as a first approximation that n may change by 0 or by ± 1. In the former case the emission or absorption of radiation is due to the constant term μ_0, and in the latter case to the term $\epsilon(r - r_0)$, the integrals being then similar to the harmonic oscillator integrals. The values of $\mu_{nn'}$ are

$$\mu_{nn} = \mu_0 \tag{40-24a}$$

and

$$\mu_{n,n-1} = \epsilon\sqrt{\frac{n}{2\alpha}}, \tag{40-24b}$$

in which $\alpha = 4\pi^2\mu\nu_0/h$ (μ being here the reduced mass for the molecule). The selection rules and intensity factors for l and m are as given in the preceding section.

It is found experimentally that dipole molecules such as the hydrogen halides absorb and emit pure rotation and oscillation-rotation bands in accordance with these equations. In all these bands the selection rule $\Delta l = \pm 1$ is obeyed, and Zeeman-effect measurements have shown similar agreement with the selection rule for m. The intensities of lines in the pure rotation bands show rough quantitative agreement with Equation 40–24a, using the dielectric constant value of μ_0, although because of experimental difficulties in the far infrared the data are as yet not very reliable. Measurements of absorption intensities for $\Delta n = 1$ have been used to calculate ϵ. As seen from the following table, ϵ is of the order of magnitude of μ_0/r_0, so that these molecules may be considered roughly as equivalent to two particles of charges $+\epsilon$ and $-\epsilon$.

TABLE 40-1

	μ_0 (dielectric constant)	r_0	μ_0/r_0	ϵ^1
HCl	1.034×10^{-18} e.s.u.	1.28 Å	$0.169\ e$	$0.086\ e$
HBr	0.788	1.42	$.116$	$.075$
HI	0.382	1.62	$.049$	$.033$

[1] E. BARTHOLOMÉ, *Z. phys. Chem.* **B 23**, 131 (1933).

It is also observed that oscillation-rotation bands with $\Delta n = 2, 3$, etc. occur; this is to be correlated with the deviation of the potential function $V(r)$ from a simple quadratic function.

In the foregoing discussion we have assumed the electronic state of the molecule to be unchanged by the transitions. The selection rule for n and the intensities are different in case there is a change in the electronic state, being then determined, according to the Franck-Condon principle,[1] mainly by the nature of the electronic potential functions for the two electronic states. As we have seen in Section 34, there is little interaction between the electronic motion and the nuclear motion in a molecule, and during an electronic transition the internuclear distance and nuclear velocities will not change very much. Let us consider the two electronic states A and B, represented by the potential curves of Figure 40-1, in which the oscillational levels are also shown. If the molecule is in the lowest oscillational level $n' = 0$ of the upper state, the probability distribution function for r is large only for r close to r_{e_A}. We would then expect an electronic transition to state B to leave the molecule at about the point P_1 on the potential curve, the nuclei having only small kinetic energy; these conditions correspond to the levels $n'' = 7$ or 8 for the lower state.

This simple argument is justified by wave-mechanical considerations. Let us consider that the wave functions for the upper electronic state may be written as $\psi_{\sigma'}\psi_{n'}$, in which $\psi_{n'}$ represents the nuclear oscillational part of the wave function, described by the quantum number n', and $\psi_{\sigma'}$ the rest of the wave function (electronic and nuclear rotational), the symbol σ' representing all other quantum numbers. Similarly, we write $\psi_{\tau''}\psi_{n''}$ for the wave functions for the lower electronic state.

[1] J. FRANCK, *Trans. Faraday Soc.*, **21**, 536 (1926); E. U. CONDON, *Phys. Rev.* **28**, 1182 (1926); **32**, 858 (1928).

The electric dipole moment integrals $\mu_{x_{\sigma'n'\sigma''n''}}$, $\mu_{y_{\sigma'n'\sigma''n''}}$, and $\mu_{z_{\sigma'n'\sigma''n''}}$ are of the form

$$\mu_{x_{\sigma'n'\sigma''n''}} = \int \psi_{\sigma'}^* \psi_{n'}^* \mu_x \psi_{\sigma''} \psi_{n''} d\tau. \qquad (40\text{--}25)$$

We assume that in this case, when there is a change in the

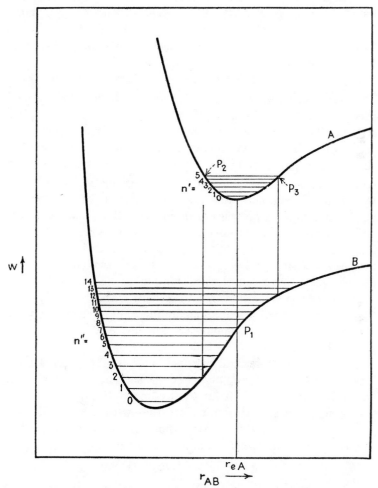

Fɪɢ. 40–1.—Energy curves for two electronic states of a molecule, to illustrate the Franck-Condon principle.

electronic state, the dipole moment function μ changes only slowly with change in the internuclear separation r, being determined essentially by the electronic coordinates. Neglecting the dependence of μ on r, we can then integrate over all coordinates except r, obtaining

$$\mu_{x_{\sigma'n'\sigma''n''}} = \mu_{x_{\sigma'\sigma''}} \int \psi_{n'}^* \psi_{n''} r^2 dr. \qquad (40\text{-}26)$$

The integral in r, determining the relative intensities of the various $n'n''$ bands, is seen to have the form of an orthogonality integral in r. Hence if the two potential functions V_A and V_B were identical except for an additive constant the integral would vanish except for $n' = n''$, the selection rule for n then being $\Delta n = 0$. In the case represented by Figure 40–1, the wave function $\psi_{n'}$ with $n' = 0$ is large only in the neighborhood of $r = r_{e_A}$. The wave functions $\psi_{n''}$ with $n'' = 7$ or 8 have large values in this region, so that the bands $n' = 0 \rightarrow n'' = 7$ or 8 will be strong. The intensity of the bands for smaller or larger values of n'' will fall off. For smaller values of n'' the functions $\psi_{n''}$ show the rapid exponential decrease in the region near r_{e_A} (corresponding to the fact that the classical motion of the nuclei would not extend into this region); whereas for larger values of n'' the functions $\psi_{n''}$ show a rapid oscillation between positive and negative values, causing the integral with the positive function $\psi_{n'}$ with $n' = 0$ to be small (the oscillation of $\psi_{n''}$ between positive and negative values corresponding to large nuclear velocities in the classical motion).

Similarly the transitions from the level with $n' = 5$, the wave function for which has its maximum values near the points P_2 and P_3, will occur mainly to the levels $n'' = 2$ or 3 and $n'' = 11$ or 12.[1]

40f. Selection Rules and Intensities for the Hydrogen Atom.— The selection rule for l, discussed in Section 40d, allows only transitions with $\Delta l = \pm 1$ for the hydrogen atom. The lines of the Lyman series, with lower state that with $n = 1$ and $l = 0$, are in consequence due to transitions from upper states with $l = 1$. The radial electric dipole moment integral

$$\mu_{n'n''} = \int R_{n'l'}^*(r) r R_{n''l''}(r) r^2 dr$$

has been evaluated by Pauli[2] for several special cases. For

---

[1] For a more complete discussion of this subject the reader is referred to the papers of Condon and to the discussions in Condon and Morse, "Quantum Mechanics," Chap. V, and RUARK and UREY, "Atoms, Molecules and Quanta," Chap. XII.

[2] Communicated in Schrödinger's third 1926 paper.

$n'' = 1$, $l'' = 0$ its value corresponds to the total intensity, aside from a constant factor, of

$$I_{n'1} = \frac{(n' - 1)^{2n'-1}}{n'(n' + 1)^{2n+1}}.$$

This has a non-vanishing value for all values of n' greater than 1. Hence there is no selection rule for n for the Lyman series, all transitions being allowed. It is similarly found that there is no selection rule for n for spectral series in general.

For the Balmer series, with lower state that with $n = 2$ and $l = 0$ or 1, the selection rule for l permits the transitions $0 \rightarrow 1$, $1 \rightarrow 0$, and $2 \rightarrow 1$. The total intensity corresponding to these transitions from the level $n = n'$ to $n = 2$ is, except for a constant factor,

$$I_{n'2} = \frac{(n' - 2)^{2n'-3}}{n'(n' + 2)^{2n'+3}}(3n'^2 - 4)(5n'^2 - 4).$$

The operation of the selection rule for l for hydrogen and hydrogenlike ions can be seen by the study of the fine structure of the lines. The phenomena are complicated, however, by the influence of electron spin.[1] In alkali atoms the levels with given n and varying l are widely separated, and the selection rule for l plays an important part in determining the nature of their spectra. Theoretical calculations have also been made of the intensities of lines in these spectra with the use of wave functions such as those described in Chapter IX, leading to results in approximate agreement with experiment.

40g. Even and Odd Electronic States and Their Selection Rules.—The wave functions for an atom can all be classified as either *even* or *odd*. An even wave function of N electrons is one such that $\psi(x_1, y_1, z_1, x_2, \cdots, z_N)$ is equal to $\psi(-x_1, -y_1, -z_1, -x_2, \cdots, -z_N)$; that is, the wave function is unchanged on changing the signs of all of the positional coordinates of the electrons. An odd wave function is one such that $\psi(x_1, y_1, z_1, x_2, \cdots, z_N)$ is equal to $-\psi(-x_1, -y_1, -z_1, -x_2, \cdots, -z_N)$.

Now we can show that the only transitions accompanied by the emission or absorption of dipole radiation which can occur are those between an even and an odd state (an even state being one represented by an even wave function, etc.). The electric

[1] See PAULING and GOUDSMIT, "The Structure of Line Spectra," Sec. 16.

moment component functions $\sum_{i=1}^{N} ex_i$, etc. change sign in case that the electronic coordinates are replaced by their negatives. Consequently an electric-moment integral such as $\int \psi_{n'}^* \sum_i ex_i \psi_{n''} d\tau$ will vanish in case that both $\psi_{n'}$ and $\psi_{n''}$ are either even or odd, but it is not required to vanish in case that one is even and the other odd. We thus have derived the very important selection rule that *transitions with the emission or absorption of dipole radiation are allowed only between even and odd states.* Because of the practical importance of this selection rule, it is customary to distinguish between even and odd states in the term symbol, by adding a superscript ° for odd states. Thus various even states are written as ^1S, ^3P, ^2D, etc., and odd states as $^1S°$, $^3P°$, $^2D°$, etc.

In case that the electronic configuration underlying a state is known, the state can be recognized as even or odd. The one-electron wave functions are even for $l = 0, 2, 4$, etc. (s, d, g, etc., orbitals) and odd for $l = 1, 3, 5$, etc. (p, f, h, etc., orbitals). Hence the configuration leads to odd states if it contains an odd number of electrons in orbitals with l odd, and otherwise to even states. For example, the configuration $1s^2 2s^2 2p^2$ leads to the even states ^1S, ^1D, and ^3P, and the configuration $1s^2 2p3d$ to the odd states $^1P°$, $^1D°$, $^1F°$, $^3P°$, $^3D°$, and $^3F°$.

Even and odd states also occur for molecules, and the selection rule is also valid here. A further discussion of this point will be given in Section 48.

Problem 40–7. Show that the selection rules forbid a hydrogen atom in a rectangular box to radiate its translational kinetic energy. Extend the proof to any atom in any kind of box.

41. THE RESONANCE PHENOMENON

The concept of *resonance* played an important part in the discussion of the behavior of certain systems by the methods of classical mechanics. Very shortly after the discovery of the new quantum mechanics it was noticed by Heisenberg that a quantum-mechanical treatment analogous to the classical treatment of resonating system can be applied to many problems, and that the results of the quantum-mechanical discussion in these cases can be given a simple interpretation as corresponding to a *quantum-mechanical resonance phenomenon.* It is not required that this interpretation be made; it has been found, however, that it is a very valuable aid to the student in the development of a reliable and productive intuitive understanding of the equations of quantum mechanics and the results of their application. In the following sections we shall discuss first classical resonance and then resonance in quantum mechanics.

41a. Resonance in Classical Mechanics.—A striking phenomenon is shown by a classical mechanical system consisting of two parts between which there is operative a small interaction, the two parts being capable of executing harmonic oscillations with the same or nearly the same frequency. It is observed that the total oscillational energy fluctuates back and forth

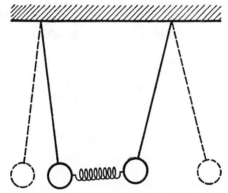

Fig. 41–1.—A system of coupled pendulums, illustrating the phenomenon of resonance.

between the two parts, one of which at a given time may be oscillating with large amplitude, and at a later time with small amplitude, while the second part has changed in the opposite direction. It is customary to say that the two parts of the system are resonating. A familiar example of such a system is composed of two similar tuning forks attached to the same base. After one fork is struck, it gradually ceases to oscillate, while at the same time the other begins its oscillation. Another example is two similar pendulums connected by a weak spring, or attached to a common support in such a way that interaction of the two occurs by way of the support (Fig. 41–1). It is observed that if only one pendulum is set to oscillating, it will gradually die down and stop, while the other begins to oscillate, ultimately reaching the amplitude of oscillation initially given the first (neglecting the frictional dissipation of energy); and that this process of transfer of energy from one pendulum to the other is repeated over and over.

It is illuminating to consider this system in greater detail. Let x_1 and x_2 be the coordinates for two oscillating particles each of mass m (such as the bobs of two pendulums restricted to small amplitudes, in order that their motion be harmonic), and let ν_0 be their oscillational frequency. We assume for the total potential energy of the system the expression

$$V(x_1, x_2) = 2\pi^2 m\nu_0^2 x_1^2 + 2\pi^2 m\nu_0^2 x_2^2 + 4\pi^2 m\lambda x_1 x_2, \quad (41\text{--}1)$$

in which $4\pi^2 m\lambda x_1 x_2$ represents the interaction of the two oscillators. This simple form corresponds to a Hooke's-law type of interaction. The solution of the equations of motion is easily accomplished by introducing the new variables[1]

$$\left.\begin{aligned} \xi &= \frac{1}{\sqrt{2}}(x_1 + x_2), \\ \eta &= \frac{1}{\sqrt{2}}(x_1 - x_2). \end{aligned}\right\} \quad (41\text{--}2)$$

In terms of these, the potential energy becomes

$$V(\xi, \eta) = 2\pi^2 m(\nu_0^2 + \lambda)\xi^2 + 2\pi^2 m(\nu_0^2 - \lambda)\eta^2,$$

while the kinetic energy has the form

$$T = \tfrac{1}{2}m\dot{x}_1^2 + \tfrac{1}{2}m\dot{x}_2^2 = \tfrac{1}{2}m\dot{\xi}^2 + \tfrac{1}{2}m\dot{\eta}^2.$$

These expressions correspond to pure harmonic oscillation of the two variables ξ and η (Sec. 1a), each oscillating with constant amplitude, ξ with the frequency $\sqrt{\nu_0^2 + \lambda}$ and η with the frequency $\sqrt{\nu_0^2 - \lambda}$, according to the equations

$$\left.\begin{aligned} \xi &= \xi_0 \cos\left(2\pi\sqrt{\nu_0^2 + \lambda}\, t + \delta_\xi\right), \\ \eta &= \eta_0 \cos\left(2\pi\sqrt{\nu_0^2 - \lambda}\, t + \delta_\eta\right). \end{aligned}\right\} \quad (41\text{--}3)$$

From these equations we obtain the equations

$$\left.\begin{aligned} x_1 &= \frac{\xi_0}{\sqrt{2}} \cos\left(2\pi\sqrt{\nu_0^2 + \lambda}\, t\right) + \frac{\eta_0}{\sqrt{2}} \cos\left(2\pi\sqrt{\nu_0^2 - \lambda}\, t\right), \\ x_2 &= \frac{\xi_0}{\sqrt{2}} \cos\left(2\pi\sqrt{\nu_0^2 + \lambda}\, t\right) - \frac{\eta_0}{\sqrt{2}} \cos\left(2\pi\sqrt{\nu_0^2 - \lambda}\, t\right), \end{aligned}\right\} \quad (41\text{--}4)$$

[1] These are the normal coordinates of the system, discussed in Section 37

for x_1 and x_2, in which we have put the phase constants δ_ξ and δ_η equal to zero, as this does not involve any loss in generality. It is seen that for λ/ν_0^2 small the two cosine functions differ only slightly from one another and both x_1 and x_2 carry out approximate harmonic oscillation with the approximate frequency ν_0, but with amplitudes which change slowly with the time. Thus at $t \cong 0$ the cosine terms are in phase, so that x_1 oscillates with the amplitude $(\xi_0 + \eta_0)/\sqrt{2}$ and x_2 with the smaller amplitude $(\xi_0 - \eta_0)/\sqrt{2}$. At the later time $t \cong t_1$ such that

$$\sqrt{\nu_0^2 + \lambda}\, t_1 = \sqrt{\nu_0^2 - \lambda}\, t_1 + \tfrac{1}{2}$$

the cosine terms are just out of phase, x_1 then oscillating with the amplitude $(\xi_0 - \eta_0)/\sqrt{2}$ and x_2 with the amplitude $(\xi_0 + \eta_0)/\sqrt{2}$. Thus we see that the period τ of the resonance, the time required for x_1 to change from its maximum to its minimum amplitude and then back to the maximum, is given by the equation

$$\sqrt{\nu_0^2 + \lambda}\, \tau = \sqrt{\nu_0^2 - \lambda}\, \tau + 1$$

or

$$\tau = \frac{1}{\sqrt{\nu_0^2 + \lambda} - \sqrt{\nu_0^2 - \lambda}} \cong \frac{\nu_0}{\lambda}. \tag{41-5}$$

It is also seen that the magnitude of the resonance depends on the constants of integration ξ_0 and η_0, the amplitudes of motion of x_1 and x_2 varying between the limits $\sqrt{2}\xi_0$ and 0 in case that $\eta_0 = \xi_0$, and retaining the constant value $\xi_0/\sqrt{2}$ (no resonance!) in case that $\eta_0 = 0$.

The behavior of the variables x_1 and x_2 may perhaps be followed more clearly by expanding the radicals $\sqrt{\nu_0^2 + \lambda}$ and $\sqrt{\nu_0^2 - \lambda}$ in powers of λ/ν_0^2 and neglecting terms beyond the first power. After simple transformations, the expressions obtained are

$$x_1 = \frac{(\xi_0 + \eta_0)}{\sqrt{2}} \cos 2\pi \frac{\lambda}{\nu_0} t \cos 2\pi \nu_0 t - \frac{(\xi_0 - \eta_0)}{\sqrt{2}} \sin 2\pi \frac{\lambda}{\nu_0} t \sin 2\pi \nu_0 t$$

and

$$x_2 = \frac{(\xi_0 - \eta_0)}{\sqrt{2}} \cos 2\pi \frac{\lambda}{\nu_0} t \cos 2\pi \nu_0 t - \frac{(\xi_0 + \eta_0)}{\sqrt{2}} \sin 2\pi \frac{\lambda}{\nu_0} t \sin 2\pi \nu_0 t.$$

It is clear from this treatment that we speak of resonance only because it is convenient for us to retain the coordinates x_1 and

x_2 in the description of the system; that is, to speak of the motion of the pendulums individually rather than of the system as a whole. We can conceive of an arrangement of levers whereby an indicator in an adjacent room would register values of ξ, and another values of η. An observer in this room would say that the system was composed of two independent harmonic oscillators with different frequencies and constant amplitudes, and would not mention resonance at all.

Despite the fact that we are not required to introduce it, the concept of resonance in classical mechanical systems has been found to be very useful in the description of the motion of systems which are for some reason or other conveniently described as containing interacting harmonic oscillators. It is found that a similar state of affairs exists in quantum mechanics. Quantum-mechanical systems which are conveniently considered to show resonance occur much more often, however, than resonating classical systems, and the resonance phenomenon has come to play an especially important part in the applications of quantum mechanics to chemistry.

41b. Resonance in Quantum Mechanics.—In order to illustrate the resonance phenomenon in quantum mechanics, let us continue to discuss the system of interacting harmonic oscillators.[1] Using the potential function of Equation 41–1, the wave equation can be at once separated in the coordinates ξ and η and solved in terms of the Hermite functions. The energy levels are given by the expression

$$W_{n_\xi n_\eta} = (n_\xi + \tfrac{1}{2})h\sqrt{\nu_0^2 + \lambda} + (n_\eta + \tfrac{1}{2})h\sqrt{\nu_0^2 - \lambda}, \quad (41\text{–}6)$$

which for λ small reduces to the approximate expression

$$W_{n_\xi n_\eta} \cong (n + 1)h\nu_0 + (n_\xi - n_\eta)\frac{h\lambda}{2\nu_0} - \frac{(n + 1)h\lambda^2}{8\nu_0^3} + \cdots,$$

$$(41\text{–}7)$$

in which $n = n_\xi + n_\eta$. The energy levels are shown in Figure 41–2; for a given value of n there are $n + 1$ approximately equally spaced levels.

This treatment, like the classical treatment using the coordinates ξ and η, makes no direct reference to resonance. Let us

---

[1] This example was used by Heisenberg in his first papers on the resonance phenomenon, *Z. f. Phys.* **38**, 411 (1926); **41**, 239 (1927).

now apply a treatment in which the concept of resonance enters, retaining the coordinates x_1 and x_2 because of their familiar physical interpretation and applying the methods of approximate solution of the wave equation given in Chapters VI and VII; indeed, if the term in λ were of more complicated form, it would be necessary to resort to some approximate treatment. This term is conveniently considered as the perturbation function in applying the first-order perturbation theory. The unperturbed

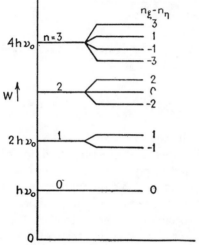

FIG. 41-2.—Energy levels for coupled harmonic oscillators; left, with $\lambda = 0$; right, with $\lambda = \nu_0^2/5$.

wave equation has as solutions products of Hermite functions in x_1 and x_2,

$$\psi^0_{n_1 n_2}(x_1, x_2) = \psi^0_{n_1}(x_1)\psi^0_{n_2}(x_2) =$$
$$N_{n_1}H_{n_1}(\sqrt{\alpha}x_1)e^{-\frac{\alpha x_1^2}{2}}N_{n_2}H_{n_2}(\sqrt{\alpha}x_2)e^{-\frac{\alpha x_2^2}{2}}, \quad (41\text{-}8)$$

corresponding to the energy values

$$W^0_{n_1 n_2} = (n_1 + n_2 + 1)h\nu_0 = (n + 1)h\nu_0,$$
$$\text{with } n = n_1 + n_2, \quad (41\text{-}9)$$

the nth level being $(n + 1)$-fold degenerate.

The perturbation energy for the non-degenerate level $n = 0$ is zero. For the level $n = 1$ the secular equation is found to be (Sec. 24)

$$\begin{vmatrix} -W' & \dfrac{h\lambda}{2\nu_0} \\ \dfrac{h\lambda}{2\nu_0} & -W' \end{vmatrix} = 0,$$

giving $W' = \pm h\lambda/2\nu_0$. A similar treatment of the succeeding degenerate levels shows that the first-order perturbation theory leads to values for the energy expressed by the first two terms of Equation 41–7.

The correct zeroth-order wave functions for the two levels with $n = 1$ are found to be

$$\psi_S = \frac{1}{\sqrt{2}} \{ \psi_1^0(x_1)\psi_0^0(x_2) + \psi_0^0(x_1)\psi_1^0(x_2) \}$$

and

$$\psi_A = \frac{1}{\sqrt{2}} \{ \psi_1^0(x_1)\psi_0^0(x_2) - \psi_0^0(x_1)\psi_1^0(x_2) \},$$

ψ_S corresponding to the lower of the two levels and ψ_A to the upper. The subscripts S and A are used to indicate that the functions are respectively symmetric and antisymmetric in the coordinates x_1 and x_2. We see that we are not justified in describing the system in either one of these stationary states as consisting of the first oscillator in the state $n_1 = 1$ and the second in the state $n_2 = 0$, or the reverse. Instead, the wave functions $n_1 = 1$, $n_2 = 0$ and $n_1 = 0$, $n_2 = 1$ contribute equally to each of the stationary states. It will be shown in Section 41c that if the perturbation is small we are justified in saying that there is resonance between these two states of motion analogous to classical resonance, one oscillator at a given time oscillating with large amplitude, corresponding to $n_1 = 1$, and at a later time with small amplitude, corresponding to $n_1 = 0$. The frequency with which the oscillators interchange their oscillational states, that is, the frequency of the resonance, is found to be λ/ν_0, which is just equal to the separation of the two energy levels divided by h. This is also the frequency of the classical resonance (Eq. 41–5).

In discussing the stationary states of the system of two interacting harmonic oscillators we have seen above that it is convenient to make use of certain wave functions $\psi_{n_1}(x_1)$, etc. which are not correct wave functions for the system, the latter being given by or approximated by linear combinations of the initially chosen functions, as found by perturbation or variation methods; and various points of analogy between this treatment and the classical treatment of the resonating system have been indicated (see also the following section). In discussing more complicated systems it is often convenient to make use of similar methods of approximate solution of the wave equation, involving the formation of linear combinations of certain initially chosen functions. The custom has arisen of describing this formation of linear combinations in certain cases as corresponding to resonance in the system. In a given stationary state the system is said to resonate among the states or structures corresponding to those initially chosen wave functions which contribute to the wave function for this stationary state, and the difference between the energy of the stationary state and the energy corresponding to the initially chosen wave functions is called *resonance energy*.[1] It is evident that any perturbation treatment for a degenerate level in which the initial wave functions are not the correct zeroth-order wave functions might be described as involving the resonance phenomenon. Whether this description would be applied or not would depend on how important the initial wave functions seem to the investigator, or how convenient this description is in his discussion.[2]

The resonance phenomenon, restricted in classical mechanics to interacting harmonic oscillators, is of much greater importance in quantum mechanics, this being, indeed, one of the most striking differences between the old and the new mechanics. It arises, for example, whenever the system under discussion contains two or more identical particles, such as two electrons or two protons; and it is also convenient to make use of the terminology in describing the approximate treatment given the structure of polyatomic molecules. The significance of the phenomenon for many-electron atoms has been seen from the discussion of the structure of the helium atom given in Chapter VIII; it was there pointed out (Sec. 29a) that the splitting of levels due to the K

---

[1] There is no close classical analogue of resonance energy.

[2] The same arbitrariness enters in the use of the word resonance in describing classical systems, inasmuch as if the interaction of the classical oscillators is increased the motion ultimately ceases to be even approximately represented by the description of the first paragraph of Section 41a.

integrals was given no satisfactory explanation until the development of the concept of quantum-mechanical resonance. The procedure which we have followed of delaying the discussion of resonance until after the complete treatment of the helium atom emphasizes the fact that the resonance phenomenon does not involve any new postulate or addition to the equations of wave mechanics but rather only a convenient method of classifying and correlating the results of wave mechanics and a basis for the development of a sound intuitive conception of the theory.

41c. A Further Discussion of Resonance.—It is illuminating to apply the perturbation method of variation of constants in order to discuss the behavior of a resonating system. Let us consider a system for which we have two wave functions, say Ψ_A^0 and Ψ_B^0, corresponding to an energy level of the unperturbed system with two-fold degeneracy. These might, for example, correspond to the sets of quantum numbers $n_1 = 1$, $n_2 = 0$ and $n_1 = 0$, $n_2 = 1$ for the system of two coupled harmonic oscillators treated in the previous section. If the perturbation were small, we could carry out an experiment at the time $t = 0$ to determine whether the system is in state A or in state B; for example, we could determine the energy of the first oscillator with sufficient accuracy to answer this question. Let us assume that at the time $t = 0$ the system is found to be in the state A. We now ask the following question: On carrying out the investigating experiment at a later time t, what is the probability that we would find that the system is in state A, and what is the probability that we would find it in state B? In answering this question we shall see that the physical interpretation of quantum-mechanical resonance is closely similar to that of classical resonance.

If the perturbation is small, with all the integrals $H'_{mn}(m \neq n)$ small compared with $W_n^0 - W_m^0$ except H'_{AB} and H'_{BA} (for which $W_A^0 = W_B^0$), we may assume as an approximation that the quantities $a_m(t)$ remain equal to zero except for a_A and a_B. From Equation 39–6 we see that these two are given by the equations

$$\left.\begin{aligned}
\dot{a}_A &= -\frac{2\pi i}{h}(H'_{AA}a_A + H'_{AB}a_B), \\
\dot{a}_B &= -\frac{2\pi i}{h}(H'_{AB}a_A + H'_{AA}a_B),
\end{aligned}\right\} \tag{41–10}$$

in which we have taken H'_{BB} equal to H'_{AA} and H'_{BA} equal to H'_{AB}

(the system being assumed to consist of two similar parts). The equations are easily solved by first forming their sum and difference. The solution which makes $a_A = 1$ and $a_B = 0$ at $t = 0$ is

$$a_A = e^{-\frac{2\pi i}{h}H'_{AA}t} \cos\left(\frac{2\pi H'_{AB}}{h}t\right), \atop a_B = -ie^{-\frac{2\pi i}{h}H'_{AA}t} \sin\left(\frac{2\pi H'_{AB}}{h}t\right). \right\} \tag{41-11}$$

The probabilities $a_A^* a_A$ and $a_B^* a_B$ of finding the system in state A and state B, respectively, at time t are hence

$$a_A^* a_A = \cos^2\left(\frac{2\pi H'_{AB}}{h}t\right), \atop a_B^* a_B = \sin^2\left(\frac{2\pi H'_{AB}}{h}t\right). \right\} \tag{41-12}$$

We see that these probabilities vary harmonically between the values 0 and 1. The period of a cycle (from $a_A^* a_A = 1$ to 0 and back to 1 again) is seen to be $h/2H'_{AB}$, and the frequency $2H'_{AB}/h$, this being, as stated in Section 41b, just $1/h$ times the separation of the levels due to the perturbation.

Let us now discuss in greater detail the sequence of conceptual experiments and calculations which leads us to the foregoing interpretation of our equations. Let us assume that we have a system composed of two coupled harmonic oscillators with coordinates x_1 and x_2, respectively, such that we can at will (by throwing a switch, say) disengage the coupling, thus causing the two oscillators to be completely independent. Let us now assume that for a period of time previous to $t = 0$ the oscillators are independent. During this period we carry out a set of two experiments consisting in separate measurements of the energy of the oscillators and in this way determine the stationary state of each oscillator. Suppose that by one such set of experiments we have found that the first oscillator is in the state $n_1 = 1$ and the second in the state $n_2 = 0$. The complete system is then in the physical situation which we have called state A in the above paragraphs, and so long as the system is left to itself it will remain in this state.

Now let us switch in the coupling at the time $t = 0$, and then switch it out again at the time $t = t'$. We now, at times later than t', investigate the system to find what the values of the quantum numbers n_1 and n_2 are. The result of this investigation will be the same, in a given case, no matter at what time later than t' the set of experiments is carried out, inasmuch as the two oscillators will remain in the definite stationary states in which they were left at time t' so long as the system is left unperturbed.

This sequence of experiments can be repeated over and over, each time starting with the system in the state $n_1 = 1$, $n_2 = 0$ and allowing the coupling to be operative for the length of time t'. In this way we can find experimentally the probability of finding the system in the various states $n_1 = 1$, $n_2 = 0$; $n_1 = 0$, $n_2 = 1$; $n_1 = 0$, $n_2 = 0$; etc.; after the perturbation has been operative for the length of time t'.

The same probabilities are given directly by our application of the method of variation of constants. The probability of transition to states of considerably different energy as the result of a small perturbation acting for a short time is very small, and we have neglected these transitions. Our calculation shows that the probability of finding the system in the state B depends on the value of t' in the way given by Equation 41–12, varying harmonically between the limits 0 and 1.

Now in case that we allow the coupling to be operative continuously, the complete system can exist in various stationary states, which we can distinguish from one another by the measurement of the energy of the system. Two of these stationary states have energy values very close to the energy for the states $n_1 = 1$, $n_2 = 0$ and $n_1 = 0$, $n_2 = 1$ of the system with the coupling removed. It is consequently natural for us to draw on the foregoing argument and to describe the coupled system in these stationary states as resonating between states A and B, with the resonance frequency $2H_{AB}/h$.

Even when it is not possible to remove the coupling interaction, it may be convenient to use this description. Thus in our discussion of the helium atom we found certain stationary states to be approximately represented by wave functions formed by linear combination of the wave functions $1s(1)\,2s(2)$ and $2s(1)\,1s(2)$. These we identify with states A and B above, saying that each electron resonates between a $1s$ and a $2s$ orbit, the two electrons changing places with the frequency $1/h$ times the separation of the energy levels $1s2s$ ^1S and $1s2s$ ^3S. It is obvious that we cannot verify this experimentally, for three reasons: we cannot remove the coupling, we cannot distinguish electron 1 from electron 2, and the interaction is so large that our calculation (based on neglect of all other unperturbed states) is very far from accurate. These limitations to the physical verification of resonance must be borne in mind; but they need not prevent us from making use of the nomenclature whenever it is convenient (as it often is in the discussion of molecular structure given in the following chapter).

CHAPTER XII

THE STRUCTURE OF SIMPLE MOLECULES

Of the various applications of wave mechanics to specific problems which have been made in the decade since its origin, probably the most satisfying to the chemist are the quantitatively successful calculations regarding the structure of very simple molecules. These calculations show that we now have at hand a theory which can be confidently applied to problems of molecular structure. They provide us with a sound conception of the interactions causing atoms to be held together in a stable molecule, enabling us to develop a reliable intuitive picture of the chemical bond. To a considerable extent the contribution of wave mechanics to our understanding of the nature of the chemical bond has consisted in the independent justification of postulates previously developed from chemical arguments, and in the removal of their indefinite character. In addition, wave-mechanical arguments have led to the development of many essentially new ideas regarding the chemical bond, such as the three-electron bond, the increase in stability of molecules by resonance among several electronic structures, and the hybridization of one-electron orbitals in bond formation. Some of these topics will be discussed in this chapter and the following one.

In Sections 42 and 43 we shall describe the accurate and reliable wave-mechanical treatments which have been given the hydrogen molecule-ion and hydrogen molecule. These treatments are necessarily rather complicated. In order to throw further light on the interactions involved in the formation of these molecules, we shall preface the accurate treatments by a discussion of various less exact treatments. The helium molecule-ion, He_2^+, will be treated in Section 44, followed in Section 45 by a general discussion of the properties of the one-electron bond, the electron-pair bond, and the three-electron bond.

42. THE HYDROGEN MOLECULE-ION

The simplest of all molecules is the hydrogen molecule-ion, H_2^+, composed of two hydrogen nuclei and one electron. This molecule was one of the stumbling blocks for the old quantum theory, for, like the helium atom, it permitted the treatment to be carried ¹hrough (by Pauli[1] and Niessen[2]) to give results in disagreement ₋vith experiment. It was accordingly very satisfying that within a year after the development of wave mechanics a discussion of the normal state of the hydrogen molecule-ion in complete agreement with experiment was carried out by Burrau by numerical integration of the wave equation. This treatment, together with somewhat more refined treatments due to Hylleraas

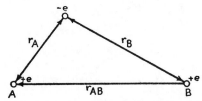

Fig. 42-1.—Coordinates used for the hydrogen molecule-ion.

and Jaffé, is described in Section 42c. Somewhat simpler and less accurate methods are described in Sections 42a and 42b, for the sake of the ease with which they can be interpreted.

42a. A Very Simple Discussion.[3]—Following the discussion of Section 34, the first step in the treatment of the complete wave equation is the solution of the wave equation for the electron alone in the field of two stationary nuclei. Using the symbols of Figure 42-1, the electronic wave equation is

$$\nabla^2\psi + \frac{8\pi^2 m_0}{h^2}\left(W + \frac{e^2}{r_A} + \frac{e^2}{r_B} - \frac{e^2}{r_{AB}}\right)\psi = 0, \qquad (42\text{-}1)$$

in which ∇^2 refers to the three coordinates of the electron and m_0 is the mass of the electron.[4]

[1] W. Pauli, *Ann. d. Phys.* **68**, 177 (1922).

[2] K. F. Niessen, Dissertation, Utrecht, 1922.

[3] L. Pauling, *Chem. Rev.* **5**, 173 (1928).

[4] We have included the mutual energy of the two nuclei e^2/H_{AB} in this equation. This is not necessary, inasmuch as the term appears unchanged in the final expression for W, and the same result would be obtained by omitting it in this equation and adding it later.

If r_{AB} is very large, the normal state of the system has the energy value $W = W_H = -Rhc$, the corresponding wave functions being u_{1s_A} or u_{1s_B}, hydrogen-atom wave functions about nucleus A or nucleus B (Sec. 21), or any two independent linear combinations of these. In other words, for large values of r_{AB} the system in its normal state is composed of a hydrogen ion A and a normal hydrogen atom B or of a normal hydrogen atom A and a hydrogen ion B.

This suggests that as a simple variation treatment of the system for smaller values of r_{AB} we make use of the same wave functions u_{1s_A} and u_{1s_B}, forming the linear combinations given by solution of the secular equation as discussed in Section 26d. The secular equation is

$$\begin{vmatrix} H_{AA} - W & H_{AB} - \Delta W \\ H_{BA} - \Delta W & H_{BB} - W \end{vmatrix} = 0, \qquad (42\text{--}2)$$

in which

$$H_{AA} = \int u_{1s_A} H u_{1s_A} d\tau,$$
$$H_{AB} = \int u_{1s_A} H u_{1s_B} d\tau,$$

and

$$\Delta = \int u_{1s_A} u_{1s_B} d\tau.$$

Δ represents the lack of orthogonality of u_{1s_A} and u_{1s_B}. Because of the equivalence of the two functions, the relations $H_{AA} = H_{BB}$ and $H_{AB} = H_{BA}$ hold. The solutions of the secular equation are hence

$$W_S = \frac{H_{AA} + H_{AB}}{1 + \Delta} \qquad (42\text{--}3)$$

and

$$W_A = \frac{H_{AA} - H_{AB}}{1 - \Delta}. \qquad (42\text{--}4)$$

These correspond respectively to the wave functions

$$\psi_S = \frac{1}{\sqrt{2 + 2\Delta}}(u_{1s_A} + u_{1s_B}) \qquad (42\text{--}5)$$

and

$$\psi_A = \frac{1}{\sqrt{2 - 2\Delta}}(u_{1s_A} - u_{1s_B}). \qquad (42\text{--}6)$$

The subscripts S and A represent the words symmetric and antisymmetric, respectively (Sec. 29a); the wave function ψ_S is symmetric in the positional coordinates of the two nuclei A and B, and ψ_A is antisymmetric in these coordinates.

Introducing W_H by use of the equation

$$-\frac{h^2}{8\pi^2 m_0}\nabla^2 u_{1s_A} - \frac{e^2}{r_A}u_{1s_A} = W_H u_{1s_A}$$

(which is the wave equation for u_{1s_A}), we obtain for the integral H_{AA} the expression

$$H_{AA} = \int u_{1s_A}\left(W_H - \frac{e^2}{r_B} + \frac{e^2}{r_{AB}}\right)u_{1s_A}d\tau = W_H + J + \frac{e^2}{a_0 D}, \quad (42\text{-}7)$$

in which

$$J = \int u_{1s_A}\left(-\frac{e^2}{r_B}\right)u_{1s_A}d\tau = \frac{e^2}{a_0}\left\{-\frac{1}{D} + e^{-2D}\left(1 + \frac{1}{D}\right)\right\}. \quad (42\text{-}8)$$

In this expression we have introduced in place of r_{AB} the variable

$$D = \frac{r_{AB}}{a_0}. \quad (42\text{-}9)$$

H_{BA} and H_{AB} are similarly given by the expression

$$H_{BA} = \int u_{1s_B}\left(W_H - \frac{e^2}{r_B} + \frac{e^2}{r_{AB}}\right)u_{1s_A}d\tau = \Delta W_H + K + \frac{\Delta e^2}{a_0 D}, \tag{42-10}$$

in which Δ is the orthogonality integral, with the value

$$\Delta = e^{-D}(1 + D + \tfrac{1}{3}D^2), \quad (42\text{-}11)$$

and K is the integral

$$K = \int u_{1s_B}\left(-\frac{e^2}{r_B}\right)u_{1s_A}d\tau = -\frac{e^2}{a_0}e^{-D}(1 + D). \quad (42\text{-}12)$$

It is seen that J represents the Coulomb interaction of an electron in a $1s$ orbital on nucleus A with nucleus B. K may be called a resonance or exchange integral, since both functions u_{1s_A} and u_{1s_B} occur in it.

Introducing these values in Equations 42–3 and 42–4, we obtain

$$W_S = W_H + \frac{e^2}{a_0 D} + \frac{J + K}{1 + \Delta} \quad (42\text{-}13)$$

and

$$W_A = W_H + \frac{e^2}{a_0 D} + \frac{J - K}{1 - \Delta}. \quad (42\text{-}14)$$

Curves showing these two quantities as functions of r_{AB} are given in Figure 42-2. It is seen that ψ_S corresponds to attraction, with the formation of a stable molecule-ion, whereas ψ_A corresponds to repulsion at all distances. There is rough agreement between observed properties of the hydrogen molecule-ion in its normal state and the values calculated in this simple way. The dissociation energy, calculated to be 1.77 v.e., is actually 2.78 v.e., and the equilibrium value of r_{AB}, calculated as 1.32 Å, is observed to be 1.06 Å.

The nature of the interactions involved in the formation of

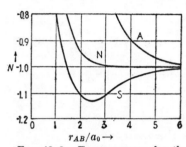

FIG. 42-2.—Energy curves for the hydrogen molecule-ion (in units $e^2/2a_0$), calculated for undistorted hydrogen atom wave functions.

this stable molecule (with a one-electron bond) is clarified by the discussion of a hypothetical case. Let us assume that our system is composed of a hydrogen atom A and a hydrogen ion B, and that even for small values of r_{AB} the electron remains attached to nucleus A, the wave function being u_{1s_A}. The energy of the system would then be H_{AA}, and the difference between this and

W_H, namely $\dfrac{e^2}{a_0}e^{-2D}\left(1 + \dfrac{1}{D}\right)$, would be the energy of interaction of a normal hydrogen atom and a hydrogen ion. The curve representing this energy function, which before the discovery of the resonance phenomenon was supposed to correspond to the hydrogen molecule-ion, is shown in Figure 42-2 with the symbol N. It is seen that it does not correspond to the formation of a stable bond but instead to repulsion at all distances. The difference between this curve and the other two is that in this case we have neglected the resonance of the electron between the two nuclei A and B. It is this resonance which causes the actual hydrogen molecule-ion to be stable—the energy of the one-electron bond is in the main the energy of resonance of the electron between the two nuclei. (Other interactions, such as polarization of the atom in the field of the ion, also contribute to some extent to the stability of the bond. An attempt to answer the question of the magnitude of this contribution will be given in the next section.)

It is seen from the figure that the resonance interaction sets in at considerably larger distances than the Coulomb interaction of atom and ion. This results from the exponential factor e^{-2D} in H_{AA}, as compared with e^{-D} in the resonance integral K. For values of r_{AB} larger than 2 Å the energy functions W_S and W_A are closely approximated by the values $W_H + K$ and $W_H - K$, respectively. In accordance with the argument of Section 41b, the resonance energy $\pm K$ corresponds to the electron's jumping back and forth between the nuclei with the frequency $2K/h$.

Problem 42–1. Verify the expressions given for H_{AA}, H_{AB}, and Δ in Equations 42–7 to 42–12.

42b. Other Simple Variation Treatments.—We can easily improve the preceding treatment by introducing an effective nuclear charge $Z'e$ in the hydrogenlike $1s$ wave functions u_{1s_A} and u_{1s_B}. This was done by Finkelstein and Horowitz.[1] On minimizing the energy W_S relative to Z' for various values of r_{AB}, they obtained a curve for W_S similar to that of Figure 42–2, but with a lower minimum displaced somewhat to the left. They found for the equilibrium value of r_{AB} the value 1.06 Å, in complete agreement with experiment. The value of the effective atomic number Z' at this point is 1.228, and the energy of the system (neglecting oscillational and rotational energy) is -15.78 v.e., as compared with the correct value -16.31 v.e.; the value of the dissociation energy $D_e = 2.25$ v.e. differing from the correct value 2.78 v.e. by 0.53 v.e. The variation of the effective atomic number from the value 1 has thus reduced the error by one-half.

The energy of the bond for this function too is essentially resonance energy. Dickinson[2] introduced an additional term, dependent on two additional parameters, in order to take polarization into account. He wrote for the (not yet normalized) variation function

$$\psi = u_{1s_A}(Z') + u_{1s_B}(Z') + \sigma\{u_{2p_A}(Z'') + u_{2p_B}(Z'')\},$$

in which the first two terms represent as before $1s$ hydrogenlike wave functions with effective nuclear charge $Z'e$ and the remain-

[1] B. N. FINKELSTEIN and G. E. HOROWITZ, Z. f. Phys. **48**, 118 (1928).
[2] B. N. DICKINSON, J. Chem. Phys. **1**, 317 (1933).

ing two terms functions such as $2p_z$ as described in Section 21,

$$u_{2p_A} = \frac{1}{4\sqrt{2\pi}}\left(\frac{Z''}{a_0}\right)^{3/2} \cdot \frac{Z''}{a_0}r_A e^{-\frac{Z''}{2a_0}r_A}\cos\vartheta,$$

in which ϑ is taken relative to a z axis extending from nucleus A toward nucleus B (and the reverse for u_{2p_B}). The parameter σ determines the extent to which these functions enter. The interpretation of the effect of these functions as representing polarization of one atom by the other follows from their nature. The function $u_{1s_A} + \sigma u_{2p_A}$ differs from u_{1s_A} by a positive amount on the side nearer B and a negative amount on the farther side, in this way being concentrated toward B in the way expected for polarization.[1]

On minimizing the energy relative to the three parameters and to r_{AB}, Dickinson found for the equilibrium distance the value 1.06 Å, and for the energy -16.26 v.e., the parameters having the values $Z' = 1.247$, $Z'' = 2.868$, and $\sigma = 0.145$.[2] The energy calculated for this function differs by only 0.05 v.e. from the correct value, so that we may say, speaking somewhat roughly, that the energy of the one-electron bond is due almost entirely to resonance of the electron between the two nuclei and to polarization of the hydrogen atom in the field of the hydrogen ion, with resonance making the greater contribution (about 2.25 v.e., as given by Finkelstein and Horowitz's function) and polarization the smaller (about 0.5 v.e.).

It was found by Guillemin and Zener[3] that another variation function containing only two parameters provides a very good value for the energy, within 0.01 v.e. of the correct value, the equilibrium separation of the nuclei being 1.06 Å, as for all functions discussed except the simple one of the preceding section. This function is

$$e^{-Z'\frac{r_A}{a_0}}e^{-Z''\frac{r_B}{a_0}} + e^{-Z''\frac{r_A}{a_0}}e^{-Z'\frac{r_B}{a_0}},$$

[1] The introduction of such a function to take care of polarization was first made (for the hydrogen molecule) by N. Rosen, *Phys. Rev.* **38**, 2099 (1931).

[2] It will be noted that Z'' is approximately twice Z'. Dickinson found that the error in the energy is changed only by 0.02 v.e. by placing Z'' equal to $2Z'$, the best values of the parameters then being $Z' = 1.254$, $\sigma = 0.1605$.

[3] V. GUILLEMIN, JR., and C. ZENER, *Proc. Nat. Acad. Sci.* **15**, 314 (1929).

the best values of the parameters being $Z' = 1.13$, $Z'' = 0.23$. The interpretation of this function is not obvious; we might say, however, that each of the two terms of the function represents a polarized hydrogen atom, the first term, for example, being large only in the neighborhood of nucleus A, and being there polarized in the direction of nucleus B by the factor $e^{-Z''\frac{r_B}{a_0}}$ multiplying the hydrogenlike function $e^{-Z'\frac{r_A}{a_0}}$, the entire wave function then differing from Dickinson's mainly in the way in which the polarization is introduced. The value of the principal effective atomic number $Z' = 1.13$ is somewhat smaller than Dickinson's value 1.247.

A still more simple variation function giving better results has been recently found by James.[1] This function is

$$e^{-\delta\xi}(1 + c\eta^2),$$

in which ξ and η are the confocal elliptic coordinates defined in the following section (Eq. 42–15), and δ and c are parameters, with best values $\delta = 1.35$ and $c = 0.448$. The value of the dissociation energy given by this function is $D_e = 2.772$ v.e., the correct value being 2.777 v.e.

42c. The Separation and Solution of the Wave Equation.—It was pointed out by Burrau[2] that the wave equation for the hydrogen molecule-ion, Equation 42–1, is separable in confocal elliptic coordinates ξ and η and the azimuthal angle φ. The coordinates ξ and η are given by the equations

$$\left. \begin{aligned} \xi &= \frac{r_A + r_B}{r_{AB}}, \\ \eta &= \frac{r_A - r_B}{r_{AB}}. \end{aligned} \right\} \qquad (42\text{--}15)$$

On introduction of these coordinates (for which the Laplace operator is given in Appendix IV), the wave equation becomes

$$\frac{\partial}{\partial\xi}\left\{(\xi^2 - 1)\frac{\partial\psi}{\partial\xi}\right\} + \frac{\partial}{\partial\eta}\left\{(1 - \eta^2)\frac{\partial\psi}{\partial\eta}\right\} + \left(\frac{1}{\xi^2 - 1} + \frac{1}{1 - \eta^2}\right)\frac{\partial^2\psi}{\partial\varphi^2} +$$
$$\frac{8\pi^2 m_0 r_{AB}^2}{h^2}\left\{\frac{W'}{4}(\xi^2 - \eta^2) + \frac{e^2}{r_{AB}}\xi\right\}\psi = 0, \quad (42\text{--}16)$$

[1] H. M. JAMES, private communication to the authors.
[2] ØYVIND BURRAU, *Det. Kgl. Danske Vid. Selskab.* **7**, 1 (1927).

in which we have made use of the relation

$$\frac{e^2}{r_A} + \frac{e^2}{r_B} = \frac{4e^2\xi}{r_{AB}(\xi^2 - \eta^2)}$$

and have multiplied through by $\dfrac{r_{AB}^2(\xi^2 - \eta^2)}{4}$. The quantity W', given by

$$W' = W - \frac{e^2}{r_{AB}}, \tag{42-17}$$

is the energy of the electron in the field of the two nuclei, the mutual energy of the two nuclei being added to this to give the total energy W.

It is seen that on replacing $\psi(\xi, \eta, \varphi)$ by the product function

$$\psi(\xi, \eta, \varphi) = \Xi(\xi)H(\eta)\Phi(\varphi) \tag{42-18}$$

this equation is separable[1] into the three differential equations

$$\frac{d^2\Phi}{d\varphi^2} = -m^2\Phi, \tag{42-19}$$

$$\frac{d}{d\eta}\left\{(1 - \eta^2)\frac{dH}{d\eta}\right\} + \left(\lambda\eta^2 - \frac{m^2}{1 - \eta^2} - \mu\right)H = 0, \tag{42-20}$$

and

$$\frac{d}{d\xi}\left\{(\xi^2 - 1)\frac{d\Xi}{d\xi}\right\} + \left(-\lambda\xi^2 + 2D\xi - \frac{m^2}{\xi^2 - 1} + \mu\right)\Xi = 0, \tag{42-21}$$

in which

$$\lambda = -\frac{2\pi^2 m_0 r_{AB}^2 W'}{h^2} \tag{42-22}$$

and

$$D = \frac{r_{AB}}{a_0}. \tag{42-23}$$

The range of the variable ξ is from 1 to ∞, and of η from -1 to $+1$. The surfaces $\xi =$ constant are confocal ellipsoids of revolution, with the nuclei at the foci, and the surfaces $\eta =$ constant are confocal hyperboloids. The parameters m, λ, and μ must assume characteristic values in order that the equations possess acceptable solutions. The familiar φ equation possesses such solutions for $m = 0, \pm 1, \pm 2, \cdots$. The subsequent procedure of solution consists in finding the relation which must exist

---

[1] The equation is also separable for the case that the two nuclei have different charges.

between λ and μ in order that the η equation possess a satisfactory solution, and, using this relation, in then finding from the ξ equation the characteristic values of λ and hence of the energy. This procedure was carried out for the normal state of the hydrogen molecule-ion by Burrau in 1927 by numerical integration of the ξ and η equations. More accurate treatments have since been given by Hylleraas[1] and by Jaffé.[2] (The simple treatment of Guillemin and Zener, described in the preceding section, approaches Burrau's in accuracy.) We shall not describe these treatments in detail but shall give a brief discussion of one of them (that of Hylleraas) after first presenting the results.

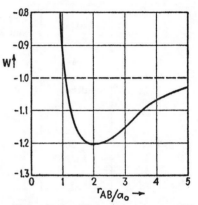

FIG. 42–3.—The energy of the normal hydrogen molecule-ion (in units $e^2/2a_0$) as a function of r_{AB}.

The energy values calculated by the three authors are given in Table 42–1 and shown graphically in Figure 42–3. It is seen that the curve is qualitatively similar to that given by the very simple treatment of Section 42a (Fig. 42–2). The three treatments agree in giving for the equilibrium value[3] of r_{AB} 2.00 a_0 or 1.06 Å, as was found for the variation functions of the preceding section also. This is in complete agreement with the band-spectral value. Spectroscopic data have not been obtained for the hydrogen molecule-ion itself but rather for various excited states of the hydrogen molecule. It is believed that these are states involving a normal hydrogen molecule-ion as core, with a highly excited outer electron in a large orbit, having little effect

[1] E. A. HYLLERAAS, *Z. f. Phys.* **71**, 739 (1931).

[2] G. JAFFÉ, *Z. f. Phys.* **87**, 535 (1934).

[3] The average value of r_{AB} for various oscillational states as determined from band-spectral data is found to be a function of the vibrational quantum number v, usually increasing somewhat with increasing v. The value for $v = 0$ is represented by the symbol r_0, and the extrapolated value corresponding to the minimum of the electronic energy function by the symbol r_e. The vibrational frequencies are similarly represented by ν_0 and ν_e (or by ω_0 and ω_e, which have found favor with band spectroscopists) and the energies of dissociation by D_0 and D_e.

on the potential function for the nuclei; this belief being supported by the constancy of the values of r_e and ν_e (the oscillational frequency) shown by them. The values of r_e were extrapolated by Birge[1] and Richardson[2] to give 1.06 Å for the molecule-ion.

TABLE 42–1.—ELECTRONIC ENERGY VALUES FOR THE HYDROGEN
MOLECULE-ION

r_{AB}/a_0	$W_{H_2^+}$ (in units Rhc)			
	Burrau	Guillemin and Zener	Hylleraas	Jaffé
0	∞	∞	∞	∞
0.5	....	....	0.5302	0.5300
1.0	−0.896	−0.903	−0.9046	−0.9035
1.25	....	....	−1.0826	
1.5	....	....	−1.1644	
1.75	−1.195*	−1.198†	−1.1980	
2.00	−1.204	−1.205	−1.20527	−1.20528
2.25	−1.198*	−1.197†	−1.1998	
2.5	....	....	−1.1878	
2.75	....	....	−1.1716	
3.0	....	−1.154	−1.1551	−1.1544
∞	−1.000	−1.000	−1.0000	−1.0000

* Interpolated between adjacent values calculated by Burrau, who estimated his accuracy in the neighborhood of the minimum as ± 0.002 Rhc.
† Interpolated values.

The value $-1.20528Rhc$ for the energy of the molecule-ion is also substantiated by experiment; the discussion of this comparison is closely connected with that for the hydrogen molecule, and we shall postpone it to Section 43d. The behavior of the minimum, however, can be compared with experiment by way of the vibrational energy levels. By matching a Morse curve to his calculated points and applying Morse's theory (Sec. 35d), Hylleraas found for the energy of the molecule ion in successive vibrational levels given by the quantum number v the expression

$$W_v = -1.20527 + 0.0206(v + \tfrac{1}{2}) - 0.00051(v + \tfrac{1}{2})^2, \quad (42\text{--}24)$$

in units Rhc. This agrees excellently with the expressions obtained by Birge[1] and Richardson[2] by extrapolation of the observed vibrational levels for excited states of the hydrogen

[1] R. T. BIRGE, Proc. Nat. Acad. Sci. 14, 12 (1928).
[2] O. W. RICHARDSON, Trans. Faraday Soc. 25, 686 (1929).

molecule, their coefficients in these units being 0.0208 and −0.00056, and 0.0210 and −0.00055, respectively.

The value $W_e = -1.20527R_Hhc$ corresponds to an electronic energy of the normal hydrogen molecule-ion of −16.3073 v.e. (using $R_Hhc = 13.5300$ v.e.) and an electronic dissociation energy into H + H$^+$ of $D_e = 2.7773$ v.e., this value being shown to be accurate to 0.0001 v.e. by the agreement between the calculations of Hylleraas and Jaffé. The value of D_0, the dissociation energy of the molecule-ion in its lowest vibrational state, differs from this by the correction terms given in Equation 42–24. These

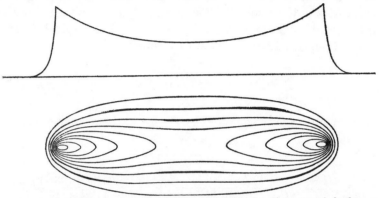

Fig. 42–4.—The electron distribution function for the normal hydrogen molecule-ion (*Burrau*). The upper curve shows the value of the function along the line passing through the two nuclei, and the lower figure shows contour lines for values 0.9, 0.8, · · · , 0.1 times the maximum value.

terms are not known so accurately, either theoretically or experimentally. Hylleraas's values lead to a correction of 0.138 v.e., Birge's to 0.139 v.e., and Richardson's to 0.140 v.e. If we accept the theoretical value 0.138 ± 0.002 v.e. we obtain

$$D_0 = 2.639 \pm 0.002 \text{ v.e.}$$

as the value of the dissociation energy of the normal hydrogen molecule-ion.

The wave function for the normal molecule-ion as evaluated by Burrau corresponds to the electron distribution function represented by Figure 42–4. It is seen that the distribution is closely concentrated about the line between the two nuclei, the electron remaining most of the time in this region.

Let us now return to a brief discussion of one of the accurate treatments of this system, that of Hylleraas, which illustrates the method of approximate solution of the wave equation discussed in Section 27a.

The variable η extends through the range -1 to $+1$, which is the range traversed by the argument $z = \cos \vartheta$ of the associated Legendre functions $P_l^{|m|}$ of Section 19. With Hylleraas we expand the function $H(\eta)$ in terms of these functions, writing

$$H(\eta) = \sum_{l=|m|}^{\infty} c_l P_l^{|m|}(\eta), \qquad (42\text{-}25)$$

in which the coefficients c_l are constants. Substituting this expression in Equation 42-20, and simplifying with the aid of the differential equation satisfied by the associated Legendre functions, Equation 19-9, we obtain the equation

$$\sum_{l=|m|}^{\infty} c_l\{\lambda\eta^2 - \mu - l(l+1)\}P_l^{|m|}(\eta) = 0. \qquad (42\text{-}26)$$

We can eliminate the factor η^2 by the use of the recursion formula

$$\eta^2 P_l^{|m|}(\eta) = \frac{(l-|m|+1)(l-|m|+2)}{(2l+1)(2l+3)}P_{l+2}^{|m|}$$
$$+ \left\{\frac{(l-|m|+1)(l+|m|+1)}{(2l+1)(2l+3)} + \frac{(l-|m|)(l+|m|)}{(2l-1)(2l+1)}\right\}P_l^{|m|}$$
$$+ \frac{(l+|m|)(l+|m|-1)}{(2l-1)(2l+1)}P_{l-2}^{|m|}, \qquad (42\text{-}27)$$

which is easily obtained by successive application of the ordinary recursion formula 19-16. On introducing this in Equation 42-26, it becomes a simple series in the functions $P_l^{|m|}(\eta)$ with coefficients independent of η. Because of the orthogonality of these functions, their coefficients must vanish independently in order that the sum vanish (Sec. 22). This gives the condition

$$\frac{(l-|m|-1)(l-|m|)}{(2l-3)(2l-1)}\lambda c_{l-2} + \left[\left\{\frac{(l-|m|+1)(l+|m|+1)}{(2l+1)(2l+3)}\right.\right.$$
$$+ \frac{(l-|m|)(l+|m|)}{(2l-1)(2l+1)}\right\}\lambda - \mu - l(l+1)\bigg]c_l$$
$$+ \frac{(l+|m|+2)(l+|m|+1)}{(2l+3)(2l+5)}\lambda c_{l+2} = 0, \qquad (42\text{-}28)$$

which is a three-term recursion formula in the coefficients c_l.

We now consider the set of equations 42–28 for different values of l as a set of simultaneous linear homogeneous equations in the unknown quantities c_l. In order that the set may possess a non-trivial solution, the determinant formed by the coefficients of the c_l's must vanish. This gives a determinantal equation involving λ and μ, from which we determine the relation between them.
We are interested in the normal state of the system, with $m = 0$ and l even. The determinantal equation for this case is

$$\begin{vmatrix} \frac{1}{3}\lambda - \mu & \frac{2}{15}\lambda & 0 & 0 & \cdots \\ \frac{2}{3}\lambda & \frac{11}{21}\lambda - 6 - \mu & \frac{4}{21}\lambda & 0 & \cdots \\ 0 & \frac{12}{35}\lambda & \frac{39}{77}\lambda - 20 - \mu & \cdots & \cdots \\ 0 & 0 & 0 & \cdots & \cdots \\ \cdots & \cdots & \cdots & \cdots & \cdots \end{vmatrix} = 0. \quad (42\text{–}29)$$

The only non-vanishing terms are in the principal diagonal and the immediately adjacent diagonals. As a rough approximation (to the first degree in λ) we can neglect the adjacent diagonals; the roots of the equation are then $\mu = \frac{1}{3}\lambda$, $\mu = \frac{11}{21}\lambda - 6$, $\mu = \frac{39}{77}\lambda - 20$, etc. We are interested in the first of these. In order to obtain it more accurately, we solve the equation again, including the first two non-diagonal terms, and replacing μ in the second diagonal term by $\frac{1}{3}\lambda$. This equation,

$$\begin{vmatrix} \frac{1}{3}\lambda - \mu & \frac{2}{15}\lambda \\ \frac{2}{3}\lambda & \frac{11}{21}\lambda - 6 - \frac{1}{3}\lambda \end{vmatrix} = 0,$$

has the solution

$$\mu = \tfrac{1}{3}\lambda + \tfrac{2}{135}\lambda^2 + \tfrac{4}{8505}\lambda^3,$$

in which powers of λ higher than the third are neglected. Hyller-aas carried the procedure one step farther, obtaining

$$\mu = \tfrac{1}{3}\lambda + \tfrac{2}{135}\lambda^2 + \tfrac{4}{8505}\lambda^3 - 0.000013\lambda^4 - 0.0000028\lambda^5.$$

This equation expresses the functional dependency of μ on λ for the normal state, as determined by the η equation. The next step is to introduce this in the ξ equation, eliminating μ, and then to solve this equation to obtain the characteristic values of λ and hence of the energy as a function of r_{AB}. Because of their more difficult character, we shall not discuss the methods of solution of this equation given by Hylleraas and Jaffé.

42d. Excited States of the Hydrogen Molecule-ion.—We have discussed (Sec. 42a) one of the excited electronic states of the hydrogen molecule-ion, with a nuclear-antisymmetric wave function formed from normal hydrogen-atom functions. This is not a stable state of the molecule-ion, inasmuch as the potential function for the nuclei does not show a minimum.

Calculations of potential functions for other excited states, many of which correspond to stable states of the molecule-ion, have been made by various investigators,[1] among whom Teller, Hylleraas, and Jaffé deserve especial mention.

43. THE HYDROGEN MOLECULE

43a. The Treatment of Heitler and London.—The following simple treatment of the hydrogen molecule (closely similar to that of the hydrogen molecule-ion described in Section 42a) does not differ essentially from that given by Heitler and London[2] in 1927, which marked the inception (except for Burrau's earlier paper on the molecule-ion) of the application of wave mechanics to problems of molecular structure and valence theory. Heitler and London's work must be considered as the greatest single contribution to the clarification of the chemist's conception of valence which has been made since G. N. Lewis's suggestion in 1916 that the chemical bond between two atoms consists of a pair of electrons held jointly by the two atoms.

Let us first consider our problem with neglect of the spin of the electrons, which we shall then discuss toward the end of the section. The system comprises two hydrogen nuclei, A and B, and two electrons, whose coordinates we shall designate by the symbols 1 and 2. Using the nomenclature of Figure 43–1, the wave equation for the two electrons corresponding to fixed positions of the two nuclei is

[1] P. M. MORSE and E. C. G. STUECKELBERG, *Phys. Rev.* **33**, 932 (1929); E. A. HYLLERAAS, *Z. f. Phys.* **51**, 150 (1928); **71**, 739 (1931); J. E. LENNARD-JONES, *Trans. Faraday Soc.* **24**, 668 (1929); E. TELLER, *Z. f. Phys.* **61**, 458 (1930); G. JAFFÉ, *Z. f. Phys.* **87**, 535 (1934).

[2] W. HEITLER and F. LONDON, *Z. f. Phys.* **44**, 455 (1927).

$$\nabla_1^2\psi + \nabla_2^2\psi + \frac{8\pi^2 m_0}{h^2}\left\{W + \frac{e^2}{r_{A1}} + \frac{e^2}{r_{B1}} + \frac{e^2}{r_{A2}} + \frac{e^2}{r_{B2}} - \frac{e^2}{r_{12}} - \right.$$
$$\left. \frac{e^2}{r_{AB}}\right\}\psi = 0. \quad (43\text{-}1)$$

For very large values of r_{AB} we know that in its normal state the system consists of two normal hydrogen atoms. Its wave functions (the state having two-fold degeneracy) are then $u_{1s_A}(1)$ $u_{1s_B}(2)$ and $u_{1s_B}(1)\,u_{1s_A}(2)$ or any two independent linear combinations of these two (the wave function $u_{1s_A}(1)$ representing a hydrogenlike $1s$ wave function for electron 1 about nucleus A,

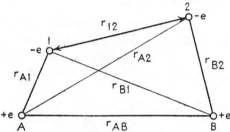

Fig. 43-1.—Coordinates used for the hydrogen molecule, represented diagrammatically.

etc., as given in Section 21). This suggests that for smaller values of r_{AB} we use as variation function a linear combination of these two product functions. We find as the secular equation corresponding to this linear variation function (Sec. 26d)

$$\begin{vmatrix} H_{I\,I} - W & H_{I\,II} - \Delta^2 W \\ H_{II\,I} - \Delta^2 W & H_{II\,II} - W \end{vmatrix} = 0, \quad (43\text{-}2)$$

in which

$$H_{I\,I} = \int\int\psi_I H\psi_I d\tau_1 d\tau_2,$$
$$H_{I\,II} = \int\int\psi_I H\psi_{II} d\tau_1 d\tau_2,$$

and

$$\Delta^2 = \int\int\psi_I\psi_{II} d\tau_1 d\tau_2,$$

with

$$\psi_I = u_{1s_A}(1)u_{1s_B}(2) \quad \text{and} \quad \psi_{II} = u_{1s_B}(1)u_{1s_A}(2).$$

It is seen that Δ is the orthogonality integral introduced in Section 42a, and given by Equation 42-11. With $H_{I\,I} = H_{II\,II}$ and $H_{I\,II} = H_{II\,I}$, the equation can be immediately solved to give

$$W_S = \frac{H_{1\,1} + H_{1\,11}}{1 + \Delta^2} \qquad (43\text{-}3)$$

and

$$W_A = \frac{H_{1\,1} - H_{1\,11}}{1 - \Delta^2}, \qquad (43\text{-}4)$$

corresponding to the wave functions

$$\psi_S = \frac{1}{\sqrt{2 + 2\Delta^2}}\{u_{1s_A}(1)u_{1s_B}(2) + u_{1s_B}(1)u_{1s_A}(2)\} \qquad (43\text{-}5)$$

and

$$\psi_A = \frac{1}{\sqrt{2 - 2\Delta^2}}\{u_{1s_A}(1)u_{1s_B}(2) - u_{1s_B}(1)u_{1s_A}(2)\}. \qquad (43\text{-}6)$$

ψ_S is symmetric in the positional coordinates of the two electrons and also in the positional coordinates of the two nuclei, whereas ψ_A is antisymmetric in both of these sets of coordinates.

On evaluation we find for $H_{1\,1}$ the expression

$$H_{1\,1} = \int\int u_{1s_A}(1)u_{1s_B}(2)\left(2W_H - \frac{e^2}{r_{B1}} - \frac{e^2}{r_{A2}} + \frac{e^2}{r_{12}} + \frac{e^2}{r_{AB}}\right)$$
$$u_{1s_A}(1)u_{1s_B}(2)d\tau_1 d\tau_2$$
$$= 2W_H + 2J + J' + \frac{e^2}{r_{AB}}, \qquad (43\text{-}7)$$

in which J is the integral of Equation 42–8 and J' is

$$J' = e^2\int\int \frac{\{u_{1s_A}(1)u_{1s_B}(2)\}^2}{r_{12}}d\tau_1 d\tau_2$$
$$= \frac{e^2}{a_0}\left\{\frac{1}{D} - e^{-2D}\left(\frac{1}{D} + \frac{11}{8} + \frac{3}{4}D + \frac{1}{6}D^2\right)\right\}, \qquad (43\text{-}8)$$

with D as before equal to r_{AB}/a_0.

Similarly we find for $H_{1\,11}$ the expression

$$H_{1\,11} = \int\int u_{1s_A}(1)u_{1s_B}(2)\left(2W_H - \frac{e^2}{r_{A1}} - \frac{e^2}{r_{B2}} + \frac{e^2}{r_{12}}\right.$$
$$\left.\frac{e^2}{r_{AB}}\right)u_{1s_B}(1)u_{1s_A}(2)d\tau_1 d\tau_2$$
$$= 2\Delta^2 W_H + 2\Delta K + K' + \Delta^2\frac{e^2}{r_{AB}}, \qquad (43\text{-}9)$$

in which K is the integral of Equation 42–12 and K' is

$$K' = e^2 \int \int \frac{u_{1s_A}(1)u_{1s_B}(2)u_{1s_B}(1)u_{1s_A}(2)}{r_{12}} d\tau_1 d\tau_2$$

$$= \frac{e^2}{5a_0}\left[-e^{-2D}\left(-\frac{25}{8} + \frac{23}{4}D + 3D^2 + \frac{1}{3}D^3 \right) \right.$$

$$\left. + \frac{6}{D}\{\Delta^2(\gamma + \log D) + \Delta'^2 Ei(-4D) - 2\Delta\Delta' Ei(-2D)\} \right],$$

$$\text{(43-10)}$$

in which $\gamma = 0.5772 \cdots$ is Euler's constant and

$$\Delta' = e^D(1 - D + \tfrac{1}{3}D^2).$$

Ei is the function known as the integral logarithm.[1] (The integral K' was first evaluated by Sugiura,[2] after Heitler and London had developed an approximate expression for it.) J' represents the Coulomb interaction of an electron in a $1s$ orbital on nucleus A with an electron in a $1s$ orbital on nucleus B, and K' is the corresponding resonance or exchange integral.

Substituting these values in Equations 43–3 and 43–4, we obtain

$$W_S = 2W_H + \frac{e^2}{r_{AB}} + \frac{2J + J' + 2\Delta K + K'}{1 + \Delta^2} \qquad \text{(43-11)}$$

and

$$W_A = 2W_H + \frac{e^2}{r_{AB}} + \frac{2J + J' - 2\Delta K - K'}{1 - \Delta^2}. \qquad \text{(43-12)}$$

Curves representing W_S and W_A as functions of r_{AB} are shown in Figure 43–2. It is seen that W_A corresponds to repulsion at all distances, there being no equilibrium position of the nuclei. The curve for W_S corresponds to attraction of the two hydrogen atoms with the formation of a stable molecule, the equilibrium value calculated for r_{AB} being 0.80 Å, in rough agreement with the experimental value 0.740 A. The energy of dissociation of the molecule into atoms (neglecting the vibrational energy of the nuclei) is calculated to be 3.14 v.e., a value somewhat smaller than the correct value 4.72 v.e. The curvature of the potential function near its minimum corresponds to a vibrational frequency for the nuclei of 4800 cm^{-1}, the band-spectral value being 4317.9 cm^{-1}.

It is seen that even this very simple treatment of the problem leads to results in approximate agreement with experiment.

---

[1] See, for example, Jahnke and Emde, "Funktionentafeln."

[2] Y. Sugiura, *Z. f. Phys.* **45**, 484 (1927).

It may be mentioned that the accuracy of the energy calculation is greater than appears from the values quoted for D_e, inasmuch as the energy of the electrons in the field of the two nuclei (differing from W_S by the term e^2/r_{AB}) at $r_{AB} = 1.5a_0$ is calculated to be $2W_H - 18.1$ v.e., and the error of 1.5 v.e. is thus only a few per cent of the total electronic interaction energy.

It is interesting and clarifying for this system also (as for the hydrogen molecule-ion) to discuss the energy function for a hypothetical case. Let us suppose that the wave function for the system were $\psi_I = u_{1s_A}(1)\, u_{1s_B}(2)$ alone. The energy of the system would then be $H_{I\,I}$, which is shown as curve N in Figure

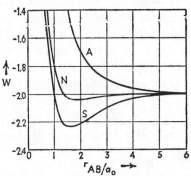

43-2. It is seen that this curve gives only a small attraction between the two atoms, with a bond energy at equilibrium only a few per cent of the observed value. The wave function ψ_S differs from this function in the interchange of the coordinates of the electrons, and we consequently say that the energy of the bond in the hydrogen molecule is in the main resonance or interchange energy.

Fig. 43-2.—Energy curves for the hydrogen molecule (in units $e^2/2a_0$).

So far we have not taken into consideration the spins of the electrons. On doing this we find, exactly as for the helium atom, that in order to make the complete wave functions antisymmetric in the electrons, as required by Pauli's principle, the orbital wave functions must be multiplied by suitably chosen spin functions, becoming

$$\psi_S \cdot \frac{1}{\sqrt{2}}\{\alpha(1)\,\beta(2) - \beta(1)\,\alpha(2)\}$$

and

$$\psi_A \cdot \alpha(1)\,\alpha(2),$$
$$\psi_A \cdot \frac{1}{\sqrt{2}}\{\alpha(1)\,\beta(2) + \beta(1)\,\alpha(2)\},$$
$$\psi_A \cdot \beta(1)\,\beta(2).$$

There are hence three repulsive states A for one attractive state S; the chance is $\frac14$ that two normal hydrogen atoms

will interact with one another in the way corresponding to the
formation of a stable molecule. It is seen that the normal state
of the hydrogen molecule is a singlet state, the spins of the
two electrons being opposed, whereas the repulsive state A is a
triplet state.

43b. Other Simple Variation Treatments.—The simple step
of introducing an effective nuclear charge $Z'e$ in the $1s$ hydrogenlike wave functions of 43–5 was taken by Wang,[1] who found that
this improved the energy somewhat, giving $D_e = 3.76$ v.e.,
and that it brought the equilibrium internuclear distance r_e
down to 0.76 Å, only slightly
greater than the experimental
value 0.740 Å. The effective
nuclear charge at the equilibrium distance was found to be
$Z'e = 1.166e$.

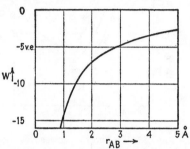

FIG. 43–3.—The mutual Coulomb
energy of two ions with charges $+e$
and $-e$ as a function of r_{AB}.

There exists the possibility
that wave functions corresponding to the ionic structures H^-H^+
and H^+H^- might also make
an appreciable contribution to
the wave function for the nor-
mal state of the molecule. These ionic functions are $u_{1s_A}(1)\, u_{1s_A}(2)$
and $u_{1s_B}(1)\, u_{1s_B}(2)$, the corresponding spin function allowed
by Pauli's principle being $\dfrac{1}{\sqrt{2}}\{\alpha(1)\,\beta(2) - \beta(1)\,\alpha(2)\}$, as for
ψ_S. It is true that for large values of r_{AB} the energy of the ionic
functions is 12.82 v.e. greater than that for the atomic functions,
this being the difference of the ionization potential and the
electron affinity (Sec. 29c) of hydrogen; but, as r_{AB} is decreased,
the Coulomb interaction of the two ions causes the energy for the
ionic functions to decrease rapidly, as shown in Figure 43–3,
the difference of 12.82 v.e. being counteracted at 1.12 Å. This
rough calculation suggests that the bond in the hydrogen molecule
may have considerable ionic character, the structures H^-H^+ and
H^+H^- of course contributing equally. The wave function

$$u_{1s_A}(1)\, u_{1s_B}(2) + u_{1s_B}(1)\, u_{1s_A}(2) + c\{u_{1s_A}(1)\, u_{1s_A}(2) +$$
$$u_{1s_B}(1)\, u_{1s_B}(2)\} \quad (43\text{--}13)$$

[1] S. C. Wang, *Phys. Rev.* **31**, 579 (1928).

was considered by Weinbaum,[1] using an effective nuclear charge
$Z'e$ in all the 1s hydrogenlike functions. On varying the parameters, he found the minimum of the energy curve (shown in
Figure 43–4) to lie at $r_{AB} = 0.77$ Å, and to correspond to the
value 4.00 v.e. for the dissociation energy D_e of the molecule.
This is an appreciable improvement, of 0.24 v.e., over the value
given by Wang's function. The parameter values minimizing
the energy[2] were found to be
$Z' = 1.193$ and $c = 0.256$.

It may be of interest to
consider the hydrogen-mole-
cule problem from another
point of view. So far we have
attempted to build a wave function for the molecule from
atomic orbital functions, a procedure which is justified as a
first approximation when r_{AB}
is large. This procedure, as
generalized to complex molecules, is called the *method of
valence-bond wave functions*, the
name sometimes being used in
the restricted sense of implying

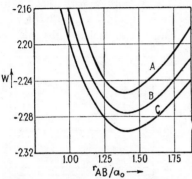

Fig. 43–4.—Energy curves for the
hydrogen molecule (in units $e^2/2a_0$): A,
for an extreme molecular-orbital wave
function; B, for an extreme valence-
bond wave function; and C, for a
valence-bond function with partial ionic
character (*Weinbaum*).

neglect of the ionic terms. Another way of considering the
structure of complex molecules, called the *method of molecular
orbitals*,[3] can be applied to the hydrogen molecule in the following
way. Let us consider that for small values of r_{AB} the interaction
of the two electrons with each other is small compared with their
interaction with the two nuclei. Neglecting the term e^2/r_{12}
in the potential energy, the wave equation separates into equations for each electron in the field of the two nuclei, as in the
hydrogen-molecule-ion problem, and the unperturbed wave
function for the normal state of the molecule is seen to be the

[1] S. WEINBAUM, *J. Chem. Phys.* **1**, 593 (1933).

[2] Weinbaum also considered a more general function with different
effective nuclear charges for the atomic and the ionic terms and found that
this reduced to 43–13 on variation.

[3] F. HUND, *Z. f. Phys.* **51**, 759 (1928); **73**, 1 (1931); etc.; R. S. MULLIKEN,
Phys. Rev. **32**, 186, 761 (1928); **41**, 49 (1932); etc.; M. DUNKEL, *Z. f. phys.
Chem.* **B7**, 81; **10**, 434 (1930); E. HÜCKEL, *Z. f. Phys.* **60**, 423 (1930); etc.

product of normal hydrogen-molecule-ion wave functions for the two electrons. Inasmuch as the function $u_{1s_A}(1) + u_{1s_B}(1)$ is a good approximation to the wave function for the electron in the normal hydrogen molecule-ion, the molecular-orbital treatment corresponds to the wave function

$$\{u_{1s_A}(1) + u_{1s_B}(1)\}\{u_{1s_A}(2) + u_{1s_B}(2)\} \qquad (43\text{–}14)$$

for the normal hydrogen molecule. It is seen that this is identical with Weinbaum's function 43–13 with $c = 1$; that is, with the ionic terms as important as the atomic terms.

If the electric charges of the nuclei were very large, the inter-electronic interaction term would actually be a small perturbation, and the molecular-orbital wave function 43–14 would be a good approximation to the wave function for the normal state of the system. In the hydrogen molecule, however, the nuclear charges are no larger than the electronic charges, and the mutual repulsion of the two electrons may well be expected to tend to keep the electrons in the neighborhood of different nuclei, as in the simple Heitler-London-Wang treatment. It would be difficult to predict which of the two simple treatments is the better. On carrying out the calculations[1] for the molecular-orbital function 43–14, introducing an effective atomic number Z', the potential curve A of Figure 43–4 is obtained, corresponding to $r_e = 0.73$ Å, $D_e = 3.47$ v.e., and $Z' = 1.193$. It is seen that the extreme atomic-orbital treatment (the Wang curve) is considerably superior to the molecular-orbital treatment for the hydrogen molecule.[2] This is also shown by the results for the more general function 43–13 including ionic terms with a coefficient c; the value of c which minimizes the energy is 0.256, which is closer to the atomic-orbital extreme ($c = 0$) than to the molecular-orbital extreme ($c = 1$).

For the doubly charged helium molecule-ion, He_2^{++}, a treatment based on the function 43–13 has been carried through,[3] leading to the energy curve shown in Figure 43–5. It is seen that at large distances the two normal He^+ ions repel each other with the force e^2/r^2. At about 1.3 Å the effect of the resonance integrals

---

[1] For this treatment we are indebted to Dr. S. Weinbaum.

[2] Similar conclusions are reached also when Z' is restricted to the value 1 (Heitler-London treatment).

[3] L. PAULING, *J. Chem. Phys.* **1**, 56 (1933).

becomes appreciable, leading to attraction of the two ions and a minimum in the energy curve at the predicted internuclear equilibrium distance $r_e = 0.75$ Å (which is very close to the value for the normal hydrogen molecule). At this distance the values of the parameters which minimize the energy are $Z' = 2.124$ and $c = 0.435$. This increase in the value of c over that for the hydrogen molecule shows that as a result of the larger nuclear charges the ionic terms become more important than for the hydrogen molecule.

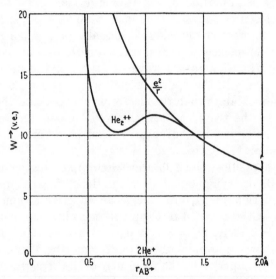

FIG. 43–5.—The energy curve for normal He_2^{++}.

We have discussed the extension of the extreme atomic-orbital treatment by the inclusion of ionic terms. A further extension could be made by adding terms corresponding to excited states of the hydrogen atoms. Similarly the molecular-orbital treatment could be extended by adding terms corresponding to excited states of the hydrogen molecule-ion. With these extensions the treatments ultimately become identical.[1] In the applications to complex molecules, however, it is usually practicable to carry through only the extremely simple atomic-orbital and molecular-orbital treatments; whether the slight superiority indicated by the above considerations for the atomic-

[1] See J. C. SLATER, *Phys. Rev.* **41**, 255 (1932).

orbital treatment is retained also for molecules containing atoms of larger atomic number remains an open question.

So far we have not considered polarization of one atom by the other in setting up the variation function. An interesting attempt to do this was made by Rosen,[1] by replacing $u_{1s_A}(1)$ in the Heitler-London-Wang function by $u_{1s_A}(1) + \sigma u_{2p_A}(1)$ (with a similar change in the other functions), as in Dickinson's treatment of the hydrogen molecule-ion (which was suggested by Rosen's work). The effective nuclear charges $Z'e$ in u_{1s} and $Z''e$ in u_{2p} were assumed to be related, with $Z'' = 2Z'$. It was found that this leads to an improvement of 0.26 v.e. in the value of D_e over Wang's treatment, the minimum in the energy corresponding to the values $r_e = 0.77$ Å, $D_e = 4.02$ v.e., $Z' = 1.19$, and $\sigma = 0.10$.

A still more general function, obtained by adding ionic terms (as in 43-13) to the Rosen function, was discussed by Weinbaum, who obtained $D_e = 4.10$ v.e., $Z' = 1.190$, $\sigma = 0.07$, and $c = 0.176$.

The results of the various calculations described in this section are collected in Table 43-1, together with the final results of James and Coolidge (see following section).

TABLE 43-1.—RESULTS OF APPROXIMATE TREATMENTS OF THE NORMAL HYDROGEN MOLECULE

	D_e	r_e	ν_e	Z'
Heitler-London-Sugiura..........	3.14 v.e.	0.80 Å	4800 cm^{-1}	
Molecular-orbital treatment......	3.47	0.73	....	1.193
Wang.........................	3.76	0.76	4900	1.166
Weinbaum (ionic)..............	4.00	0.77	4750	1.193
Rosen (polarization).............	4.02	0.77	4260	1.19
Weinbaum (ionic-polarization)....	4.10	......	....	1.190
James-Coolidge.................	4.722	0.74		
Experiment....................	4.72	0.7395	4317.9	

43c. The Treatment of James and Coolidge.—In none of the variation functions discussed in the preceding section does the interelectronic interaction find suitable expression. A major advance in the treatment of the hydrogen molecule was made by James and Coolidge[2] by the explicit introduction of the

[1] N. ROSEN, *Phys. Rev.* **38**, 2099 (1931).

[2] H. M. JAMES and A. S. COOLIDGE, *J. Chem. Phys.* **1**, 825 (1933).

interelectronic separation r_{12} in the variation function (the similar step by Hylleraas having led to the ultimate solution of the problem of the normal helium atom). Using the elliptic coordinates (Sec. 42c)

$$\xi_1 = \frac{r_{A1} + r_{B1}}{r_{AB}}, \qquad \xi_2 = \frac{r_{A2} + r_{B2}}{r_{AB}},$$

$$\eta_1 = \frac{r_{A1} - r_{B1}}{r_{AB}}, \qquad \eta_2 = \frac{r_{A2} - r_{B2}}{r_{AB}},$$

and the new coordinate

$$u = \frac{2r_{12}}{r_{AB}},$$

James and Coolidge chose as the variation function the expression

$$\psi = \frac{1}{2\pi} e^{-\delta(\xi_1 + \xi_2)} \sum_{mnjkp} c_{mnjkp} (\xi_1^m \xi_2^n \eta_1^j \eta_2^k u^p + \xi_1^n \xi_2^m \eta_1^k \eta_2^j u^p), \quad (43\text{-}15)$$

the summation to include zero and positive values of the indices, with the restriction that $j + k$ be even, which is required to make the function symmetric in the coordinates of the nuclei.

Calculations were first made for $r_{AB} = 1.40a_0$ (the experimental value of r_e) and $\delta = 0.75$; with these fixed values the variation of the parameters can be carried out by the solution of a determinantal equation (Sec. 26d). It was found that five terms alone lead to an energy value much better than any that had been previously obtained,[1] the improvement being due in the main to the inclusion of one term involving u (Tables 43-2 and 43-3). It is seen from Table 43-2 that the eleven-term and thirteen-term functions lead to only slightly different energy values, and the authors' estimate that the further terms will contribute only a small amount, making $D_e = 4.722 \pm 0.013$ v.e., seems not unreasonable.

Using the eleven-term function, James and Coolidge investigated the effects of varying δ and r_{AB}, concluding that the values previously assumed minimize the energy, corresponding to agreement between the theoretical and the experimental value of r_e, and that the energy depends on r_{AB} in such a way as to correspond closely to the experimental value of ν_e.

---

[1] It is of interest that the best value found by including only terms with $p = 0$ is $D_e = 4.27$ v.e., which is only slightly better than the best values of the preceding section.

This must be considered as a thoroughly satisfactory treatment of the normal hydrogen molecule, the only improvement which we may look forward to being the increase in accuracy by the inclusion of further terms.

TABLE 43–2.—SUCCESSIVE APPROXIMATIONS WITH THE JAMES-COOLIDGE WAVE FUNCTION FOR THE HYDROGEN MOLECULE

Number of terms	Total energy	D_e
1	$-2.189\,R_H hc$	2.56 v.e.
5	-2.33290	4.504
11	-2.34609	4.685
13	-2.34705	4.698

TABLE 43–3.—VALUES OF COEFFICIENTS c_{mnjkp} FOR NORMALIZED WAVE FUNCTIONS FOR THE HYDROGEN MOLECULE*

Term $mnjkp$	Values of c_{mnjkp}			
	1 term	5 terms	11 terms	13 terms
00000	1.69609	2.23779	2.29326	2.22350
00020	....	0.80483	1.19526	1.19279
10000	....	−0.60985	−0.86693	−0.82767
00110	....	−0.27997	−0.49921	−0.45805
00001	....	0.19917	0.33977	0.35076
10200	....	....	−0.13656	−0.17134
10110	....	....	0.14330	0.12394
10020	....	....	−0.07214	−0.12101
20000	....	....	0.06621	0.08323
00021	....	....	....	0.07090
10001	....	....	....	−0.03987
00002	....	....	−0.02456	−0.01197
00111	....	....	−0.03143	−0.01143

*In a later note, *J. Chem. Phys.* **3**, 129 (1935), James and Coolidge state that these values are about 0.05 per cent too large.

43d. Comparison with Experiment.—The theoretical values for the energy of dissociation of the hydrogen molecule and molecule-ion discussed in the preceding sections can be compared with experiment both directly and indirectly. The value

$$D_0 = 2.639 \pm 0.002 \text{ v.e.}$$

for H_2^+ is in agreement with the approximate value 2.6 ± 0.1 v.e. found from the extrapolated vibrational frequencies for excited

352   *THE STRUCTURE OF SIMPLE MOLECULES*   [XII-43d

states of H_2. For the hydrogen molecule the energy calculations of James and Coolidge with an estimate of the effect of further terms and corrections for zero-point vibration (using a Morse function) and for the rapid motion of the nuclei (corresponding to the introduction of a reduced mass of electron and proton) lead to the value[1] 4.454 ± 0.013 v.e. for the dissociation energy D_0. This is in entire agreement with the most accurate experi-

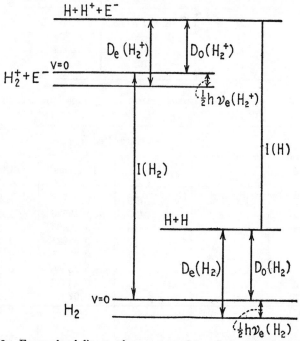

Fig. 43–6.—Energy-level diagram for a system of two electrons and two protons.

mental value, 4.454 ± 0.005 v.e., obtained by Beutler[2] by the extrapolation of observed vibrational levels.

Another test of the values can be made in the following way. From the energy-level diagram for a system of two electrons and

[1] H. M. JAMES and A. S. COOLIDGE, *J. Chem. Phys.* **3**, 129 (1935). We are indebted to Drs. James and Coolidge for the personal communication of this and other results of their work.

[2] H. BEUTLER, *Z. phys. Chem.* **B27**, 287 (1934). A direct thermochemical determination by F. R. Bichowsky and L. C. Copeland, *Jour. Am. Chem. Soc.* **50**. 1315 (1928), gave the value 4.55 ± 0.15 v.e.

two protons shown in Figure 43–6 we see that the relation

$$I(H_2) + D_0(H_2^+) = I(H) + D_0(H_2) \qquad (43\text{-}16)$$

holds between the various ionization energies and dissociation energies. With the use of the known values of $I(H)$ and $D_0(H_2^+)$ (the latter being the theoretical value) and of the extrapolated spectroscopic value of $I(H_2)$, $D_0(H_2)$ is determined[1] as

$$4.448 \pm 0.005 \text{ v.e.},$$

again in excellent agreement with the value given by James and Coolidge.

43e. Excited States of the Hydrogen Molecule.—Several excited states of the hydrogen molecule have been treated by perturbation and variation methods,[2] with results in approximate agreement with experiment.

Instead of discussing these results, let us consider the simple question as to what wave functions for the hydrogen molecule can be built from $1s$ hydrogenlike functions u_A and u_B alone. There are four product functions of this type, $u_A(1)\,u_B(2)$, $u_B(1)\,u_A(2)$, $u_A(1)\,u_A(2)$, and $u_B(1)\,u_B(2)$. The equivalence of the two electrons and of the two nuclei requires that the wave functions obtained from these by solution of the secular equation be either symmetric or antisymmetric in the positional coordinates of the two electrons and also either symmetric or antisymmetric in the two nuclei. These functions are

$$\left.\begin{array}{c} \text{I} \\ \text{II} \end{array}\right\} \{u_A(1)\,u_B(2) + u_B(1)\,u_A(2)\},\ \{u_A(1)\,u_A(2) + u_B(1)\,u_B(2)\}$$

$$\begin{cases} S^N S^E \ {}^1\Sigma_g^+ \\ S^N S^E \ {}^1\Sigma_g^+, \end{cases}$$

$$\text{III} \quad u_A(1)\,u_B(2) - u_B(1)\,u_A(2),\, A^N A^E \ {}^3\Sigma_u^+,$$
$$\text{IV} \quad u_A(1)\,u_A(2) - u_B(1)\,u_B(2),\, A^N S^E \ {}^1\Sigma_u^+,$$

functions I and II being formed by linear combination of the two indicated functions. One of these (I, say) represents the

---

[1] Personal communication to Dr. James from Prof. O. W. Richardson.

[2] E. C. KEMBLE and C. ZENER, *Phys. Rev.* **33**, 512 (1929); C. ZENER and V. GUILLEMIN, *Phys. Rev.* **34**, 999 (1929); E. A. HYLLERAAS, *Z. f. Phys.* **71,**-739 (1931); E. MAJORANA, *Atti Accad. Lincei* **13**, 58 (1931); J. K. L. MAC DONALD, *Proc. Roy. Soc.* **A136**, 528 (1932). The method of James and Coolidge has been applied to several excited states of the hydrogen molecule by R. D. Present, *J Chem. Phys.* **3**, 122 (1935), and by H. M. James, A. S. Coolidge, and R. D. Present, in a paper to be published soon.

normal state of the molecule (Sec. 43*b*, Weinbaum), and the other
an excited state.   The term symbol $^1\Sigma_g^+$ for these states contains
the letter Σ to show that there is no component of electronic
orbital angular momentum along the nuclear axis; the superscript
1 to show that the molecule is in a singlet state, as shown also
by the symbol S^E, meaning symmetric in the positional coordinates of the two electrons, Pauli's principle then requiring that
the electron-spin function be the singlet function

$$\alpha(1)\ \beta(2) - \beta(1)\ \alpha(2);$$

and the superscript $+$ to show that the electronic wave function
is symmetric in the two nuclei, as shown also by S^N.   In
addition the subscript g (German *gerade*) is given to show that
the electronic wave function is an even function of the electronic
coordinates.   Functions III and IV are both antisymmetric in
the nuclei, as indicated by the symbol A^N and the superscript $-$,
and are odd functions, as shown by the subscript u (German
ungerade), III being a triplet and IV a singlet function.   A
further discussion of these symmetry properties will be given in
the next section and in Section 48.

Function III represents the repulsive interaction of two
normal hydrogen atoms, as mentioned in Section 43*a*.   Function
II is mainly ionic in character and function IV completely so,
representing the interaction of H^+ and H^-.   Of these IV corresponds to a known state, the first electronically excited state
of the molecule.   As might have been anticipated from the
ionic character of the wave function, the state differs in its properties from the other known excited states, having $r_e = 1.29$ Å
and $\nu_e = 1358$ cm^{-1}, whereas the others have values of r_e and ν_e
close to those for the normal hydrogen molecule-ion, 1.06 Å
and 2250 cm^{-1}.   The calculations of Zener and Guillemin and
of Hylleraas have shown that at the equilibrium distance the
wave function for this state involves some contribution from
wave functions for one normal and one excited atom (with
$n = 2, l = 1$), and with increase in r_{AB} this contribution increases,
the molecule in this state dissociating into a normal and an
excited atom.

The state corresponding to II has not yet been identified.

Problem 43–1.   Construct a wave function of symmetry type $A^N S^g$
from $1s$ and $2p$ functions.

43f. Oscillation and Rotation of the Molecule. Ortho and Para Hydrogen.—In accordance with the discussion of the preceding sections and of Chapter X, we can represent the complete wave function for the hydrogen molecule as the product of five functions, one describing the orbital motion of the electrons, the second the orientation of electron spins, the third the oscillational motion of the nuclei, the fourth the rotational motion of the nuclei, and the fifth the orientation of nuclear spins (assuming them to exist):

$$\psi = \begin{pmatrix} \text{electronic} \\ \text{orbital} \\ \text{motion} \end{pmatrix} \begin{pmatrix} \text{electronic-} \\ \text{spin} \\ \text{orientation} \end{pmatrix} \begin{pmatrix} \text{nuclear} \\ \text{oscillation} \end{pmatrix} \begin{pmatrix} \text{nuclear} \\ \text{rotation} \end{pmatrix} \begin{pmatrix} \text{nuclear-} \\ \text{spin} \\ \text{orientation} \end{pmatrix}.$$

For the normal electronic state the first of these is symmetric in the two electrons, the second antisymmetric, and the remaining three independent of the electrons (and hence symmetric), making the entire function antisymmetric in the two electrons, as required by Pauli's principle. Let us now consider the symmetry character of these functions with respect to the nuclei. The first we have seen to be symmetric in the nuclei. The second is also symmetric, not being dependent on the nuclear coordinates. The third is also symmetric for all oscillational states, inasmuch as the variable r which occurs in the oscillational wave function is unchanged by interchanging the nuclei. The rotational function, however, may be either symmetric or antisymmetric. Interchanging the two nuclei converts the polar angle ϑ into $\pi - \vartheta$ and φ into $\pi + \varphi$; the consideration of the rotational wave functions (Secs. 35a and 21) shows that this causes a change in sign if the rotational quantum number K is odd, and leaves the function unchanged if K is even. Hence the rotational wave function is symmetric in the nuclei for even rotational states and antisymmetric for odd rotational states. The nuclear-spin function can be either symmetric or antisymmetric, providing that the nuclei possess spins.

By an argument identical with that given in Section 29b for the electrons in the helium atom we know that a system containing two identical protons can be represented either by wave functions which are symmetric in the protons or by wave functions which are antisymmetric in the protons. Let us assume that the protons possessed no spins and that the symmetric functions existed in nature. Then only the even rotational states of the normal hydrogen molecule would occur (and only the odd rotational states of the A^N electronic state IV of the preceding section). Similarly, if the antisymmetric functions existed in nature, only the odd rotational states of the normal molecule would occur. If, on the other hand, the protons possessed spins of $\frac{1}{2}$ (this being the value of the nuclear-spin quantum number I), both even and odd rotational states would occur, in the ratio of 3 to 1 if the complete wave function were symmetric or 1 to 3 if it were antisymmetric, inasmuch as there are for $I = \frac{1}{2}$ three symmetric nuclear-spin wave functions,

$$\alpha(A)\ \alpha(B),$$

$$\frac{1}{\sqrt{2}}\{\alpha(A)\ \beta(B) + \beta(A)\ \alpha(B)\},$$

and

$$\beta(A)\ \beta(B),$$

and one antisymmetric one,

$$\frac{1}{\sqrt{2}}\{\alpha(A)\ \beta(B) - \beta(A)\ \alpha(B)\}.$$

In this case, then, we would observe alternating intensities in the rotational fine structure of the hydrogen bands, with the ratio of intensities 3:1 or 1:3, depending on the symmetry character of protons. Similar alternating intensities result from larger values of I, the ratio being[1] $I + 1$ to I. It is seen that

---

[1] Thus for $I = 1$ there are three spin functions for one particle, α, β, and γ, say, corresponding to $m_I = +1$, 0, -1. From these we can build the following wave functions for two particles, giving the ratio 2:1.

Symmetric	Antisymmetric
$\alpha(A)\ \alpha(B)$	
$\beta(A)\ \beta(B)$	
$\gamma(A)\ \gamma(B)$	
$\frac{1}{\sqrt{2}}\{\alpha(A)\ \beta(B) + \beta(A)\ \alpha(B)\}$	$\frac{1}{\sqrt{2}}\{\alpha(A)\ \beta(B) - \beta(A)\ \alpha(B)\}$
$\frac{1}{\sqrt{2}}\{\alpha(A)\ \gamma(B) + \gamma(A)\ \alpha(B)\}$	$\frac{1}{\sqrt{2}}\{\alpha(A)\ \gamma(B) - \gamma(A)\ \alpha(B)\}$
$\frac{1}{\sqrt{2}}\{\beta(A)\ \gamma(B) + \gamma(A)\ \beta(B)\}$	$\frac{1}{\sqrt{2}}\{\beta(A)\ \gamma(B) - \gamma(A)\ \beta(B)\}$

from the observation and analysis of band spectra of molecules containing two identical nuclei the symmetry character and the spin of the nuclei can be determined.

It was found by Dennison[1] (by a different method—the study of the heat capacity of the gas, discussed in Section 49e) that protons (like electrons) have a spin of one-half, and that the allowed wave functions are completely antisymmetric in the proton coordinates (positional plus spin). This last statement is the exact analogue of the Pauli exclusion principle.[2]

Each of the even rotational wave functions for the normal hydrogen molecule is required by this exclusion principle to be combined with the antisymmetric spin function, whereas each of the odd rotational wave functions can be associated with the three symmetric spin functions, giving three complete wave functions. Hence on the average there are three times as many complete wave functions for odd rotational states as for even, and at high temperatures three times as many molecules will be in odd as in even rotational states (Sec. 49e). Moreover, a molecule in an odd rotational state will undergo a transition to an even rotational state (of the normal molecule) only extremely rarely, for such a transition would result only from a perturbation involving the nuclear spins, and these are extremely small in magnitude. Hence (as was assumed by Dennison) under ordinary circumstances we can consider hydrogen as consisting of two distinct molecular species, one, called *para hydrogen*, having the nuclear spins opposed and existing only in even rotational states (for the normal electronic state), and the other, called *ortho hydrogen*, having the nuclear spins parallel and existing only in the odd rotational states. Ordinary hydrogen is one-quarter para and three-quarters ortho hydrogen.

On cooling to liquid-air temperatures the molecules of para hydrogen in the main go over to the state with $K = 0$ and those of ortho hydrogen to the state with $K = 1$, despite the fact that at thermodynamic equilibrium almost all molecules would be in the state with $K = 0$, this metastable condition being retained for months. It was discovered by Bonhoeffer

---

[1] D. M. DENNISON, *Proc. Roy. Soc.* **A115**, 483 (1927).

[2] The spins and symmetry nature for other nuclei must at present be determined experimentally; for example, it is known that the deuteron has $I = 1$ and symmetric wave functions.

and Harteck,[1] however, that a catalyst such as charcoal causes thermodynamic equilibrium to be quickly reached, permitting the preparation of nearly pure para hydrogen. It is believed that under these conditions the ortho-para conversion is due to a magnetic interaction with the nuclear spins,[2] and not to dissociation into atoms and subsequent recombination, inasmuch as the reaction $H_2 + D_2 \rightleftarrows 2HD$ is not catalyzed under the same conditions. The conversion is catalyzed by paramagnetic substances[3] (oxygen, nitric oxide, paramagnetic ions in solution), and a theoretical discussion of the phenomenon has been published.[4] At higher temperatures the conversion over solid catalysts seems to take place through dissociation and recombination.

44. THE HELIUM MOLECULE-ION He_2^+ AND THE INTERACTION OF TWO NORMAL HELIUM ATOMS

In the preceding sections we have discussed systems of two nuclei and one or two electrons. Systems of two nuclei and three or four electrons, represented by the helium molecule-ion He_2^+ and by two interacting helium atoms, respectively, are treated in the following paragraphs. A discussion of the results obtained for systems of these four types and of their general significance in regard to the nature of the chemical bond and to the structure of molecules will then be presented in Section 45.

44a. The Helium Molecule-ion He_2^+.—In treating the system of two helium nuclei and three electrons by the variation method let us first construct electronic wave functions by using only hydrogenlike $1s$ orbital wave functions for the two atoms, which we may designate as u_A and u_B, omitting the subscripts $1s$ for the sake of simplicity. Four completely antisymmetric wave functions can be built from these and the spin functions α and β. These are (before normalization)

$$\psi_I = \begin{vmatrix} u_A(1)\ \alpha(1) & u_A(1)\ \beta(1) & u_B(1)\ \alpha(1) \\ u_A(2)\ \alpha(2) & u_A(2)\ \beta(2) & u_B(2)\ \alpha(2) \\ u_A(3)\ \alpha(3) & u_A(3)\ \beta(3) & u_B(3)\ \alpha(3) \end{vmatrix} \quad (44\text{-}1)$$

---

[1] K. F. BONHOEFFER and P. HARTECK, *Z. f. phys. Chem.* **B4**, 113 (1929).

[2] K. F. BONHOEFFER, A. FARKAS, and K. W. RUMMEL, *Z. f. phys. Chem.* **B31**, 225 (1933).

[3] L. FARKAS and H. SACHSSE, *Z. f. phys. Chem.* **B23**, 1, 19 (1933).

[4] E. WIGNER, *Z. f. phys. Chem.* **B23**, 28 (1933).

and

$$\psi_{II} = \begin{vmatrix} u_B(1) \; \alpha(1) & u_B(1) \; \beta(1) & u_A(1) \; \alpha(1) \\ u_B(2) \; \alpha(2) & u_B(2) \; \beta(2) & u_A(2) \; \alpha(2) \\ u_B(3) \; \alpha(3) & u_B(3) \; \beta(3) & u_A(3) \; \alpha(3) \end{vmatrix}, \qquad (44\text{-}2)$$

and two other functions, ψ_{III} and ψ_{IV}, obtained by replacing α by β in the last column of these functions. It is seen that the function ψ_I represents a pair of electrons with opposed spins on nucleus A (as in the normal helium atom) and a single electron with positive spin on nucleus B; this we might write as He: ·He$^+$. Function ψ_{II} similarly represents the structure He·$^+$:He, the nuclei having interchanged their roles. It is evident that this system shows the same degeneracy as the hydrogen molecule-ion, and that the solution of the secular equation for ψ_I and ψ_{II} will lead to the functions ψ_S and ψ_A, the nuclear-symmetric and nuclear-antisymmetric combinations of ψ_I and ψ_{II} (their sum and difference), as the best wave functions given by this approximate treatment. The other wave functions ψ_{III} and ψ_{IV} lead to the same energy levels.

The results of the energy calculation[1] (which, because of its similarity to those of the preceding sections, does not need to be given in detail) are shown in Figure 44–1. It is seen that the nuclear-antisymmetric wave function ψ_A corresponds to repulsion at all distances, whereas the nuclear-symmetric function ψ_S leads to attraction and the formation of a stable molecule-ion. That this attraction is due to resonance between the structures He: ·He$^+$ and He·$^+$:He is shown by comparison with the energy curve for ψ_I or ψ_{II} alone, given by the dashed line in Figure 44–1. We might express this fact by writing for the normal helium molecule-ion the structure He···He$^+$, and saying that its stability is due to the presence of a *three-electron bond* between the two atoms.

The function ψ_S composed of $1s$ hydrogenlike orbital wave functions with effective nuclear charge $2e$ leads to a minimum in the energy curve at $r_e = 1.01$ Å and the value 2.9 v.e. for the energy of dissociation D_e into He + He$^+$. A more accurate treatment[2] can be made by minimizing the energy for each value

[1] L. PAULING, *J. Chem. Phys.* **1**, 56 (1933).

[2] L. PAULING, *loc. cit.* The same calculation with Z' given the fixed value 1.8 was made by E. MAJORANA, *Nuovo Cim.* **8**, 22 (1931).

of r_{AB} with respect to an effective nuclear charge $Z'e$. This leads to $r_e = 1.085$ Å, $D_e = 2.47$ v.e., and the vibrational frequency $\nu_e = 1950$ cm^{-1}, with Z' equal to 1.833 at the equilibrium distance. A still more reliable treatment can be made by introducing two effective nuclear charges $Z'e$ and $Z''e$, one for the helium atom and one for the ion, and minimizing the energy with respect to Z' and Z''. This has been done by Weinbaum,[1]

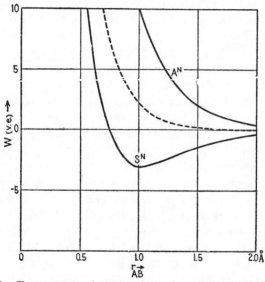

Fig. 44-1.—Energy curves for attractive and repulsive states of He$_2^+$. The dashed curve corresponds to a non-existent state, resonance between He$\cdot$ $\cdot$He$^+$ and He$\cdot^+$ $\cdot$He being excluded.

who obtained the values $r_e = 1.097$ A, $D_e = 2.22$ v.e., $Z' = 1.734$, and $Z'' = 2.029$. The results of these calculations are in good agreement with the experimental values given by excited states of the diatomic helium molecule (consisting of the normal molecule-ion and an outer electron), which are $r_e = 1.09$ Å, $D_e = 2.5$ v.e., and $\nu_e = 1650$ cm^{-1}.

It is of interest that the system of a helium nucleus and a hydrogen nucleus and three electrons does not show the degeneracy of functions ψ_I and ψ_{II}, and that in consequence the interaction of a normal helium atom and a normal hydrogen

[1] S. WEINBAUM, J. Chem. Phys. 3, 547 (1935).

atom corresponds to repulsion, as has been verified by approximate calculations.[1]

44b. The Interaction of Two Normal Helium Atoms.—We may write for the wave function for the normal state of a system consisting of two nuclei and four electrons the expression

$$\psi = N \begin{vmatrix} u_A(1)\,\alpha(1) & u_A(1)\,\beta(1) & u_B(1)\,\alpha(1) & u_B(1)\,\beta(1) \\ u_A(2)\,\alpha(2) & u_A(2)\,\beta(2) & u_B(2)\,\alpha(2) & u_B(2)\,\beta(2) \\ u_A(3)\,\alpha(3) & u_A(3)\,\beta(3) & u_B(3)\,\alpha(3) & u_B(3)\,\beta(3) \\ u_A(4)\,\alpha(4) & u_A(4)\,\beta(4) & u_B(4)\,\alpha(4) & u_B(4)\,\beta(4) \end{vmatrix}, \quad (44\text{-}3)$$

in which u_A and u_B represent $1s$ wave functions about nuclei A and B, respectively, and N is a normalizing factor. This wave function satisfies Pauli's principle, being completely antisymmetric in the four electrons. It is the only wave function of this type which can be constructed with the use of the one-electron orbital functions u_A and u_B alone.

It was mentioned by Heitler and London in their first paper[1] that rough theoretical considerations show that two normal helium atoms repel each other at all distances. The evaluation of the energy for the wave function ψ of Equation 44-3 with u_A and u_B hydrogenlike $1s$ wave functions with effective atomic number $Z' = \frac{27}{16}$ was carried out by Gentile.[2] A more accurate calculation based on a helium-atom wave function not given by a single algebraic expression has been made by Slater,[3] who found that the interaction energy is given by the approximate expression

$$W - W^0 = 7.70 \cdot 10^{-10} e^{-\frac{2.43R}{a_0}} \text{ ergs.} \quad (44\text{-}4)$$

This represents the repulsion which prevents the helium atoms from approaching one another very closely. The weak attractive forces which give rise to the constant a of the van der Waals equation of state cannot be treated by a calculation of this type based on unperturbed helium-atom wave functions. It will be shown in Section 47b that the van der Waals attraction is given approximately by the energy term $-1.41e^2\dfrac{a_0^5}{R^6}$ or $-0.607\dfrac{a_0^6}{R^6}$

[1] W. HEITLER and F. LONDON, *Z. f. Phys.* **44**, 455 (1927).
[2] G. GENTILE, *Z. f. Phys.* **63**, 795 (1930).
[3] J. C. SLATER, *Phys. Rev.* **32**, 349 (1928).

ergs. The equilibrium interatomic distance corresponding to this attraction term and the repulsion term of Equation 44–4 is 3.0 Å, in rough agreement with the experimental value of about 3.5 Å for solid helium, showing that the theoretical calculations are of the correct order of magnitude.

45. THE ONE-ELECTRON BOND, THE ELECTRON-PAIR BOND, AND THE THREE-ELECTRON BOND

In the preceding sections we have discussed systems containing two nuclei, each with one stable orbital wave function (a $1s$ function), and one, two, three, or four electrons. We have found that in each case an antisymmetric variation function of the determinantal type constructed from atomic orbitals and spin functions leads to repulsion rather than to attraction and the formation of a stable molecule. For the four-electron system only one such wave function can be constructed, so that two normal helium atoms, with completed K shells, interact with one another in this way. For the other systems, on the other hand, more than one function of this type can be set up (the two corresponding to the structures H· H$^+$ and H$^+$ ·H for the hydrogen molecule-ion, for example); and it is found on solution of the secular equation that the correct approximate wave functions are the sum and difference of these, and that in each case one of the corresponding energy curves leads to attraction of the atoms and the formation of a stable bond. We call the bonds involving two orbitals (one for each nucleus) and one, two, and three electrons the *one-electron bond*, the *electron-pair bond*, and the *three-electron bond*, respectively.

The calculations for the hydrogen molecule-ion, the hydrogen molecule, and the helium molecule-ion show that for these systems the electron-pair bond is about twice as strong a bond (using the dissociation energy as a measure of the strength of a bond) as the one-electron bond or the three-electron bond.[1] This fact alone provides us with some explanation of the great importance of the electron-pair bond in molecular structure in general and the subsidiary roles played by the one-electron bond and the three-electron bond.[2]

[1] See, however, the treatment of Li$_2^+$ by H. M. JAMES, *J. Chem. Phys.* **3**, 9 (1935).

[2] L. PAULING, *J. Am. Chem. Soc.* **53**, 3225 (1931).

There is a still more cogent reason for the importance of the electron-pair bond. This is the nature of the dependence of the energy of the bond on the similarity or dissimilarity of the two nuclei (or the two orbitals) involved. Using only two orbitals, u_A and u_B, we can construct for the one-electron system only the two wave functions

$$\psi_{\mathrm{I}} = u_A(1)\ \alpha(1)$$

and

$$\psi_{\mathrm{II}} = u_B(1)\ \alpha(1)$$

(together with two others involving $\beta(1)$ which do not combine with these and which lead to the same energy curves). These correspond to the electronic structures $A \cdot B^+$ and $A^+ \cdot B$. If A and B are identical (or if ψ_{I} and ψ_{II} correspond to the same energy because of an accidental relation between the orbitals and the nuclear charges) there is degeneracy, and the interaction of ψ_{I} and ψ_{II} causes the formation of a stable one-electron bond. If this equality of the energy does not obtain, the bond is weakened, the bond energy falling to zero as the energy difference for ψ_{I} and ψ_{II} becomes very large.

The three-electron bond behaves similarly. The wave functions (Eqs. 44-1 and 44-2) are closely related to those for the one-electron system, and the bond energy similarly decreases rapidly in magnitude as the energy difference for the two wave functions increases. Hence, in general, we expect strong one-electron bonds and three-electron bonds not to be formed between unlike atoms.

The behavior of the electron-pair bond is entirely different. The principal degeneracy leading to bond formation is that between the wave functions

$$\psi_{\mathrm{I}} = \begin{vmatrix} u_A(1)\ \alpha(1) & u_B(1)\ \beta(1) \\ u_A(2)\ \alpha(2) & u_B(2)\ \beta(2) \end{vmatrix}$$

and

$$\psi_{\mathrm{II}} = \begin{vmatrix} u_A(1)\ \beta(1) & u_B(1)\ \alpha(1) \\ u_A(2)\ \beta(2) & u_B(2)\ \alpha(2) \end{vmatrix}$$

These correspond to the same energy value even when A and B are not identical; hence there is just the same resonance stabilizing an electron-pair bond between unlike atoms as between like atoms. Moreover, the influence of the ionic terms is such as to introduce still greater stability as the nuclei become more unlike. One of the ionic wave functions

$$\psi_{\mathrm{III}} = u_A(1)\, u_A(2) \begin{vmatrix} \alpha(1) & \beta(1) \\ \alpha(2) & \beta(2) \end{vmatrix}$$

and

$$\psi_{\mathrm{IV}} = u_B(1)\, u_B(2) \begin{vmatrix} \alpha(1) & \beta(1) \\ \alpha(2) & \beta(2) \end{vmatrix},$$

corresponding to the ionic structures $A:^- B^+$ and $A^+ :B^-$, becomes more and more important (contributing more and more to the normal state of the molecule) as one of the atoms becomes more electronegative than the other, in consequence of the lowering of the energy for that ionic function. This phenomenon causes electron-pair bonds between unlike atoms to be, in general, somewhat stronger than those between like atoms. The discussion of this subject has been in the main empirical.[1]

It has been found possible to apply quantum-mechanical methods such as those described in this chapter in the detailed discussion of the electronic structure of polyatomic molecules and of valence and chemical bond formation in general. Only in a very few cases has the numerical treatment of polyatomic molecules been carried through with much accuracy; the most satisfactory calculation of this type which has been made is that of Coolidge[2] for the water molecule. General arguments have been presented[3] which provide a sound formal justification for the postulates previously made by the chemist regarding the nature of valence. It can be shown, for example, that one bond of the types discussed in this section can be formed by an atom for each stable orbital of the atom. Thus the first-row elements of the periodic system can form as many as four bonds, by using the four orbitals of the L shell, but not more. This result and other results[4] regarding the relative orientation of the bond axes provide the quantum-mechanical basis for the conception of the tetrahedral carbon atom. Special methods for the

[1] L. PAULING, J. Am. Chem. Soc. 54, 3570 (1932).

[2] A. S. COOLIDGE, Phys. Rev. 42, 189 (1932).

[3] W. HEITLER, Z. f. Phys. 47, 835 (1928), etc.; F. LONDON, Z. f. Phys. 50, 24 (1928), etc.; M. BORN, Z. f. Phys. 64, 729 (1930); J. C. SLATER, Phys. Rev. 38, 1109 (1931).

[4] J. C. SLATER, Phys. Rev. 34, 1293 (1929); L. PAULING, J. Am. Chem. Soc. 53, 1367 (1931); J. H. VAN VLECK, J. Chem. Phys. 1, 177 (1933), etc.

approximate treatment of the stability of very complex molecules such as the aromatic hydrocarbons[1] have also been developed and found to be useful in the discussion of the properties of these substances. The already very extensive application of wave mechanics to these problems cannot be adequately discussed in the small space which could be allowed it in this volume.

[1] E. Hückel, *Z. f. Phys.* **70**, 204 (1931), etc.; G. Rumer, *Göttinger Nachr.* p. 337, 1932; L. Pauling, *J. Chem. Phys.* **1**, 280 (1933); L. Pauling and G. W. Wheland, *ibid.* **1**, 362 (1933); L. Pauling and J. Sherman, *ibid.* **1**, 679 (1933), etc.

CHAPTER XIII

THE STRUCTURE OF COMPLEX MOLECULES

In carrying out the simple treatments of the hydrogen molecule-ion, the hydrogen molecule, the helium molecule-ion, and the system composed of two normal helium atoms discussed in the last chapter, we encountered no difficulty in constructing a small number of properly antisymmetric approximate wave functions out of one-electron orbital functions for the atoms of the molecule. The same procedure can be followed for more complex molecules; it is found, however, that it becomes so complicated as to be impracticable for any but the simplest molecules, unless some method of simplifying and systematizing the treatment is used. A treatment of this type, devised by Slater,[1] is described in the following sections, in conjunction with the discussion of a special application (to the system of three hydrogen atoms). Slater's treatment of complex molecules has been the basis of most of the theoretical work which has been carried on in this field in the last three years.

46. SLATER'S TREATMENT OF COMPLEX MOLECULES

In the last chapter we have seen that a good approximation to the wave function for a system of atoms at a considerable distance from one another is obtained by using single-electron orbital functions $u_a(1)$, etc., belonging to the individual atoms, and combining them with the electron-spin functions α and β in the form of a determinant such as that of Equation 44-3. Such a function is antisymmetric in the electrons, as required by Pauli's principle, and would be an exact solution of the wave equation for the system if the interactions between the electrons and those between the electrons of one atom and the nuclei of the other atoms could be neglected. Such determinantal

[1] J. C. SLATER. *Phys. Rev.* **38**, 1109 (1931).

functions are exactly analogous to the functions[1] used in Section 30a in the treatment of the electronic structure of atoms.

It may be possible to construct for a complex molecule many such functions with nearly the same energy, all of which would have to be considered in any satisfactory approximate treatment. Thus if we consider one atom to have the configuration $1s^2 2s^2 2p$, we must consider the determinantal functions involving all three $2p$ functions for that atom. A system of this type, in which there are a large number of available orbitals, is said to involve *orbital degeneracy*. Even in the absence of orbital degeneracy, the number of determinantal functions to be considered may be large because of the variety of ways in which the spin functions α and β can be associated with the orbital functions. This *spin degeneracy* has been encountered in the last chapter; in the simple treatment of the hydrogen molecule we considered the two functions corresponding to associating positive spin with the orbital u_A and negative spin with u_B, and then negative spin with u_A and positive spin with u_B (Sec. 45). The four wave functions described in Section 44a for the helium molecule-ion might be represented by the scheme of Table 46–1. The plus

TABLE 46–1.—WAVE FUNCTIONS FOR THE HELIUM MOLECULE-ION, He_2^+

Function	u_A	u_B	Σm_s
I	+ −	+	$+\frac{1}{2}$
II	+	+ −	$+\frac{1}{2}$
III	+ −	−	$-\frac{1}{2}$
IV	−	+ −	$-\frac{1}{2}$

and minus signs show which spin function α or β is to be associated with the orbital functions u_A and u_B (in this case $1s$ functions on the atoms A and B, respectively) in building up the determinantal wave functions. Thus row 1 of Table 46–1 corresponds to the function ψ_I given in Equation 44–1.

The column labeled Σm_s has the same meaning as in the atomic problem; namely, it is the sum of the z-components of the spin angular momentum of the electrons (with the factor $h/2\pi$). Just as in the atomic case, the wave functions which have different

---

[1] In Section 30a the convention was adopted that the symbol $u_\alpha(i)$ should include the spin function $\alpha(i)$ or $\beta(i)$. In this section we shall not use the convention, instead writing the spin function α or β explicitly each time.

values of Σm_s do not combine with one another, so that we were justified in Section 44a in considering only ψ_I and ψ_{II}.

Problem 46–1. Set up tables similar to Table 46–1 for the hydrogen molecule using the following choices of orbital functions: (a) 1s orbitals on the two atoms, allowing only one electron in each. (b) The same orbitals but allowing two electrons to occur in a single orbital also; i.e., allowing ionic functions. (c) The same as (a) with the addition of functions $2p_z$ on each atom. (d) The molecular orbital (call it u) obtained by the accurate treatment of the normal state of the hydrogen molecule-ion.

46a. Approximate Wave Functions for the System of Three Hydrogen Atoms.—In the case of three hydrogen atoms we can set up a similar table, restricting ourselves to the three 1s functions u_a, u_b, and u_c on three atoms a, b, and c, respectively, and neglecting ionic structures (Table 46–2).

TABLE 46–2.—WAVE FUNCTIONS FOR THE SYSTEM OF THREE HYDROGEN ATOMS

Function	u_a	u_b	u_c	Σm_s
I	+	+	+	$+\frac{3}{2}$
II	+	+	−	$+\frac{1}{2}$
III	+	−	+	$+\frac{1}{2}$
IV	−	+	+	$+\frac{1}{2}$
V	+	−	−	$-\frac{1}{2}$
VI	−	+	−	$-\frac{1}{2}$
VII	−	−	+	$-\frac{1}{2}$
VIII	−	−	−	$-\frac{3}{2}$

The wave function corresponding to row II of Table 46–2 is, for illustration,

$$\psi_{II} = \frac{1}{\sqrt{3!}} \begin{vmatrix} u_a(1)\ \alpha(1) & u_b(1)\ \alpha(1) & u_c(1)\ \beta(1) \\ u_a(2)\ \alpha(2) & u_b(2)\ \alpha(2) & u_c(2)\ \beta(2) \\ u_a(3)\ \alpha(3) & u_b(3)\ \alpha(3) & u_c(3)\ \beta(3) \end{vmatrix} . \quad (46\text{--}1)$$

Each of the functions described in Table 46–2 is an approximate solution of the wave equation for three hydrogen atoms; it is therefore reasonable to consider the sum of them with undetermined coefficients as a linear variation function. The determination of the coefficients and the energy values then requires the solution of a secular equation (Sec. 26d) of eight rows and columns, a typical element of which is

$$H_{I\ II} - \Delta_{I\ II} W \qquad (46\text{--}2)$$

where

$$H_{I\,II} = \int \psi_I^* H \psi_{II} d\tau, \qquad (46\text{--}3)$$

and

$$\Delta_{I\,II} = \int \psi_I^* \psi_{II} d\tau, \qquad (46\text{--}4)$$

H being the complete Hamiltonian operator for the system.

Problem 46–2. Make a table similar to Table 46–2 but including all ionic functions that can be made with the use of u_a, u_b, and u_c.

46b. Factoring the Secular Equation.—In the discussion of the electronic structure of atoms (Sec. 30c) we found that the secular equation could be factored to a considerable extent because integrals involving wave functions having different values of Σm_s or different values of Σm_l (the quantum numbers of the components of spin and orbital angular momentum, respectively) are zero. In the molecular case the orbital angular momentum component is no longer a constant of the motion (Sec. 52), so that only the spin quantum numbers are useful in factoring the secular equation.

In the case of the system under discussion, we see from Table 46–2 that the secular equation factors into two linear factors ($\Sigma m_s = \frac{3}{2}$ and $-\frac{3}{2}$) and two cubic factors ($\Sigma m_s = \frac{1}{2}$ and $-\frac{1}{2}$). On the basis of exactly the same reasoning as used in Section 30c for the atomic case, we conclude that the roots of the two linear factors will be equal to each other and also to one of the roots of each of the cubic factors.[1] The four corresponding wave functions are therefore associated with a quartet energy level, which on the vector picture corresponds to the parallel orientation of the three spin vectors, the four states differing only in the orientation of the resultant vector.

The two remaining energy levels will occur twice, once in each of the cubic factors. Each of them is, therefore, a doublet level. The straightforward way of obtaining their energy values would be to solve the cubic equation; but this is unnecessary, inasmuch as by taking the right linear combinations of II, III, and IV it is possible to factor the cubic equation into a linear factor and a quadratic factor, the linear factor yielding the energy of the quartet level. Such combinations are

---

[1] These statements can easily be verified by direct comparison of the roots obtained, using the expressions for the integrals given in Section 46c.

$$A = \frac{1}{\sqrt{2}}(\text{II} - \text{III}), \tag{46-5}$$

$$B = \frac{1}{\sqrt{2}}(\text{III} - \text{IV}), \tag{46-6}$$

$$C = \frac{1}{\sqrt{2}}(\text{IV} - \text{II}), \tag{46-7}$$

and

$$D = \frac{1}{\sqrt{3}}(\text{II} + \text{III} + \text{IV}). \tag{46-8}$$

Since these four functions are constructed from only three linearly independent functions II, III, and IV, they cannot be linearly independent; in fact, it is seen that $A + B + C = 0$. The factoring of the secular equation will be found to occur when it is set up in terms of D and any two of the functions A, B, and C.

The energy of the quartet level can be obtained from either of the linear factors; it is given by the relation

$$W = \frac{H_{\text{II}}}{\Delta_{\text{II}}}. \tag{46-9}$$

The values of the energy of the two doublet levels are obtained from the quadratic equation

$$\begin{vmatrix} H_{AA} - \Delta_{AA}W & H_{AB} - \Delta_{AB}W \\ H_{BA} - \Delta_{BA}W & H_{BB} - \Delta_{BB}W \end{vmatrix} = 0, \tag{46-10}$$

in which

$$\left. \begin{aligned} H_{AB} &= \int A^*HB d\tau, \\ \Delta_{AB} &= \int A^*B d\tau, \\ &\cdots\cdots\cdots \end{aligned} \right\} \tag{46-11}$$

Problem 46-3. Indicate how the secular equation for each of the cases of Problem 46-1 will factor by drawing a square with rows and columns labeled by the wave functions which enter the secular equation, and indicating by zeros in the proper places in the square the vanishing matrix elements.

46c. Reduction of Integrals.—Before discussing the conclusions which can be drawn from these equations, let us reduce somewhat further the integrals $H_{\text{II III}}$, etc. The wave function II can be written in the form (Sec. 30a)

$$\psi_{\text{II}} = \frac{1}{\sqrt{3!}} \sum_P (-1)^P P u_a(1)\alpha(1)u_b(2)\alpha(2)u_c(3)\beta(3), \tag{46-12}$$

in which P represents a permutation of the functions $u_a\alpha$, etc., among the electrons. A typical integral can thus be expressed in the form

$$H_{\text{II III}} = \frac{1}{3!}\sum_{P'}\sum_{P}(-1)^{P'+P}\int P'u_a^*(1)\alpha(1)u_b^*(2)\alpha(2)u_c^*(3)\beta(3)H$$

$$Pu_a(1)\alpha(1)u_b(2)\beta(2)u_c(3)\alpha(3)d\tau. \quad (46\text{--}13)$$

Following exactly the argument of Section 30d for the atomic case, we can reduce this to the form

$$H_{\text{II III}} = \sum_{P}(-1)^P\int u_a^*(1)\alpha(1)u_b^*(2)\alpha(2)u_c^*(3)\beta(3)HPu_a(1)\alpha(1)$$

$$u_b(2)\beta(2)u_c(3)\alpha(3)d\tau. \quad (46\text{--}14)$$

As in the atomic case, the integral vanishes unless the spins match, and there can be no permutation P which matches the spins unless Σm_s is the same for II and III. In this case we see that the spins are matched for the permutations P which permute 123 into 132 or 231 so that only these terms contribute to the sum. When the spins match in an integral, the integration over the spin can be carried out at once, yielding the factor unity. We thus have the result

$$\int u_a^*(1)\,\alpha(1)\,u_b^*(2)\,\alpha(2)\,u_c^*(3)\,\beta(3)\,Hu_a(1)\,\alpha(1)\,u_b(3)\,\beta(3)\,u_c(2)\,\alpha(2)$$
$$d\tau = \int u_a^*(1)\,u_b^*(2)\,u_c^*(3)\,Hu_a(1)\,u_c(2)\,u_b(3)\,d\tau = (abc|H|acb),$$
$$(46\text{--}15)$$

in which we have introduced a convenient abbreviation, $(abc|H|acb)$.

In this way we obtain the following expressions:

$$\left.\begin{aligned}
H_{\text{I I}} &= (abc|H|abc) - (abc|H|bac) - (abc|H|acb) \\
&\quad - (abc|H|cba) + (abc|H|bca) + (abc|H|cab), \\
H_{\text{II II}} &= (abc|H|abc) - (abc|H|bac), \\
H_{\text{III III}} &= (abc|H|abc) - (abc|H|cba), \\
H_{\text{IV IV}} &= (abc|H|abc) - (abc|H|acb), \\
H_{\text{II III}} &= (abc|H|cab) - (abc|H|acb), \\
H_{\text{III IV}} &= (abc|H|cab) - (abc|H|bac), \\
H_{\text{II IV}} &= (abc|H|bca) - (abc|H|cba).
\end{aligned}\right\} \quad (46\text{--}16)$$

The expressions for the Δ's are the same with H replaced by unity. The integral $(abc|H|abc)$ is frequently called the *Coulomb integral*, because it involves the Coulomb interaction of two distributions of electricity determined by u_a, u_b, and u_c. The other integrals such as $(abc|H|bac)$ are called *exchange integrals*. If only one pair of orbitals has been permuted, the integral is called a *single exchange integral;* if more than one, a *multiple exchange integral.* If the orbital functions u_a, u_b, and u_c were mutually orthogonal, many of these integrals would vanish, but it is seldom convenient to utilize orthogonal orbital functions in molecular calculations. Nevertheless, the deviation from orthogonality may not be great, in which case many of the integrals can be neglected.

46d. Limiting Cases for the System of Three Hydrogen Atoms. The values of the integrals $H_{II III}$, etc., depend on the distances between the atoms a, b, and c, and therefore the energy values and wave functions will also depend on these distances. It is interesting to consider the limiting case in which a is a large distance from b and c, which are close together. It is clear that the wave function u_a will not overlap appreciably with either u_b or u_c, so that the products $u_a u_b$ and $u_a u_c$ will be essentially zero for all values of the coordinates. Such integrals as $(abc|H|bac)$ will therefore be practically zero, and we can write

$$H_{II II} = H_{III III} = (abc|H|abc),$$
$$H_{III IV} = H_{II IV} = 0,$$
$$H_{IV IV} = (abc|H|abc) - (abc|H|acb),$$

and

$$H_{II III} = -(abc|H|acb),$$

thus obtaining the further relations

$$H_{AA} = (abc|H|abc) + (abc|H|acb),$$
$$H_{BB} = (abc|H|abc) - \tfrac{1}{2}(abc|H|acb),$$

and

$$H_{AB} = -\tfrac{1}{2}(abc|H|abc) - \tfrac{1}{2}(abc|H|acb).$$

If we insert these values into the secular equation 46–10 we obtain as one of the roots the energy value

$$W = \frac{H_{AA}}{\Delta_{AA}}, \tag{46–17}$$

and we find that the corresponding wave function is just the function A itself.

It is found on calculation that exchange integrals involving orbitals on different atoms are usually negative in sign. In case that such an integral occurs in the energy expression with a positive coefficient, it will contribute to stabilizing the molecule by attracting the atoms toward one another. Thus the expression for H_{AA} includes the Coulomb integral $(abc|H|abc)$ and the exchange integral $(abc|H|acb)$ with positive coefficient. Hence atoms b and c will attract one another, in the way corresponding to the formation of an electron-pair bond between them (exactly as in the hydrogen molecule alone). Similarly the function B represents the structure in which atoms a and c are bonded, and C that in which a and b are bonded.

When we bring the three atoms closer together, so that all the interactions are important, none of these functions alone is the correct combination; they must be combined to give a wave function which represents the state of the system. Therefore when three hydrogen atoms are near together, it is not strictly correct to say that a certain two of them are bonded, while the third is not.

We can, however, make some statements regarding the interaction of a hydrogen molecule and a hydrogen atom on the basis of the foregoing considerations. We have seen that when atom a is far removed from atoms b and c (which form a normal hydrogen molecule), the wave function for the system is function A. As a approaches b and c the wave function will not differ much from A, so long as the ab and ac distances are considerably larger than the bc distance. An approximate value for the interaction energy will thus be H_{AA}/Δ_{AA}, with

$$H_{AA} = \tfrac{1}{2}(H_{\mathrm{II\,II}} + H_{\mathrm{III\,III}} - 2H_{\mathrm{II\,III}})$$
$$= (abc|H|abc) + (abc|H|acb)$$
$$- \tfrac{1}{2}(abc|H|bac) - \tfrac{1}{2}(abc|H|cba) - (abc|H|cab),$$

and a similar expression for Δ_{AA}. It is found by calculation that in general the single exchange integrals become important at distances at which the Coulomb integral and the orthogonality integral have not begun to change appreciably, and at which the multiple exchange integrals $[(abc|H|cab)$ in this case] are still negligible. Thus we see that the interaction energy of a hydrogen atom and a hydrogen molecule at large distances is

$$- \tfrac{1}{2}(abc|H|bac) - \tfrac{1}{2}(abc|H|cba).$$

Each of these terms corresponds to repulsion, showing that the molecule will repel the atom.

Approximate discussions of the interaction of a hydrogen atom and hydrogen molecule have been given by Eyring and Polanyi,[1] and a more accurate treatment for some configurations has been carried out by Coolidge and James.[2]

46e. Generalization of the Method of Valence-bond Wave Functions.—The procedure which we have described above for discussing the interaction of three hydrogen atoms can be generalized to provide an analogous treatment of a system consisting of many atoms.   Many investigators have contributed to the attack on the problem of the electronic structure of complex molecules, and several methods of approximate treatment have been devised.   In this section we shall outline a method of treatment (due in large part to Slater) which may be called the *method of valence-bond wave functions*, without giving proofs of the pertinent theorems.   The method is essentially the same as that used above for the three-hydrogen-atom problem.

Let us now restrict our discussion to the singlet states of molecules with spin degeneracy only.   For a system involving $2n$ electrons and $2n$ stable orbitals (such as the $1s$ orbitals in $2n$ hydrogen atoms), there are $(2n)!/2^n n!$ different ways in which valence bonds can be drawn between the orbitals in pairs.   Thus for the case of four orbitals a, b, c, and d the bonds can be drawn in three ways, namely,

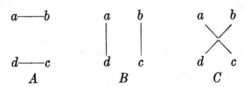

There are, however, only $\dfrac{(2n)!}{n!(n+1)!}$ independent singlet wave functions which can be constructed from the $2n$ orbitals with one electron assigned to each orbital (that is, with neglect of ionic structures).   It was shown by Slater that wave functions can be set up representing structures A, B, and C, and that only two of them are independent.   The situation is very closely analogous to that described in Section 46b.

[1] H. EYRING and M. POLANYI, *Naturwissenschaften*, **18**, 914 (1930); *Z. f. phys. Chem.* **B12**, 279 (1931).

[2] A. S. COOLIDGE and H. M. JAMES, *J. Chem. Phys.*, **2**, 811 (1934).

The very important observation was made by Rumer[1] that if the orbitals a, b, etc. are arranged in a ring or other closed concave curve (which need have no relation to the nuclear configuration of the molecule), and lines are drawn between orbitals bonded together (the lines remaining within the closed curve), the structures represented by diagrams in which no lines intersect are independent. These structures are said to form a *canonical set*. Thus in the above example the canonical set (corresponding to the order a, b, c, d) comprises structures A and B. For six orbitals there are five independent structures, as shown in Figure 46-1.

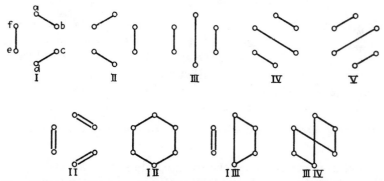

Fig. 46-1.—The five canonical valence-bond structures for six orbitals, and some of their superposition patterns.

The wave function corresponding to the structure in which orbitals a and b, c and d, etc. are bonded is

$$\psi = \frac{1}{2^{n/2}} \sum_R (-1)^R R \left[\frac{1}{\{(2n)!\}^{1/2}} \sum_P (-1)^P \right.$$

$$\left. Pa(1)\ \beta(1)\ b(2)\ \alpha(2)\ c(3)\ \beta(3)\ d(4)\ \alpha(4)\ \cdots \right], \quad (46\text{–}18)$$

in which P is the permutation operator described above (Sec. 46c), and R is the operation of interchanging the spin functions α and β of bonded orbitals, such as a and b. The factor $(-1)^R$ equals $+1$ for an even number of interchanges and -1 for an odd number. The convention is adopted of initially assigning the spin function β to orbital a, α to b, etc.

[1] G. Rumer, *Göttinger Nachr.*, p. 377, 1932.

A simple method has been developed[1] of calculating the coefficients of the Coulomb and exchange integrals in setting up the secular equation. To find the coefficient of the Coulomb integral for two structures, superimpose the two bond diagrams, as shown in Figure 46–1. The superposition pattern consists of closed polygons or *islands*, each formed by an even number of bonds. The coefficient of the Coulomb integral is $\frac{1}{2}^{n-i}$, where i is the number of islands in the superposition diagram. Thus we obtain $H_{II} = Q + \cdots$, $H_{I\,II} = \frac{1}{4}Q + \cdots$, etc., in which Q represents the Coulomb integral $(abcd \cdots |H|abcd \cdots)$.

The coefficient of a single exchange integral such as

$$(ab) = (abcd \cdots |H|bacd \cdots)$$

is equal to $f/2^{n-i}$, in which f has the value $-\frac{1}{2}$ if the two orbitals involved (a and b) are in different islands of the superposition pattern; $+1$ if they are in the same island and separated by an odd number of bonds (along either direction around the polygon); and -2 if they are in the same island and separated by an even number of bonds. Thus we see that

$$H_{II} = Q - \tfrac{1}{2}(ac) + (ab) + \cdots, \quad H_{I\,II} = Q - 2(ac) + (ab) + \cdots, \text{ etc.}$$

Let us now discuss the energy integral for a particular valence-bond wave function, in order to justify our correlation of valence-bond distribution and wave function as given in Equation 46–18. The superposition pattern for a structure with itself, as shown by I I in Figure 46–1, consists of n islands, each consisting of two bonded orbitals. We see that

$$W_I = \frac{H_{II}}{\Delta_{II}} = \frac{1}{\Delta_{II}}\left\{ Q + \sum \binom{\text{single exchange integrals for bonded}}{\text{pairs of orbitals}} \right.$$

$$- \tfrac{1}{2}\sum \binom{\text{single exchange integrals for non-bonded}}{\text{pairs of orbitals}}$$

$$\left. + \text{ higher exchange integrals} \right\}. \quad (46\text{--}19)$$

It is found by calculation that the single exchange integrals are as a rule somewhat larger in magnitude than the other integrals. Moreover, the single exchange integral for two orbitals

[1] L. PAULING, *J. Chem. Phys.* **1**, 280 (1933). See also H. Eyring and G. E. Kimball, *J. Chem. Phys.* **1**, 239 (1933), for another procedure.

on different atoms is usually negative in value for the interatomic distances occurring in molecules, changing with interatomic distance in the way given by a Morse curve (Sec. 35d). Those single exchange integrals which occur with the coefficient $+1$ in Equation 46–19 consequently lead to attraction of the atoms involved in the exchange, and the other single exchange integrals (with coefficient $-\frac{1}{2}$) lead to repulsion; in other words, the wave function corresponds to attraction of bonded atoms and repulsion of non-bonded atoms and is hence a satisfactory wave function to represent the valence-bond structure under discussion.

The valence-bond method has been applied to many problems, some of which are mentioned in the following section. It has been found possible to discuss many of the properties of the chemical bond by approximate wave-mechanical methods; an especially interesting application has been made in the treatment of the mutual orientation of directed valence bonds,[1] leading to the explanation of such properties as the tetrahedral orientation and the equivalence of the four carbon valences.

46f. Resonance among Two or More Valence-bond Structures.—It is found that for many molecules a single wave function of the type given in Equation 46–18 is a good approximation to the correct wave function for the normal state of the system; that is, it corresponds closely to the lowest root of the secular equation for the spin-degeneracy problem. To each of these molecules we attribute a single valence-bond structure, or electronic structure of the type introduced by G. N. Lewis, with two electrons shared between two bonded atoms, as representing satisfactorily the properties of the molecule.

In certain cases, however, it is evident from symmetry or other considerations that more than one valence-bond wave function is important. For example, for six equivalent atoms arranged at the corners of a regular hexagon the two structures I and II of Figure 46–1 are equivalent and must contribute equally to the wave function representing the normal state of the system. It can be shown that, as an approximation, the benzene molecule can be treated as a six-electron system. Of the total of 30 valence electrons of the carbon and hydrogen atoms, 24 can be considered

---

[1] J. C. SLATER, *Phys. Rev.* **37**, 481 (1931); L. PAULING, *J. Am. Chem. Soc.* **53**, 1367 (1931); J. H. VAN VLECK, *J. Chem. Phys.* **1**, 177 (1933); R. HULTGREN, *Phys. Rev.* **40**, 891 (1932).

to be involved in the formation of single bonds between adjacent atoms, giving the structure

$$
\begin{array}{c}
\text{H} \\
| \\
\text{C}
\end{array}
$$

These single bonds use the $1s$ orbital for each of the hydrogen atoms and three of the L orbitals for each carbon atom. There remain six L orbitals for the carbon atoms and six electrons, which can be represented by five independent wave functions corresponding to the five structures of Figure 46–1. We see that structures I and II are the Kekulé structures, with three double bonds between adjacent atoms, whereas the other structures involve only two double bonds between adjacent atoms. If, as an approximation, we consider only the Kekulé structures, we obtain as the secular equation

$$
\begin{vmatrix}
H_{\mathrm{I\,I}} - \Delta_{\mathrm{I\,I}}W & H_{\mathrm{I\,II}} - \Delta_{\mathrm{I\,II}}W \\
H_{\mathrm{I\,II}} - \Delta_{\mathrm{I\,II}}W & H_{\mathrm{II\,II}} - \Delta_{\mathrm{II\,II}}W
\end{vmatrix} = 0,
$$

in which also $H_{\mathrm{II\,II}} = H_{\mathrm{I\,I}}$ and $\Delta_{\mathrm{II\,II}} = \Delta_{\mathrm{I\,I}}$.
The solutions of this are

$$
W = \frac{H_{\mathrm{I\,I}} + H_{\mathrm{I\,II}}}{\Delta_{\mathrm{I\,I}} + \Delta_{\mathrm{I\,II}}}
$$

and

$$
W = \frac{H_{\mathrm{I\,I}} - H_{\mathrm{I\,II}}}{\Delta_{\mathrm{I\,I}} - \Delta_{\mathrm{I\,II}}},
$$

the corresponding wave functions being $\psi_\mathrm{I} + \psi_\mathrm{II}$ and $\psi_\mathrm{I} - \psi_\mathrm{II}$. Thus the normal state of the system is more stable than would correspond to either structure I or structure II. In agreement with the discussion of Section 41, this energy difference is called the *energy of resonance* between the structures I and II.

As a simple example let us discuss the system of four equivalent univalent atoms arranged at the corners of a square. The two structures of a canonical set are

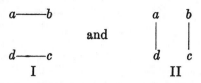

If we neglect all exchange integrals of H except the single exchange integrals between adjacent atoms, which we call α [$\alpha = (ab) = (bc) = (cd) = (da)$], and all exchange integrals occurring in Δ, the secular equation is found by the rules of Section 46e to be

$$\begin{vmatrix} Q + \alpha - W & \tfrac{1}{2}Q + 2\alpha - \tfrac{1}{2}W \\ \tfrac{1}{2}Q + 2\alpha - \tfrac{1}{2}W & Q + \alpha - W \end{vmatrix} = 0.$$

The solutions of this are $W = Q + 2\alpha$ and $W = Q - 2\alpha$, of which the former represents the normal state, α being negative in sign. The energy for a single structure (I or II) is $W_I = Q + \alpha$; hence the *resonance* between the two structures stabilizes the system by the amount α.

Extensive approximate calculations of resonance energies for molecules, especially the aromatic carbon compounds, have been made, and explanations of several previously puzzling phenomena have been developed.[1] Empirical evidence has also been advanced to show the existence of resonance among several valence-bond structures in many simple and complex molecules.[2]

It must be emphasized, as was done in Section 41, that the use of the term *resonance* implies that a certain type of approximate treatment is being used. In this case the treatment is based on the valence-bond wave functions described above, a procedure which is closely related to the systematization of molecule formation developed by chemists over a long period of years, and the introduction of the conception of resonance has permitted the valence-bond picture to be extended to include

[1] E. Hückel, *Z. f. Phys.* **70**, 204 (1931), etc.; L. Pauling and G. W. Wheland, *J. Chem. Phys.* **1**, 362 (1933); L. Pauling and J. Sherman, *ibid.* **1**, 679 (1933); J. Sherman, *ibid.* **2**, 488 (1934); W. G. Penney, *Proc. Roy. Soc.* **A146**, 223 (1934); G. W. Wheland, *J. Chem. Phys.* **3**, 230 (1935).

[2] L. Pauling, *J. Am. Chem. Soc.* **54**, 3570 (1932); *Proc. Nat. Acad. Sci.* **18**, 293 (1932); L. Pauling and J. Sherman, *J. Chem. Phys.* **1**, 606 (1933); G. W. Wheland, *ibid*, **1**, 731 (1933); L. O. Brockway and L. Pauling, *Proc. Nat. Acad. Sci.* **19**, 860 (1933).

previously anomalous cases. A further discussion of this point is given in the following section.

Problem 46-4. Set up the problem of resonance between three equivalent structures or functions ψ_I, ψ_{II}, ψ_{III}, assuming that $H_{I\,I} = H_{II\,II} = H_{III\,III}$, etc. Solve for the energy levels and correct combinations, putting $\Delta_{I\,I} = 1$ and $\Delta_{I\,II} = 0$.

Problem 46-5. Evaluate the energy of a benzene molecule, considered as a six-electron problem: (a) considering only one Kekulé structure; (b) considering both Kekulé structures; (c) considering all five structures. Neglect all exchange integrals of H except

$$(ab) = (bc) = (cd) = (de) = (ef) = (fa) = \alpha,$$

and all exchange integrals entering in Δ.

46g. The Meaning of Chemical Valence Formulas.—The structural formulas of the organic chemist have been determined over a long period of years as a shorthand notation which describes the behavior of the compound in various reactions, indicates the number of isomers, etc. It is only recently that physical methods have shown directly that they are also frequently valid as rather accurate representations of the spatial arrangement of the atoms. The electronic theory of valence attempted to burden them with the additional significance of maps of the positions of the valence electrons. With the advent of quantum mechanics, we know that it is not possible to locate the electrons at definite points in the molecule or even to specify the paths on which they move. However, the positions of maximum electron density can be calculated, and, as shown in Figure 42-4, the formation of a bond does tend to increase the electron density in the region between the bonded atoms, which therefore provides a revised interpretation of the old concept that the valence electrons occupy positions between the atoms.

The discussion of Section 46e shows that, at least in certain cases, the valence-bond picture can be correlated with an approximate solution of the wave-mechanical problem. This correlation, however, is not exact in polyatomic molecules because functions corresponding to other ways of drawing the valence bonds also enter, although usually to a lesser extent.

Thus the valence picture may be said to have a definite significance in terms of wave mechanics in those cases in which one valence-bond wave function is considerably more important than the others, but where this is not true the significance of the structural formulas is less definite. Such less definite cases are those which can be described in terms of resonance. It is notable that the deficiency of the single structural formula in such cases has long been recognized by the organic chemist, who found that no single formula was capable of describing the reactions and isomers of such a substance as benzene. In a sense, the use of the term *resonance* is an effort to extend the usefulness of the valence picture, which otherwise is found to be an imperfect way of describing the state of many molecules.

46h. The Method of Molecular Orbitals.—Another method of approximate treatment of the electronic structure of molecules, called the *method of molecular orbitals*, has been developed and extensively applied, especially by Hund, Mulliken, and Hückel.[1] This method, as usually carried out, consists in the approximate determination of the wave functions (molecular orbitals) and the associated energy values for one electron in a potential field corresponding to the molecule. The energy of the entire molecule is then considered to be the sum of the energies of all the electrons, distributed among the more stable molecular orbitals with no more than two electrons per orbital (Pauli's principle). A refinement of this method has been discussed in Section 43b in connection with the hydrogen molecule.

As an example let us consider the system of four equivalent univalent atoms at the corners of a square, discussed in the previous section by the valence-bond method. The secular equation for a one-electron wave function (molecular orbital), expressed as a linear combination of the four atomic orbitals u_a, u_b, u_c, and u_d, is

$$\begin{vmatrix} q - W & \beta & 0 & \beta \\ \beta & q - W & \beta & 0 \\ 0 & \beta & q - W & \beta \\ \beta & 0 & \beta & q - W \end{vmatrix} = 0,$$

in which q is the Coulomb integral $\int u_a(1) \, H' u_a(1) \, d\tau$ and β is the exchange integral $\int u_a(1) \, H' u_b(1) \, d\tau$ for adjacent atoms, H' being the Hamiltonian operator corresponding to the molecular

---

[1] F. HUND, *Z. f. Phys.* **73**, 1, 565 (1931–1932); R. S. MULLIKEN, *J. Chem. Phys.* **1**, 492 (1933); etc.; J. E. LENNARD-JONES. *Trans. Faraday Soc.* **25**, 668 (1929); E. HÜCKEL, *Z. f. Phys.* **72**, 310 (1931); **76**, 628 (1932); **83**, 632 (1933); *Trans. Faraday Soc.* **30**, 40 (1934).

potential function assumed. We neglect all other integrals. The roots of this equation are

$$W_1 = q + 2\beta,$$
$$W_2 = q,$$
$$W_3 = q,$$
$$W_4 = q - 2\beta.$$

Since β is negative, the two lowest roots are W_1 and W_2 (or W_3), and the total energy for four electrons in the normal state is

$$W = 2W_1 + 2W_2 = 4q + 4\beta.$$

If there were no interaction between atoms a, b and c, d (corresponding to bond formation allowed only between a and b and between c and d), the energy for four electrons would still be $4q + 4\beta$. Accordingly in this example the method of molecular orbitals leads to zero resonance energy. This is in poor agreement with the valence-bond method, which gave the resonance energy α. In most cases, however, it is found that the results of the two methods are in reasonably good agreement, provided that β be given a value equal to about $0.6\,\alpha$ (for aromatic compounds). A comparison of the two methods of treatment has been made by Wheland.[1] It is found that the valence-bond method, when it can be applied, seems to be somewhat more reliable than the molecular-orbital method. On the other hand, the latter method is the more simple one, and can be applied to problems which are too difficult for treatment by the valence-bond method.

Problem 46–6. Treat the system of Problem 46–5 by the molecular-orbital method. Note that the resonance energy given by the two methods is the same if $\beta = 0.553\,\alpha$ (using part c of Problem 46–5).

[1] G. W. WHELAND, *J. Chem. Phys.* **2**, 474 (1934).

CHAPTER XIV

MISCELLANEOUS APPLICATIONS OF QUANTUM MECHANICS

In the following three sections we shall discuss four applications of quantum mechanics to miscellaneous problems, selected from the very large number of applications which have been made. These are: the van der Waals attraction between molecules (Sec. 47), the symmetry properties of molecular wave functions (Sec. 48), statistical quantum mechanics, including the theory of the dielectric constant of a diatomic dipole gas (Sec. 49), and the energy of activation of chemical reactions (Sec. 50). With reluctance we omit mention of many other important applications, such as to the theories of the radioactive decomposition of nuclei, the structure of metals, the diffraction of electrons by gas molecules and crystals, electrode reactions in electrolysis, and heterogeneous catalysis.

47. VAN DER WAALS FORCES

The first detailed treatments of the weak forces between atoms and molecules known as *van der Waals forces* (which are responsible for the constant a of the van der Waals equation of state) were based upon the idea that these forces result from the polarization of one molecule in the field of a permanent dipole moment or quadrupole moment of another molecule,[1] or from the interaction of the permanent dipole or quadrupole moments themselves.[2] With the development of the quantum mechanics it has been recognized (especially by London[3]) that for most molecules these interactions are small compared with another interaction, namely, that corresponding to the polarization of one molecule in the rapidly changing field due to the instan-

---

[1] P. Debye, *Phys. Z.* **21**, 178 (1920); **22**, 302 (1921).

[2] W. H. Keesom, *Proc. Acad. Sci. Amsterdam* **18**, 636 (1915); *Phys. Z.* **22**, 129, 643 (1921).

[3] F. London, *Z. f. Phys.* **63**, 245 (1930).

taneous configuration of electrons and nuclei of another molecule; that is, in the main the polarization of one molecule by the time-varying dipole moment of another. In the following sections we shall discuss the approximate evaluation of the energy of this interaction by variation and perturbation methods for hydrogen atoms (Sec. 47a) and helium atoms (Sec. 47b), and then briefly mention the approximate semiempirical discussion for molecules in general (Sec. 47c).

47a. Van der Waals Forces for Hydrogen Atoms.—For large values of the internuclear distance $r_{AB} = R$ the exchange phenomenon is unimportant, and we can take as the unperturbed wave function for a system of two hydrogen atoms the simple product of two hydrogenlike 1s wave functions,

$$\psi^0 = u_{1sA}(1)\, u_{1sB}(2). \tag{47-1}$$

The perturbation for this function consists of the potential energy terms

$$H' = -\frac{e^2}{r_{B1}} - \frac{e^2}{r_{A2}} + \frac{e^2}{r_{AB}} + \frac{e^2}{r_{12}}. \tag{47-2}$$

Now this expression can be expanded in a Taylor's series in inverse powers of $R = r_{AB}$, to give (with the two atoms located on the z axis)

$$
\begin{aligned}
H' = {} & \frac{e^2}{R^3}(x_1x_2 + y_1y_2 - 2z_1z_2) + \frac{3}{2}\frac{e^2}{R^4}\{r_1^2z_2 - r_2^2z_1 \\
& + (2x_1x_2 + 2y_1y_2 - 3z_1z_2)(z_1 - z_2)\} \\
& + \frac{3}{4}\frac{e^2}{R^5}\{r_1^2r_2^2 - 5r_2^2z_1^2 - 5r_1^2z_2^2 - 15z_1^2z_2^2 \\
& + 2(x_1x_2 + y_1y_2 + 4z_1z_2)^2\} + \cdots ,
\end{aligned}
\tag{47-3}
$$

in which x_1, y_1, z_1 are coordinates of the first electron relative to its nucleus, and x_2, y_2, z_2 are coordinates of the second electron relative to its nucleus. The first term represents the interaction of the dipole moments of the two atoms, the second the dipole-quadrupole interaction, the third the quadrupole-quadrupole interaction, and so on.

Let us first consider only the dipole-dipole interaction, using the approximate second-order perturbation treatment[1] of Section

---

[1] The first-order perturbation energy is zero, as can be seen from inspection of the perturbation function.

27e. It is necessary for us to evaluate the integral

$$\int \psi^{0*}(H')^2\psi^0 d\tau,$$

with H' given by

$$H' = \frac{e^2}{R^3}(x_1x_2 + y_1y_2 - 2z_1z_2). \tag{47-4}$$

It is seen that the cross-products in $(H')^2$ vanish on integration, so that we obtain

$$(H'^2)_{00} = \frac{e^4}{R^6}\int \psi^{0*}(x_1^2x_2^2 + y_1^2y_2^2 + 4z_1^2z_2^2)\psi^0 d\tau$$

or

$$(H'^2)_{00} = \frac{2e^4}{3R^6}\int \psi^{0*}r_1^2r_2^2\psi^0 d\tau = \frac{2e^4}{3R^6}\overline{r_1^2}\,\overline{r_2^2}. \tag{47-5}$$

This expression, with $\overline{r_1^2}$ and $\overline{r_2^2}$ replaced by their value $3a_0^2$ (Sec. 21c), gives, when introduced in Equation 27-47 together with $W_0^0 = -e^2/a_0$, the value for the interaction energy

$$W_0'' = -\frac{6e^2a_0^5}{R^6}. \tag{47-6}$$

The fact that this value is also given by the variation method with the variation function $\psi^0(1 + AH')$ shows that this is an upper limit for W_0'' (a lower limit for the coefficient of $-e^2a_0^5/R^6$). Moreover, by an argument similar to that of the next to the last paragraph of Section 27e it can be shown that the value $-8\frac{e^2a_0^5}{R^6}$ is a lower limit to W_0'', so that we have thus determined the value of the dipole-dipole interaction to within about 15 per cent.

Variation treatments of this problem have been given by Slater and Kirkwood,[1] Hassé,[2] and Pauling and Beach.[3] It can be easily shown[4] that the second-order perturbation energy can be obtained by the use of a variation function of the form

$$\psi = \psi^0\{1 + H'f(r_1, r_2)\},$$

with H' given by Equation 47-4. The results of the variation

[1] J. C. SLATER and J. G. KIRKWOOD, *Phys. Rev.* **37**, 682 (1931).

[2] H. R. HASSÉ, *Proc. Cambridge Phil. Soc.* **27**, 66 (1931). A rough treatment for various states has been given by J. Podolanski, *Ann. d. Phys.* **10**, 695 (1931).

[3] L. PAULING and J. Y. BEACH, *Phys. Rev.* **47**, 686 (1935).

[4] This was first shown by Slater and Kirkwood.

treatment for different functions $f(r_1, r_2)$ are given in Table 47–1. It is seen that the coefficient of $-e^2a_0^5/R^6$ approaches a value[1] only slightly larger than 6.499; this can be accepted as very close to the correct value.

So far we have considered only dipole-dipole interactions. Margenau[2] has applied the approximate second-order perturbation method of Section 27e to the three terms of Equation 47-3, obtaining the expression

$$W_0'' \cong -\frac{6e^2a_0^5}{R^6} - \frac{135e^2a_0^7}{R^8} - \frac{1416e^2a_0^9}{R^{10}} + \cdots . \quad (47–7)$$

It is seen that the higher-order terms become important at small distances.

TABLE 47–1.—VARIATION TREATMENT OF VAN DER WAALS INTERACTION
OF TWO HYDROGEN ATOMS

Variation function $u_{1sA}(1)\ u_{1sB}(2)\left\{1 + \dfrac{e^2}{R^3}(x_1x_2 + y_1y_2 - 2z_1z_2)f(r_1, r_2)\right\}$

$f(r_1, r_2)$	$E - W^0$	Reference*
1. A	$-6.00e^2a_0^5/R^6$	H
2. $-\frac{1}{2}r_1r_2/(r_1 + r_2)$	-6.14	SK
3. $A + B(r_1 + r_2)$	-6.462	PB
4. $A + B\,r_1r_2$	-6.469	H
5. $A + B(r_1 + r_2) + C\,r_1r_2$	-6.482	PB
6. $A\,r_1^\nu r_2^\nu(\nu = 0.325)$	-6.49	SK
7. $A + B\,r_1r_2 + C\,r_1^2r_2^2$	-6.490	H
8. $A + B\,r_1r_2 + C\,r_1^2r_2^2 + D\,r_1^3r_2^3$	-6.498	H
9. Polynomial† to $r_1^2r_2^2$	-6.4984	PB
10. Polynomial to $r_1^3r_2^3$	-6.49899	PB
11. Polynomial to $r_1^4r_2^4$	-6.49903	PB

* H = Hassé, SK = Slater and Kirkwood, PB = Pauling and Beach.
† The polynomial contains all terms of degree 2 or less in r_1 and 2 or less in r_2.

[1] A straightforward but approximate application of second-order perturbation theory by R. Eisenschitz and F. London gave the value 6.47 for this coefficient [*Z. f. Phys.* **60**, 491 (1930)]. The first attack on this problem was made by S. C. Wang, *Phys. Z.* **28**, 663 (1927). The value found by him for the coefficient, $2\,4\%_8 = 8.68$, must be in error (as first pointed out by Eisenschitz and London), being larger than the upper limit 8 given above. The source of the error has been pointed out by Pauling and Beach, *loc. cit.*

[2] H. MARGENAU, *Phys. Rev.* **38**, 747 (1931). More accurate values of the coefficients have been calculated by Pauling and Beach, *loc. cit.*

47b. Van der Waals Forces for Helium.—In treating the dipole-dipole interaction of two helium atoms, the expression for H' consists of four terms like that of Equation 47–4, corresponding to taking the electrons in pairs (each pair consisting of an electron on one atom and one on the other atom). The variation function has the form

$$\psi = \psi^0 \left\{ 1 + \sum_{i,j} H'_{ij} f(r_i, r_j) \right\}.$$

Hassé[1] has considered five variation functions of this form, shown with their results in Table 47–2. The success of his similar treatment of the polarizability of helium (function 6 of Table 29–3) makes it probable that the value $-1.413 e^2 a_0^5 / R^6$ for W'' is not in error by more than a few per cent. Slater and Kirkwood[1] obtained values 1.13, 1.78, and 1.59 for the coefficient of $-e^2 a_0^5 / R^6$ by the use of variation functions based on their helium atom functions mentioned in Section 29e. An approximate discussion of dipole-quadrupole and quadrupole-quadrupole interactions has been given by Margenau.[1]

TABLE 47–2.—VARIATION TREATMENT OF VAN DER WAALS INTERACTION OF TWO HELIUM ATOMS

ψ^0	$f(r_1 r_2)$	$E - W^0$
1. $e^{-Z's}$	A	$-1.079 e^2 a_0^5 / R^6$
2. $e^{-Z's}$	$A + B r_1 r_2$	-1.225
3. $e^{-Z's}$	$A + B r_1 r_2 + C r_1^2 r_2^2$	-1.226
4. $e^{-Z's}(1 + c_1 u)$	A	-1.280
5. $e^{-Z's}(1 + c_1 u)$	$A + B r_1 r_2$	-1.413

47c. The Estimation of van der Waals Forces from Molecular Polarizabilities.—London[2] has suggested a rough method of estimating the van der Waals forces between two atoms or molecules, based on the approximate second-order perturbation treatment of Section 27e. We obtain by this treatment (see Secs. 27e and 29e) the expression

$$\alpha \cong \frac{2 n e^2 \overline{z^2}}{I} \tag{47–8}$$

---

[1] *Loc. cit.*

[2] F. LONDON, *Z. f. Phys.* **63**, 245 (1930).

for the polarizability of an atom or molecule, in which n is the number of effective electrons, $\overline{z^2}$ the average value of z^2 for these electrons (z being the coordinate of the electron relative to the nucleus in the field direction), and I the energy difference of the normal state and the effective zero point for energy, about equal in value to the first ionization energy. The van der Waals interaction energy may be similarly written as

$$W'' = -\frac{6n_A n_B e^4 \overline{z_A^2}\, \overline{z_B^2}}{R^6(I_A + I_B)}, \qquad (47\text{-}9)$$

which becomes on introduction of α_A and α_B

$$W'' = -\frac{3}{2}\frac{\alpha_A \alpha_B}{R^6}\frac{I_A I_B}{I_A + I_B}, \qquad (47\text{-}10)$$

or, in case the molecules are identical,

$$W'' = -\frac{3}{4}\frac{\alpha^2 I}{R^6}. \qquad (47\text{-}11)$$

With α in units 10^{-24} cm^3 and I in volt electrons, this is

$$W'' = -\frac{De^2 a_0^5}{R^6},$$

in which

$$D = 1.27\alpha^2 I.$$

It must be realized that this is only a very rough approximation. For hydrogen atoms it yields $D = 7.65$ (correct value 6.50) and for helium 1.31 (correct value about 1.4).

For the further discussion of the validity of London's relation between van der Waals forces and polarizabilities, and of other applications of the relation, such as to the heats of sublimation of molecular crystals and the unactivated adsorption of gases by solids, the reader is referred to the original papers.[1]

48. THE SYMMETRY PROPERTIES OF MOLECULAR WAVE FUNCTIONS

In this section we shall discuss the symmetry properties of molecular wave functions to the extent necessary for an under-

[1] F. LONDON, loc. cit.; F. LONDON and M. POLANYI, Z. f. phys. Chem. 11B, 222 (1930); M. POLANYI, Trans. Faraday Soc. 28, 316 (1932); J. E. LENNARD-JONES, ibid. 28, 333 (1932).

standing of the meaning and significance of the term symbols used for diatomic molecules by the spectroscopist.

In Section 34 it was mentioned that the nuclear and electronic parts of an approximate wave function for a molecule can be separated by referring the electronic coordinates to axes determined by the nuclear configuration. Let us now discuss this choice of coordinates for a diatomic molecule in greater detail. We first introduce the Cartesian coordinates X, Y, Z of the center of mass of the two nuclei relative to axes fixed in space, and the polar coordinates r, ϑ, φ of nucleus A relative to a point midway between nucleus A and nucleus B as origin,[1] also referred to axes

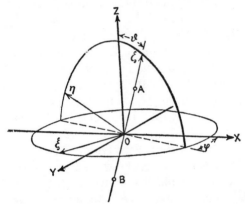

Fig. 48–1.—The relation between axes ξ, η, ζ and X, Y, Z.

fixed in space, as indicated in Figure 48–1. We next introduce the Cartesian coordinates ξ_i, η_i, ζ_i or the polar coordinates r_i, ϑ_i, φ_i of each of the electrons, measured with reference, not to axes fixed in space, but instead to axes dependent on the angular coordinates ϑ and φ determining the orientation of the nuclear axis. These axes, ξ, η, ζ, are chosen in the following way. ζ is taken along the nuclear axis OA (Fig. 48–1), and ξ lies in the XY plane, its sense being such that the Z axis lies between the η and ζ axes (ξ, η, ζ forming a left-handed system, say). It is, moreover, often convenient to refer the azimuthal angles of all electrons but one to the azimuthal angle of this electron, using the coordinates φ_1, $\varphi_2 - \varphi_1$, $\varphi_3 - \varphi_1$, $\cdots$ in place of φ_1, φ_2, φ_3, $\cdots$.

---

[1] It is convenient in this section to use these coordinates, which differ slightly from those adopted in Chapter X.

It has been shown[1] that these coordinates can be introduced in the wave equation, and that the wave functions then assume a simple form. We have discussed the wave function for the nuclear motions in detail in Chapter X. The only part of the electronic wave function which can be written down at once is that dependent on φ_1. Inasmuch as the potential energy of the system is independent of φ_1 (as a result of our subterfuge of measuring the φ's of the other electrons relative to φ_1), φ_1 is a cyclic coordinate, and occurs in the wave function only in the factor $e^{\pm i\Lambda\varphi_1}$, in which Λ can assume the values 0, 1, 2, $\cdots$. The quantum number Λ thus determines the magnitude of the component of electronic orbital angular momentum along the line joining the nuclei. [Λ is somewhat analogous to the component M_L of the resultant orbital angular momentum (or azimuthal) quantum number L for atoms.] The value of Λ is expressed by the principal character of a molecular term symbol: Σ denoting $\Lambda = 0$; Π, $\Lambda = \pm 1$; Δ, $\Lambda = \pm 2$; etc. As in the case of atomic terms, the multiplicity due to electron spin is indicated by a superscript to the left, $^1\Sigma$ indicating a singlet, $^2\Sigma$ a doublet, etc.

It may be mentioned that if we ignore the interactions of the electronic and nuclear motions the wave functions corresponding to Λ and $-\Lambda$ correspond to identical energy values. This degeneracy is removed by these interactions, however, which lead to a small splitting of energy levels for $\Lambda > 0$, called Λ-*type doubling*.[2] The correct wave functions are then the sum and difference of those corresponding to Λ and $-\Lambda$.

In the following sections we shall discuss the characteristic properties of diatomic molecules containing two identical nuclei (*symmetrical diatomic molecules*).

48a. Even and Odd Electronic Wave Functions. Selection Rules.—By the argument of Section 40e we have shown that the transition probabilities for a diatomic molecule are determined in the main by the electric-moment integrals over the electronic parts of the wave functions, taken relative to the axes ξ, η, ζ determined by the positions of the nuclei. Let us now classify the electronic wave functions of symmetrical diatomic molecules

[1] F. Hund, *Z. f. Phys.* **42**, 93 (1927); R. deL. Kronig, *ibid.* **46**, 814; **50**, 347 (1928); E. Wigner and E. E. Witmer, *ibid.* **51**, 859 (1928).

[2] See, for example, J. H. Van Vleck, *Phys. Rev.* **33**, 467 (1929).

as *even* or *odd*, introducing the subscripts g (German *gerade*) for even terms and u (*ungerade*) for odd terms in the term symbols for identification. This classification depends on the behavior of the electronic wave function with respect to the transformation $\xi_i,\ \eta_i,\ \zeta_i \to -\xi_i,\ -\eta_i,\ -\zeta_i$, that is, on inversion through the origin, even functions remaining unchanged by this operation, and odd functions changing sign. The argument of Section 40g leads to the following selection rule: *Transitions are allowed only between even and odd levels* ($g \to u,\ u \to g$).

(Although electronic wave functions for diatomic molecules containing unlike nuclei cannot be rigorously classified as even or odd, they often approach members of these classes rather closely, and obey an approximate selection rule of the above type.)

48b. The Nuclear Symmetry Character of the Electronic Wave Function.—We are now in a position to discuss the nuclear symmetry character of the electronic wave function for a diatomic molecule in which the nuclei are identical. Interchanging the two nuclei A and B converts ϑ into $\pi - \vartheta$ and φ into $\pi + \varphi$; these coordinates, however, do not occur in the electronic wave function. The interchange of the nuclei also converts the coordinates $\xi_i,\ \eta_i,\ \zeta_i$ of each electron into $-\xi_i,\ \eta_i,\ -\zeta_i$, and hence $r_i,\ \vartheta_i,\ \varphi_i$ into $r_i,\ \pi - \vartheta_i,\ \pi - \varphi_i$ [or $\varphi_i - \varphi_1$ into $-(\varphi_i - \varphi_1)$]. In case that the electronic wave function is left unchanged by this transformation, the electronic wave function is *symmetric in the nuclei;* if the factor -1 is introduced by the transformation, the electronic wave function is *antisymmetric in the nuclei.*

The nuclear symmetry character of the electronic wave function is represented in the term symbol by introducing the superscript $+$ or $-$ after taking cognizance of the presence of the subscript g or u discussed in Section 48a, the combinations $\overset{+}{g}$ and $\overset{-}{u}$ representing electronic wave functions symmetric in the nuclei, and $\overset{-}{g}$ and $\overset{+}{u}$ those antisymmetric in the nuclei. Thus we see that

$$\Sigma_g^+ \text{ and } \Sigma_u^- \text{ are } S^N$$

and

$$\Sigma_g^- \text{ and } \Sigma_u^+ \text{ are } A^N.$$

For $\Lambda \neq 0$ there is little need to represent the symmetry character in the term symbol, inasmuch as the S^N and A^N states occur in pairs corresponding to nearly the same energy value (Λ-type doubling), and in consequence the $+$ and $-$ superscripts are usually omitted.

The states with superscript $+$ are called *positive states*, and those with superscript $-$ *negative states*.

The principal use of the nuclear symmetry character is in determining the allowed values of the rotational quantum number K of the molecule. The complete wave functions for a molecule (including the nuclear-spin function) must be either symmetric or antisymmetric in the nuclei, depending on the nature of the nuclei involved. If the nuclei have no spins, then the existent functions are of one or the other of the types listed below.

I. Complete wave function S^N:

$$\overset{+}{g}, \; K \text{ even} \xleftarrow{\;\;\Delta K \text{ even}\;\;} \overset{-}{u}, \; K \text{ even}$$

$$\big\updownarrow \Delta K \text{ odd} \qquad\qquad \big\updownarrow \Delta K \text{ odd}$$

$$\overset{+}{u}, \; K \text{ odd} \xleftarrow{\;\;\Delta K \text{ even}\;\;} \overset{-}{g}, \; K \text{ odd}$$

II. Complete wave function A^N:

$$\overset{+}{g}, \; K \text{ odd} \xleftarrow{\;\;\Delta K \text{ even}\;\;} \overset{-}{u}, \; K \text{ odd}$$

$$\big\updownarrow \Delta K \text{ odd} \qquad\qquad \big\updownarrow \Delta K \text{ odd}$$

$$\overset{+}{u}, \; K \text{ even} \xleftarrow{\;\;\Delta K \text{ even}\;\;} \overset{-}{g}, \; K \text{ even}$$

It is seen that in either case the transitions allowed by the selection rule $g \leftrightarrow u$ are such that ΔK is even for $+ \rightarrow -$ or $- \rightarrow +$ transitions, and odd for $+ \rightarrow +$ or $- \rightarrow -$ transitions.

The selection rule $\Delta K = 0, \pm 1$ can be derived by the methods of Chapter XI; this becomes $\Delta K = 0$ for positive $\leftrightarrow$ negative transitions, and $\Delta K = \pm 1$ for positive $\rightarrow$ positive or negative $\rightarrow$ negative transitions.

In case that the nuclei possess spins, with spin quantum number I, both types of functions and transitions occur (the two not forming combinations), with the relative weights $(I + 1)/I$ or $I/(I + 1)$, as discussed in Section 43*f*.

Let us now consider a very simple example, in order to clarify the question; namely, the case of a molecule possessing only one electron, in the states represented by approximate wave functions which can be built from the four orbitals $u_A = s, p_z, p_x, p_y$ about nucleus A, and four similar ones u_B about nucleus B; s, p_z, p_x, p_y being real one-electron wave functions such as given in Table 21-4 for the L shell. We can combine these into eight functions of the form $s_A + s_B$, $s_A - s_B$, etc. If the functions are referred to parallel axes for the two atoms and taken as in Table 21-4

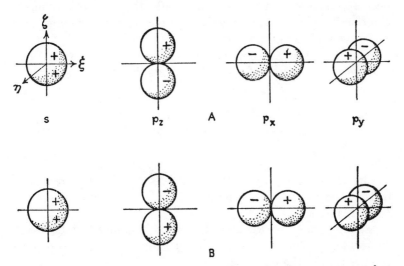

Fig. 48-2.—Positive and negative regions of wave functions s, p_z, p_x, and p_y for atoms A and B.

except for a factor -1 for p_{z_B} (introduced for convenience), then they have the general nature shown in Figure 48-2, in which the functions $u_A + u_B$ are designated, the plus and minus signs representing regions equivalent except for sign. From the inspection of this figure and a similar one for $u_A - u_B$ (in which the signs are changed for u_B), it is seen that the eight functions have the following symmetry character in the nuclei:

Function	s	p_z	p_x	p_y
$u_A + u_B$	S^N	S^N	A^N	S^N
$u_A - u_B$	A^N	A^N	S^N	A^N

By the argument given above we know that four of these are Σ states, with $\Lambda = 0$, and four are Π states. The Π states are those formed from p_x and p_y (which are the linear combinations of the complex exponential functions p_{+1} and p_{-1}). The two Π states $u_A + u_B$ are separated widely by the exchange integrals from the two $u_A - u_B$, and the Λ-type doubling will cause a further small separation of the nuclear-symmetric and nuclear-antisymmetric levels. The exchange terms similarly separate the $u_A + u_B$ s and p_z functions from the $u_A - u_B$ functions. The best approximate wave functions would then be certain linear combinations of the two nuclear-symmetric functions and also of the two nuclear-antisymmetric functions.

We can now write complete term symbols for the eight electronic wave functions of our simple example, as follows:

$$
\begin{array}{ccccc}
 & s & p_z & p_x & p_y \\
u_A + u_B & {}^2\Sigma_g^+ & {}^2\Sigma_g^+ & \{{}^2\Pi_u^+ & {}^2\Pi_u^-\} \\
u_A - u_B & {}^2\Sigma_u^+ & {}^2\Sigma_u^+ & \{{}^2\Pi_g^+ & {}^2\Pi_g^-\}.
\end{array}
$$

The identification as even or odd is easily made by inspection of Figure 48–2. The two ${}^2\Pi_u$ terms (one S^N and one A^N) are placed in brackets to show that they form a Λ-type doublet, as are the two ${}^2\Pi_g$ terms.

48c. Summary of Results Regarding Symmetrical Diatomic Molecules.—The various symmetry properties which we have considered are the following:

1. Even and odd electronic functions, indicated by subscripts g and u (Sec. 48a). Selection rule: Transitions allowed only between g and u.

2. The nuclear symmetry of the complete wave function (including rotation of the molecule but not nuclear spin). Selection rule: Symmetric-antisymmetric transitions not allowed.

3. The nuclear symmetry of the electronic wave function, represented by the superscripts $+$ and $-$, $\overset{+}{g}$ and $\overset{-}{u}$ being S^N; $\overset{-}{g}$ and $\overset{+}{u}$, A^N. Selection rule: $\Delta K = 0$ for positive-negative transitions, and $\Delta K = \pm 1$ for positive-positive and negative-negative transitions. (This is not independent of 1 and 2. In practice 1 and 3 are usually applied.)

We are now in a position to discuss the nature of the spectral lines to be expected for a symmetrical diatomic molecule. We have not treated the spin moment vector of the electrons, which combines with the angular momentum vectors Λ and K in various ways to form resultants; the details of this can be found in the treatises on molecular spectroscopy listed at the end of Chapter X. Let us now for simplicity consider transitions among $^1\Sigma$ states, assuming that the nuclei have no spins, and that the existent complete wave functions are symmetric in the nuclei (as for helium). The allowed rotational states are then those with K even for $^1\Sigma_g^+$ and $^1\Sigma_u^-$, and those with K odd for $^1\Sigma_g^-$ and $^1\Sigma_u^+$, and the transitions allowed by 1 and 3 are the following:

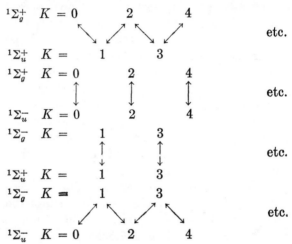

49. STATISTICAL QUANTUM MECHANICS. SYSTEMS IN THERMODYNAMIC EQUILIBRIUM

The subject of statistical mechanics is a branch of mechanics which has been found very useful in the discussion of the properties of complicated systems, such as a gas. In the following sections we shall give a brief discussion of the fundamental theorem of statistical quantum mechanics (Sec. 49a), its application to a simple system (Sec. 49b), the Boltzmann distribution law (Sec. 49c), Fermi-Dirac and Bose-Einstein statistics (Sec. 49d), the rotational and vibrational energy of molecules (Sec. 49e), and the dielectric constant of a diatomic dipole gas (Sec. 49f). The discussion in these sections is mainly descriptive and elementary; we have made no effort to carry through the difficult derivations or to enter into the refined arguments needed in a thorough and detailed treatment of the subject, but have endeavored to present an understandable general survey.

49a. The Fundamental Theorem of Statistical Quantum Mechanics.—Let us consider a large system with total energy known to lie in the range W to $W + \Delta W$. We inquire as to the properties of this system. If we knew the wave function representing the system, values of the dynamical quantities corresponding to the properties of the system could be calculated by the methods of Section 12*d*. In general, however, there will be many stationary states of the system (especially if it be a very complicated system, such as a sample of gas of measurable volume) with energy values lying in the range W to $W + \Delta W$, and our knowledge of the state of the system may not allow us to select one wave function alone as representing the system. Moreover, it might be possible for us to find a set of approximate wave functions for the system by ignoring weak interactions of parts of the system with each other or of the system and its environment; no one of these approximate wave functions would represent the state of the system over any appreciable period of time, and so we would not be justified in selecting any one of them for use in calculating values of dynamical quantities.

Under these circumstances we might make calculations regarding the properties of the system for each of the wave functions with energy between W and ΔW, and then average the various calculations to obtain predictions regarding the average expected behavior of the system. The important question immediately arises as to what weights are to be assigned the various wave functions in carrying out this averaging. The answer to this question is given by the fundamental theorem of statistical quantum mechanics, as follows: *In calculating average values of properties of a system with energy between W and ΔW, the same weight is to be assigned to every accessible wave function with energy in this range, in default of other information.* (The wave functions are of course to be normalized and mutually orthogonal.) This theorem can be derived from the equations of quantum mechanics (by methods such as the variation of constants, discussed in Chapter XI), with the aid of an additional postulate,[1] which is the quantum-mechanical analogue of the

---

[1] The postulate of *randomness of phases*. See, for example, W. Pauli, "Probleme der modernen Physik," S. Hirzel, Leipzig, 1928.

ergodic hypothesis of classical statistical mechanics. We shall not discuss this derivation.

The word *accessible* appears in the theorem for the following reason. If a system is known to be in one state at a given instant, and if it is known that it is impossible for any operative perturbation to cause a transition to a certain other state, then it is obviously wrong to include this latter state in the expression for the average. We have already met such non-combining states in our discussion of the symmetry of wave functions for collections of identical particles (Secs. 29b, 30a). It was shown that if the system is known to be represented by a wave function symmetrical in all the identical particles composing it, no perturbation can cause it to change over to a state with an antisymmetrical wave function. The nature of the wave functions which actually occur is dependent upon the nature of the system. If it is composed of electrons or protons, the wave functions must be antisymmetric; if it is composed of hydrogen atoms, thought of as entities, the wave functions must be symmetric in these atoms; etc. Moreover, we may sometimes have to take the passage of time into consideration in interpreting the word accessible. Let us consider as our system a helium atom, for example, which is known at the time $t = 0$ to be in some excited singlet state, the wave function being symmetric in the positions of the electrons and antisymmetric in their spins. Transitions to triplet states can occur only as a result of perturbations affecting the electron spins; and, since these perturbations are very small, the probability of transition to all triplet states in a short time will be very small. In predicting properties for this system for a short period after the time $t = 0$, we would accordingly be justified in considering only the singlet states as accessible.

49b. A Simple Application.—In order to illustrate the use of the fundamental theorem of statistical quantum mechanics, we shall discuss a very simple problem in detail.

Let us consider a system composed of five harmonic oscillators, all with the same characteristic frequency ν, which are coupled with one another by weak interactions. The set of product wave functions $\Psi(a)\Psi(b)\Psi(c)\Psi(d)\Psi(e)$ can be used to construct approximate wave functions for the system by the use of the method of variation of constants (Chap. XI). Here $\Psi(a)$, $\cdots$

represent the harmonic oscillator wave functions (Sec. 11), the letters a, b, c, d, e representing the coordinates of the five oscillators. For each oscillator there is a set of functions $\Psi_{n_a}(a)$ corresponding to the values 0, 1, 2, $\cdots$ for the quantum number n_a. The total unperturbed energy of the system is $W_n^0 = (n_a + \frac{1}{2})h\nu + \cdots + (n_e + \frac{1}{2})h\nu = (n + \frac{5}{2})h\nu$, in which $n = n_a + n_b + n_c + n_d + n_e$.

The application of the variation-of-constants treatment shows that if the system at one time is known to have a total energy value close to $W_{n'}^0$, where n' is a particular value of the quantum number n, then the wave function at later times can be expressed essentially as a combination of the product wave functions for $n = n'$, the wave functions for $n \neq n'$ making a negligible contribution provided that the mutual interactions of the oscillators are weak. Let us suppose that the system has an energy value close to $12\frac{1}{2}h\nu$, that is, that n' is equal to 10. The product wave functions corresponding to this value of n' are those represented by the 1001 sets of values of the quantum numbers $n_a, \cdots, n_e$ given in Table 49–1.

TABLE 49–1.—SETS OF QUANTUM NUMBERS FOR FIVE COUPLED HARMONIC OSCILLATORS WITH TOTAL QUANTUM NUMBER 10

n_a n_b n_c n_d n_e		n_a n_b n_c n_d n_e	
10 . 0 . 0 . 0 . 0 etc.*	(5)	6 . 2 . 1 . 1 . 0 etc.	(60)
9 . 1 . 0 . 0 . 0	(20)	5 . 3 . 1 . 1 . 0	(60)
8 . 2 . 0 . 0 . 0	(20)	5 . 2 . 2 . 1 . 0	(60)
7 . 3 . 0 . 0 . 0	(20)	4 . 4 . 1 . 1 . 0	(30)
6 . 4 . 0 . 0 . 0	(20)	A 4 . 3 . 2 . 1 . 0	(120)
5 . 5 . 0 . 0 . 0	(10)	4 . 2 . 2 . 2 . 0	(20)
8 . 1 . 1 . 0 . 0	(30)	3 . 3 . 3 . 1 . 0	(20)
7 . 2 . 1 . 0 . 0	(60)	3 . 3 . 2 . 2 . 0	(30)
6 . 3 . 1 . 0 . 0	(60)	6 . 1 . 1 . 1 . 1	(5)
6 . 2 . 2 . 0 . 0	(30)	5 . 2 . 1 . 1 . 1	(20)
5 . 4 . 1 . 0 . 0	(60)	4 . 3 . 1 . 1 . 1	(20)
5 . 3 . 2 . 0 . 0	(60)	4 . 2 . 2 . 1 . 1	(30)
4 . 4 . 2 . 0 . 0	(30)	3 . 3 . 2 . 1 . 1	(30)
4 . 3 . 3 . 0 . 0	(30)	3 . 2 . 2 . 2 . 1	(20)
7 . 1 . 1 . 1 . 0	(20)	2 . 2 . 2 . 2 . 2	(1)

* The other sets indicated by "etc." are in this case 0 . 10 . 0 . 0 . 0, 0 . 0 . 10 . 0 . 0, 0 . 0 . 0 . 10 . 0, and 0 . 0 . 0 . 0 . 10, a total of five, as shown by the number in parentheses.

In case that the interactions between the oscillators are of a general nature (the ab, ac, bc, $\cdots$ interactions being different), all of the product functions will be accessible, and the fundamental theorem then requires that over a long period of time the 1001 product functions will contribute equally to the wave function of the system. In calculating the contribution of oscillator a, for example, to the properties of the system, we would calculate the properties of oscillator a in the states $n_a = 0$ [using the wave function $\Psi_0(a)$], $n_a = 1, \cdots, n_a = 10$, and then

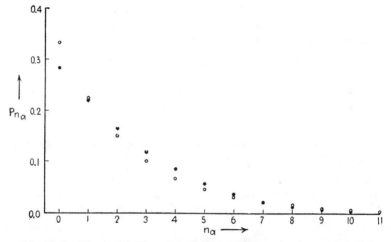

Fig. 49-1.—The probability values P_{n_a} for system-part a in a system of five coupled harmonic oscillators with total quantum number $n = 10$ (closed circles), and values calculated by the Boltzmann distribution law (open circles).

average them, using as weights the numbers of times that $n_a = 0, 1, 2, \cdots, 10$ occur in Table 49-1. These weights are given in Table 49-2. The numbers obtained by dividing by the total (1001) can be described as the probabilities that oscillator a (or b, c, $\cdots$) be in the states $n_a = 0, 1, 2, \cdots, 10$. These probability values are represented graphically in Figure 49-1.

49c. The Boltzmann Distribution Law.—We have been discussing a system composed of a small number (five) of weakly interacting parts. A similar discussion (which we shall not give because it is necessarily rather involved) of a system composed of an extremely large number of weakly interacting parts can be carried through, leading to a general expression for the probability of distribution of any one of the parts among its

TABLE 49-2.—WEIGHTS FOR STATES OF INDIVIDUAL OSCILLATORS IN COUPLED SYSTEM

n_a, etc.	Weight	Probability P_{n_a}
0	286	0.286
1	220	.220
2	165	.165
3	120	.120
4	84	.084
5	56	.056
6	35	.035
7	20	.020
8	10	.010
9	4	.004
10	1	.001
Total.........	1001	1.001

stationary states.[1]   The result of the treatment is the *Boltzmann distribution law* in its quantum-mechanical form:

If all the product wave functions $\Psi(a)\Psi(b)\cdots$ of a system composed of a very large number of weakly interacting parts a, b, $\cdots$ are accessible, then the probability of distribution of one of the parts, say a, among its states, represented by the quantum number n_a, is given by the equation

$$P_{n_a} = Ae^{-\frac{W_{n_a}}{kT}}, \qquad (49\text{-}1)$$

in which W_{n_a} is the energy of the part a in its various states and the constant A has such a value as to make

$$\sum_{n_a=0}^{\infty} P_{n_a} = 1. \qquad (49\text{-}2)$$

There is considered to be one state for every independent wave function $\Psi(a)$. The exponential factor, called the *Boltzmann exponential factor*, is the same as in the classical Boltzmann distribution law, which differs from Equation 49-1 only in the way the state of the system part is described. The constant k is the *Boltzmann constant*, with the value 1.3709×10^{-16} erg deg^{-1}. The absolute temperature T occurring in Equation 49-1

[1] That is, among the stationary states for this part of the system when isolated from the other parts.

is introduced in the derivation of this equation by methods closely similar to those of classical statistical mechanics.

Some indication of the reasonableness of this equation is given by comparing it with the results of our discussion of the system of five coupled harmonic oscillators. The open circles in Figure 49–1 represent values of P_{n_a} calculated by Equation 49–1, with kT placed equal to $\frac{5}{2}h\nu$ (this leading approximately to the average value $\frac{5}{2}h\nu$ for W_{n_a}, as assumed in the earlier discussion). It is seen that there is general agreement, the discrepancies arising from the fact that the number of parts of the system (five) is small (rather than very large, as required in order that the Boltzmann distribution law be applicable).

In Equation 49–1 each wave function is represented separately. It is often convenient to group together all wave functions corresponding to the same energy, and to write

$$P_i = A p_i e^{-\frac{W_i}{kT}}, \tag{49–3}$$

in which p_i is the degree of degeneracy or *a priori probability* or *quantum weight* of the energy level W_i.

In case that the wave functions for the part of the system under consideration are very numerous and correspond to energy values lying very close together, it is convenient to rewrite the distribution law in terms of $P(W)$, such that $P(W)dW$ is the probability that the energy of the system part lie between W and $W + dW$, in the form

$$P(W) = A p(W) e^{-\frac{W}{kT}}, \tag{49–4}$$

in which $p(W)dW$ is the number of wave functions for the system part in the energy range W to $W + dW$.

As an illustration of the use of Equation 49–4 let us consider the distribution in translational energy of the molecules of a gas (the entire gas being the system and the molecules the system parts) such that all product wave functions are accessible.[1] It is found (by the use of the results of Section 14, for example) that $p(W)$ is given by the equation

$$p(W) = \frac{4\sqrt{2}\pi m^{3/2} V}{h^3} W^{1/2}, \tag{49–5}$$

[1] We shall see in the next section that actual gases are not of this type.

in which V is the volume of the box containing the gas and m is the mass of a molecule. The *Maxwell distribution law for velocities* is obtained by substituting this in Equation 49–4 and replacing W by $\frac{1}{2}mv^2$, v being the velocity of the molecule.

Problem 49–1. Derive Equation 49–5 with the use of the results of Section 14. By equating W to the kinetic energy $\frac{1}{2}mv^2$ (v being the velocity), derive the Maxwell distribution law for velocities, and from it calculate expressions for the mean velocity and root-mean-square velocity of gas molecules.

It will be shown in the following section that the Boltzmann distribution law is usually not strictly applicable in discussing the translational motion of molecules.

49d. Fermi-Dirac and Bose-Einstein Statistics.—As stated in the foregoing section, the Boltzmann distribution law is applicable to the parts of a system for which all product wave functions are accessible. The parts of such a system are said to conform to *Boltzmann statistics*. Very often, however, we encounter systems for which not all product wave functions are accessible. We have seen before (Sec. 29, etc.) that the wave functions for a system of identical particles can be grouped into non-combining sets of different symmetry character, one set being completely symmetric in the coordinates of the particles, one completely antisymmetric, and the others of intermediate symmetry character. Only the wave functions of one symmetry character are accessible to a given system of identical particles.

Thus our simple system of five harmonic oscillators would be restricted to wave functions of one symmetry character if the interactions ab, ac, bc, $\cdots$ were equivalent, that is, if the oscillators were identical.[1] It was to avoid this that we made the explicit assumption of non-equivalence of the interactions in Section 49b. The accessible wave functions for five identical oscillators would be the completely symmetric ones, the com-

---

[1] In order for the oscillators to behave identically with respect to external perturbations as well as mutual interactions they would have to occupy the same position in space; that is, to oscillate about the same point. A system such as a crystal is often treated approximately as a set of coupled harmonic oscillators (the atoms oscillating about their equilibrium positions). The Boltzmann statistics would be used for this set of oscillators, inasmuch as the interactions depend on the positions of the oscillators in space in such a way as to make them non-identical.

pletely antisymmetric ones, or those with the various intermediate symmetry characters. It is only the two extreme types which have been observed in nature. There are 30 completely symmetric wave functions for $n = 10$; they are formed from the successive sets in Table 49–1 by addition, the first being

$$\frac{1}{\sqrt{5}}\{(10.0.0.0.0) + (0.10.0.0.0) + (0.0.10.0.0) + (0.0.0.10.0) +$$

$$(0.0.0.0.10)\}$$

and the last being (2.2.2.2.2). From these we can obtain weights for the successive values, similar to those given in Table 49–2; these weights will not be identical with those of the table, however, and so will correspond to a new statistics. This is very clearly seen for the case that only the completely antisymmetric wave functions are accessible. The only wave function with $n = 10$ which is completely antisymmetric is that formed by suitable linear combination of the 120 product functions (4.3.2.1.0), etc., marked A in Table 49–1 (the other functions violate Pauli's principle, the quantum numbers not being all different). Hence even at the lowest temperatures only one of the five oscillators could occupy the lowest vibrational state, whereas the Boltzmann distribution law would in the limit $T \to 0$ place all five in this state.

If only the completely antisymmetric wave functions are accessible to a system composed of a large number of weakly interacting parts, the system parts conform to the Fermi-Dirac statistics;[1] *if only the completely symmetric wave functions are accessible, they conform to the Bose-Einstein statistics.*[2]

The Fermi-Dirac distribution law in the forms analogous to Equations 49–1, 49–3, and 49–4 is

$$P_n = \frac{1}{A e^{\frac{W_n}{kT}} + N}, \tag{49–6}$$

---

[1] E. Fermi, *Z. f. Phys.* **36**, 902 (1926); P. A. M. Dirac, *Proc. Roy. Soc.* **A112**, 661 (1926). This statistics was first developed by Fermi, on the basis of the Pauli exclusion principle, and was discovered independently by Dirac, using antisymmetric wave functions.

[2] S. N. Bose, *Z. f. Phys.* **26**, 178 (1924); A. Einstein, *Sitzber. Preuss. Akad. Wiss.* p. 261, 1924; p. 3, 1925. Bose developed this statistics to obtain a formal treatment of a photon gas, and Einstein extended it to the case of material gases.

$$P_i = \frac{p_i}{Ae^{\frac{W_i}{kT}} + N}, \qquad (49\text{-}7)$$

and

$$P(W) = \frac{p(W)}{Ae^{\frac{W}{kT}} + N}, \qquad (49\text{-}8)$$

in each of which the constant A has such a value as to make the sum or integral of P equal to unity. Here N is the number of identical system parts in which the accessible wave functions are antisymmetric.

Problem 49-2. Show that at very low temperatures the Fermi-Dirac distribution law places one system part in each of the N lowest states.

The Fermi-Dirac distribution law for the kinetic energy of the particles of a gas would be obtained by replacing $p(W)$ by the expression of Equation 49-5 for point particles (without spin) or molecules all of which are in the same non-degenerate state (aside from translation), or by this expression multiplied by the appropriate degeneracy factor, which is 2 for electrons or protons (with spin quantum number $\frac{1}{2}$), or in general $2I + 1$ for spin quantum number I. This law can be used, for example, in discussing the behavior of a gas of electrons. The principal application which has been made of it is in the theory of metals,[1] a metal being considered as a first approximation as a gas of electrons in a volume equal to the volume of the metal.

Problem 49-3. (a) Evaluate the average kinetic energy of the valence electrons (ignoring the K electrons and the nuclei) in a crystal of lithium metal at $0°A$, and discuss the distribution of energy. (b) Calculate the number of electrons at $298°A$ with kinetic energy 0.10 v.e. greater than the maximum for $0°A$. The density of lithium is 0.53 g./cm³.

The Bose-Einstein distribution law in the forms analogous to Equations 49-6, 49-7, and 49-8 is

$$P_n = \frac{1}{Ae^{\frac{W_n}{kT}} - N}, \qquad (49\text{-}9)$$

[1] W. Pauli, *Z. f. Phys.* **41**, 81 (1927); A. Sommerffld, *Z. f. Phys.* **47**, 1, 43 (1928); etc. Review articles have been published by K. K. Darrow, *Rev. Mod. Phys.* **1**, 90 (1929); J. C. Slater, *Rev. Mod. Phys.* **6**, 209 (1934); etc.

$$P_i = \frac{p_i}{A e^{\frac{W_i}{kT}} - N}, \tag{49-10}$$

and

$$P(W) = \frac{p(W)}{A e^{\frac{W}{kT}} - N}, \tag{49-11}$$

in which the symbols retain their former significance. The Bose-Einstein statistics is to be used for photons,[1] deuterons, helium atoms, hydrogen molecules, etc.

For many systems to which Fermi-Dirac or Bose-Einstein statistics is to be applied the term $\pm N$ is negligible compared to $A e^{\frac{W}{kT}}$, and the appropriate equations are very closely approximated by the corresponding Boltzmann equations. Thus helium gas under ordinary conditions shows no deviations from the perfect gas laws (Boltzmann statistics) which can be attributed to the operation of Bose-Einstein statistics. At very low temperatures and very high pressures, deviations due to this cause should occur, however; this *degeneration*[2] has not been definitely shown to occur for material gases by experiment,[3] the principal difficulty being that real gases elude investigation under extreme conditions by condensing to a liquid or solid phase.

49e. The Rotational and Vibrational Energy of Molecules.— In the statistical discussion of any gas containing identical molecules, cognizance must be taken of the type of statistics applicable. Often, however, we are not primarily interested in the translational motion of the molecules but only in their distribution among various rotational, vibrational, and electronic states. This distribution can usually be calculated by the use of the Boltzmann distribution law, the effect of the symmetry character being ordinarily negligible (except in so far as the sym-

---

[1] With appropriate modifications to take account of the vanishing rest mass of photons.

[2] The word *degeneracy* is used in this sense (distinct from that of Section 14), the electrons in a metal being described as constituting a *degenerate electron gas*.

[3] G. E. UHLENBECK and L. GROPPER, *Phys. Rev.* **41**, 79 (1932), and references there quoted.

metry character relative to identical particles in the same molecule determines the allowed wave functions for the molecule).

In case that the energy of a molecule can be represented as the sum of several terms (such as rotational, vibrational, electronic, and translational energy), the Boltzmann factor can be written as the product of individual Boltzmann factors, and the contributions of the various energy terms to the total energy of the system in thermodynamic equilibrium and to the heat capacity, entropy, and other properties can be calculated separately. To illustrate this we shall discuss the contributions of rotational and vibrational motion to the energy content, heat capacity, and entropy of hydrogen chloride gas.

As shown in Chapter X, the energy of a hydrogen chloride molecule in its normal electronic state can be approximately represented as

$$W_{v,K} = (v + \tfrac{1}{2})h\nu + K(K + 1)\frac{h^2}{8\pi^2 I}, \qquad (49\text{-}12)$$

in which ν is the vibrational frequency, I the moment of inertia of the molecule, and v and K the vibrational and rotational quantum numbers, with allowed values $v = 0, 1, 2, \cdots$ and $K = 0, 1, 2 \cdots$. At all but very high temperatures the Boltzmann factor for excited electronic states is very small, so that only the normal electronic state need be considered. Using Equation 49-3, we write for the probability that a molecule be in the state v,K the expression

$$P_{vK} = P_v P_K \qquad (49\text{-}13)$$

in which

$$P_v = Be^{-\frac{(v + \frac{1}{2})h\nu}{kT}}, \qquad (49\text{-}14)$$

and

$$P_K = C(2K + 1)e^{-\frac{K(K+1)h^2}{8\pi^2 I kT}}, \qquad (49\text{-}15)$$

$2K + 1$ being the quantum weight of the Kth rotational state. B and C have values such that

$$\sum_{v=0}^{\infty} P_v = 1 \quad \text{and} \quad \sum_{K=0}^{\infty} P_K = 1.$$

It is seen that the average rotational and vibrational energy per molecule can hence be written as

$$\bar{W} = \sum_{v=0}^{\infty} \sum_{K=0}^{\infty} P_v P_K \left\{ (v + \tfrac{1}{2})h\nu + K(K + 1)\frac{h^2}{8\pi^2 I} \right\},$$

or, since the summation over K can be at once carried out for the first term (to give the factor 1) and that over v for the second term,

$$\bar{W} = \bar{W}_{\text{vibr.}} + \bar{W}_{\text{rot.}},$$

with

$$\bar{W}_{\text{vibr.}} = \sum_{v=0}^{\infty} (v + \tfrac{1}{2})h\nu P_v$$

and

$$\bar{W}_{\text{rot.}} = \sum_{K=0}^{\infty} K(K + 1)\frac{h^2}{8\pi^2 I} P_K;$$

that is, the average energy is separable into two parts in the same way as the energy $W_{v,K}$ (Eq. 49–12). By introducing the variables[1]

$$x = \frac{h\nu}{kT},$$
$$\sigma = \frac{h^2}{8\pi^2 I kT},$$
$$\left. \right\} \qquad (49\text{–}16)$$

these parts can be written as

$$\bar{W}_{\text{vibr.}} = kT \frac{\displaystyle\sum_{v=0}^{\infty} (v + \tfrac{1}{2})x e^{-(v+\frac{1}{2})x}}{\displaystyle\sum_{v=0}^{\infty} e^{-(v+\frac{1}{2})x}} \qquad (49\text{–}17)$$

and

$$\bar{W}_{\text{rot.}} = kT \frac{\displaystyle\sum_{K=0}^{\infty} K(K + 1)(2K + 1)\sigma e^{-K(K+1)\sigma}}{\displaystyle\sum_{K=0}^{\infty} (2K + 1)e^{-K(K+1)\sigma}}, \qquad (49\text{–}18)$$

[1] The symbol σ is conventionally used in this way as well as for the quantity $h^2/8\pi^2 I$, as in Section 35.

the sums in the denominators corresponding to the factors B and C of Equations 49–14 and 49–15. Expressions for the vibrational and rotational heat capacity $C_{\text{vibr.}}$ and $C_{\text{rot.}}$ can be obtained by differentiating with respect to T, and the contributions of vibration and rotation to the entropy can then be obtained as $S_{\text{vibr.}} = \int_0^T \dfrac{C_{\text{vibr.}}}{T} dT$ and $S_{\text{rot.}} = \int_0^T \dfrac{C_{\text{rot.}}}{T} dT$.

Problem 49–4. Considering only the first two or three excited states, calculate the molal vibrational energy, heat capacity, and entropy of hydrogen chloride at 25°C., using the vibrational wave number $\nu = 2990$ cm^{-1}.

Problem 49–5. By replacing the sums by integrals, show that the expressions 49–17 and 49–18 approach the classical value kT for large T.

Problem 49–6. Calculate the rotational energy curve (as a function of T) for hydrogen chloride at temperatures at which it begins to deviate from zero. The internuclear distance is 1.27 Å.

The treatment of ortho and para hydrogen, mentioned in Section 43f, differs from that of hydrogen chloride only in the choice of accessible rotational wave functions. For para hydrogen K can assume only the values 0, 2, 4, $\cdots$, the quantum weight being $2K + 1$. For ortho hydrogen K can have the values 1, 3, 5, $\cdots$, with quantum weight $3(2K + 1)$, the factor 3 being due to the triplet nuclear-spin functions. Ordinary hydrogen is to be treated as a mixture of one-quarter para and three-quarters ortho hydrogen, inasmuch as only the states with K even are to be considered as accessible to the para molecules, and those with K odd to the ortho molecules. In the presence of a catalyst, however, all states become accessible, and the gas is to be treated as consisting of molecules of a single species.

Problem 49–7. Discuss the thermodynamic properties (in their dependence on rotation) of the types of hydrogen mentioned above.

Problem 49–8. Similarly treat deuterium and protium-deuterium molecules (see footnote, Sec. 43f).

49f. The Dielectric Constant of a Diatomic Dipole Gas.— Under the influence of an electric field, a gas whose molecules have a permanent electric moment and in addition can have a further moment induced in them by electronic polarization becomes polarized in the direction of the field, the polarization per unit volume being

$$P = \frac{3}{4\pi}\frac{\epsilon - 1}{\epsilon + 2}F = N\overline{\overline{\mu_z}} + N\alpha F, \tag{49-19}$$

in which ϵ is the dielectric constant of the gas, F the strength of the applied field (assumed to be parallel to the z axis), N the number of molecules in unit volume, and α the polarizability of the molecule. $\overline{\overline{\mu_z}}$ represents the average value of $\overline{\mu_z}$ for all molecules in the gas, $\overline{\mu_z}$ being the average value of the z component of the permanent electric moment μ of a molecule in a given state of motion. It was shown by Debye[1] that according to classical theory $\overline{\overline{\mu_z}}$ has the value

$$\overline{\overline{\mu_z}} = \frac{\mu^2 F}{3kT}. \tag{49-20}$$

We shall now show that for the special case of a diatomic dipole gas, such as hydrogen chloride, the same expression is given by quantum mechanics.

Let us consider that the change of the permanent moment μ with change in the vibrational quantum number v can be neglected. $\overline{\overline{\mu_z}}$ is then given by the equation

$$\overline{\overline{\mu_z}} = \sum_{K,M} P_{KM}\overline{\mu_z}(KM), \tag{49-21}$$

in which[2]

$$P_{KM} = Ae^{-K(K+1)\sigma}, \tag{49-22}$$

with $\sigma = h^2/8\pi^2 IkT$, as in Equation 49-16. Our first task is hence to evaluate $\overline{\mu_z}(KM)$, which is the average value of

$$\mu_z = \mu \cos \vartheta$$

for a molecule in the rotational state described by the quantum numbers K and M, ϑ being the angle between the moment μ of the molecule (that is, the nuclear axis) and the z axis.

The value of $\overline{\mu_z}(KM)$ is given by the integral

$$\overline{\mu_z}(KM) = \int \psi_{KM}^* \mu \cos \vartheta \psi_{KM} d\tau, \tag{49-23}$$

in which ψ_{KM} is the first-order perturbed wave function for the molecule in the electric field. It is found, on application of the

[1] P. DEBYE, *Phys. Z.* **13**, 97 (1912).

[2] It is assumed at this point that the energy of interaction of the molecule and the field can be neglected in the exponent of the Boltzmann factor. An investigation shows that this assumption is valid.

usual methods of Chapters VI and VII, the perturbation function being

$$H' = -\mu F \cos \vartheta, \qquad (49\text{-}24)$$

that $\overline{\mu_z}(KM)$ has the value

$$\overline{\mu_z}(KM) = \frac{8\pi^2 I \mu^2 F}{h^2} \frac{\{3M^2 - K(K+1)\}}{(2K-1)K(K+1)(2K+3)} \qquad (49\text{-}25)$$

(see Prob. 49–9).

Inasmuch as P_{KM} is independent of the quantum number M, to the degree of approximation of our treatment, we can at once calculate the average value of $\overline{\mu_z}(KM)$ for all states with the same value of K, by summing $\overline{\mu_z}(KM)$ for $M = -K, K+1, \cdots, +K$, and dividing by $2K+1$. The only part of 49–25 which involves any difficulty is that in M^2. The value of $\sum_{M=-K}^{+K} M^2$ is $\frac{1}{3}K(K+1)(2K+1)$; using this, we see that

$$\overline{\mu_z}(K) = \frac{1}{2K+1} \sum_{M=-K}^{+K} \overline{\mu_z}(KM) = 0, \qquad K > 0. \quad (49\text{-}26)$$

Thus we have obtained the interesting result that the *only rotational state which contributes to the polarization is that with $K = 0$*. The value of $\overline{\mu_z}$ for this state is seen from Equation 49–25 to be

$$\overline{\mu_z}(0) = \frac{8\pi^2 I \mu^2 F}{3h^2}, \qquad (49\text{-}27)$$

and $\overline{\overline{\mu_z}}$ hence is given by the equation

$$\overline{\overline{\mu_z}} = \frac{\mu^2 F}{3kT} \frac{1}{\sigma \sum_{K=0}^{\infty} (2K+1)e^{-K(K+1)\sigma}}, \qquad (49\text{-}28)$$

in which the sum in the denominator corresponds to the constant A of Equation 49–22. For small values of σ (such as occur in actual experiments) this reduces to

$$\overline{\overline{\mu_z}} = \frac{\mu^2 F}{3kT}, \qquad (49\text{-}29)$$

which is identical with the classical expression 49–20. On introduction in Equation 49–19, this gives the equation

$$P = \frac{3}{4\pi}\frac{\epsilon - 1}{\epsilon + 2}F = \frac{N\mu^2 F}{3kT} + N\alpha F. \qquad (49\text{--}30)$$

Problem 49-9. Using the surface-harmonic wave functions mentioned in the footnote at the end of Section 35c, derive Equation 49–25, applying either the ordinary second-order perturbation theory or the method of Section 27a.

Problem 49-10. Discuss the approximation to Equation 49–28 provided by 49–29 for hydrogen chloride molecules ($\mu = 1.03 \times 10^{-18}$ e.s.u.) in a field of 1000 volts per centimeter.

It can be shown[1] that Equation 49–30 is not restricted to diatomic molecules in its application, but is valid in general, except for a few special cases (as, for example, for a molecule with electric moment largely dependent on the vibrational state, or on the state of rotation of one part of the molecule about a single bond, etc.). With the use of this equation the electric moments of molecules can be determined from measurements on the temperature coefficient of the dielectric constants of gases and dilute solutions and in other ways. This has been done for a very large number of substances, with many interesting structural conclusions. An illustration is the question of which of the two isomers of dichlorethylene is the *cis* and which the *trans* form, i.e., which compound is to be assigned to each of the formulas shown below:

$$
\begin{array}{ccc}
\text{H} & & \text{H} \\
& \diagdown & \diagup \\
& \text{C} = \text{C} & \\
& \diagup & \diagdown \\
\text{Cl} & & \text{Cl}
\end{array}
\qquad
\begin{array}{ccc}
\text{H} & & \text{Cl} \\
& \diagdown & \diagup \\
& \text{C} = \text{C} & \\
& \diagup & \diagdown \\
\text{Cl} & & \text{H}
\end{array}
$$

cis form   trans form

The *trans* form is symmetrical and therefore is expected to have zero electric moment. It is found experimentally that the compound which the chemists had previously selected as the *trans* form does in fact have zero moment, whereas the *cis* form has a moment of about 1.74×10^{-18} e.s.u. (The unit 10^{-18} e.s.u. is sometimes called a *Debye unit*.) Strong evidence for the plane structure of benzene is also provided by electric-moment data, and many other problems of interest to chemists have been attacked in this way.

---

[1] See the references at the end of the section, in particular Van Vleck.

An equation which is very closely related to Equation 49–30 is also applicable to the magnetic susceptibility of substances; indeed, this equation was first derived (by Langevin[1] in 1905) for the magnetic case. The temperature-dependent term in this case corresponds to paramagnetism, μ representing the magnetic moment of the molecule; and the other term, which in the magnetic case is negative, corresponds to diamagnetism. For discussions of the origin of diamagnetism, the composition of the resultant magnetic moment μ from the spin and orbital moments of electrons, etc., the reader is referred to the references mentioned below.

References on Magnetic and Electric Moments

J. H. Van Vleck: "The Theory of Electric and Magnetic Susceptibilities," Oxford University Press, 1932.

C. P. Smyth: "Dielectric Constant and Molecular Structure," Chemical Catalog Company, Inc., New York, 1931.

P. Debye: "Polar Molecules," Chemical Catalog Company, Inc., New York, 1929.

E. C. Stoner: "Magnetism and Atomic Structure," E. P. Dutton & Co., Inc., New York, 1926.

The most extensive table of values of dipole moments available at present is that given in an Appendix of the *Transactions of the Faraday Society*, 1934.

General References on Statistical Mechanics

R. C. Tolman: "Statistical Mechanics with Applications to Physics and Chemistry," Chemical Catalog Company, Inc., New York, 1927.

R. H. Fowler: "Statistical Mechanics," Cambridge University Press, 1929.

L. Brillouin: "Les Statistiques Quantiques," *Les presses universitaires de France*, Paris, 1930.

K. K. Darrow: *Rev. Mod. Phys.* **1**, 90 (1929).

R. H. Fowler and T. E. Sterne: *Rev. Mod. Phys.* **4**, 635 (1932).

50. THE ENERGY OF ACTIVATION OF CHEMICAL REACTIONS

A simple interpretation of the activation energy E of a chemical reaction such as

$$A + BC \rightarrow AB + C \tag{50-1}$$

is provided by the assumption that the molecule BC in its normal electronic state is not able to react with the atom A, and that reaction occurs only between A and an electronically excited molecule BC*, E being then the energy difference of the normal

[1] P. Langevin, *J. de phys.* **4**, 678 (1905).

and the excited molecule. A reasonable alternative to this was given in 1928 by London,[1] who suggested that such a reaction might take place without any change in the electronic state of the system (other than that accompanying the change in the internuclear distances corresponding to the reaction 50–1, as discussed in Section 34). The heat of activation would then be obtained in the following way. We consider the electronic energy $W_0(\xi)$ for the normal electronic state of the system as a

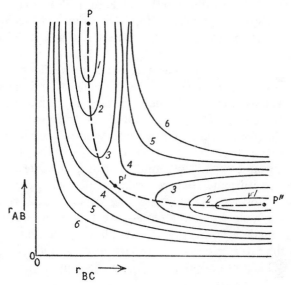

Fig. 50–1.—The electronic energy surface (showing contour lines with increasing energy 1, 2, 3, etc.) for a system of three atoms arranged linearly, as a function of the internuclear distances r_{AB} and r_{BC}.

function of the nuclear coordinates ξ. $W_0(\xi)$ will have one value for the nuclear configuration in which nuclei B and C are close together, as in the normal molecule BC, and A is far removed, and another value for the AB + C nuclear configuration. (The difference of these, corrected for the energy of oscillation and rotation of the molecules, is the energy change during the reaction.) Now in order to change from one extreme configuration to the other, the nuclei must pass through intermediate configurations, as atom A approaches B and C recedes from it, and the electronic energy $W_0(\xi)$ would change with change in

[1] F. LONDON in the Sommerfeld Festschrift, "Probleme der modernen Physik," p. 104, S. Hirzel, Leipzig, 1928.

configuration, perhaps as shown in Figure 50–1. The change from A + BC, represented by the configuration point P, to AB + C, represented by the configuration point P'', could take place most easily along the path shown by the dotted line. We have seen in Section 34 that the electronic energy can be treated as a potential function for the nuclei; it is evident that in order for reaction to take place the nuclei must possess initially enough kinetic energy to carry them over the high point P' of the saddle of the potential function of Figure 50–1. The energy difference $W_0(P') - W_0(P)$, after correction for zero-point oscillational energy, etc., would be interpreted as the activation energy E.

No thoroughly satisfactory calculation of activation energies in this way has yet been made. The methods of treatment discussed for the hydrogen molecule in Section 43, in particular the method of James and Coolidge, could of course be extended to a system of three protons and three electrons to provide a satisfactory treatment of the reaction $H + H_2 \rightarrow H_2 + H$. This calculation would be difficult and laborious, however, and has not been carried out. Several rough calculations, providing values of E for comparison with the experimental value[1] of about 6 kcal/mole (from the ortho-para hydrogen conversion), have been made. In Section 46d we have seen that at large distances the interaction of a hydrogen atom A and a hydrogen molecule BC is given approximately by the expression

$$- \tfrac{1}{2}(abc\,|H|\,bac) + \tfrac{1}{2}(abc\,|H|\,cba),$$

the first term corresponding to repulsion of A by B and the second to repulsion of A by C. It is reasonable then that the easiest path for the reaction would correspond to a linear arrangement ABC, the repulsion of A and C then being a minimum for given values of r_{AB} and r_{BC}. Eyring and Polanyi[2] calculated energy surfaces for linear configurations by neglecting higher exchange integrals and making other simplifying assumptions, the values

[1] A. Farkas, *Z. f. phys. Chem.* **B10**, 419 (1930); P. Harteck and K. H. Geib, *ibid.* **B15**, 116 (1931).

[2] H. Eyring and M. Polanyi, *Naturwissenschaften* **18**, 914 (1930); *Z. f. phys. Chem.* **B12**, 279 (1931); H. Eyring, *Naturwissenschaften* **18**, 915 (1930), *J. Am. Chem. Soc.* **53**, 2537 (1931); H. Pelzer and E. Wigner, *Z. f. phys. Chem.* **B15**, 445 (1932).

of the Coulomb and single exchange integrals being taken from the simple Heitler-London-Sugiura treatment of the hydrogen molecule or estimated from the empirical potential function for this molecule. These approximate treatments led to values in the neighborhood of 10 to 15 kcal for the activation energy. Coolidge and James[1] have recently pointed out that the approximate agreement with experiment depends on the cancellation of large errors arising from the various approximations.

The similar discussion of the activation energies of a number of more complicated reactions has been given by Eyring and collaborators.[2]

[1] A. S. COOLIDGE and H. M. JAMES, *J. Chem. Phys.* **2**, 811 (1934).

[2] H. EYRING, *J. Am. Chem. Soc.* **53**, 2537 (1931); G. E. KIMBALL and H. EYRING, *ibid.* **54**, 3876 (1932); A. SHERMAN and H. EYRING, *ibid.* **54**, 2661 (1932); R. S. BEAR and H. EYRING, *ibid.* **56**, 2020 (1934); H. EYRING, A. SHERMAN, and G. E. KIMBALL, *J. Chem. Phys.* **1**, 586 (1933); A. SHERMAN, C. E. SUN, and H. EYRING, *ibid.* **3**, 49 (1935).

CHAPTER XV

GENERAL THEORY OF QUANTUM MECHANICS

The branch of quantum mechanics to which we have devoted our attention in the preceding chapters, based on the Schrödinger wave equation, can be applied in the discussion of most questions which arise in physics and chemistry. It is sometimes convenient, however, to use somewhat different mathematical methods; and, moreover, it has been found that a thoroughly satisfactory general theory of quantum mechanics and its physical interpretation require that a considerable extension of the simple theory be made. In the following sections we shall give a brief discussion of matrix mechanics (Sec. 51), the properties of angular momentum (Sec. 52), the uncertainty principle (Sec. 53), and transformation theory (Sec. 54).

51. MATRIX MECHANICS

In the first paper written on the quantum mechanics[1] Heisenberg formulated and successfully attacked the problem of calculating values of the frequencies and intensities of the spectral lines which a system could emit or absorb; that is, of the energy levels and the electric-moment integrals which we have been discussing. He did not use wave functions and wave equations, however, but instead developed a formal mathematical method for calculating values of these quantities. The mathematical method is one with which most chemists and physicists are not familiar (or were not, ten years ago), some of the operations involved being surprisingly different from those of ordinary algebra. Heisenberg invented the new type of algebra as he needed it; it was immediately pointed out by Born and Jordan,[2] however, that in his new quantum mechanics Heisenberg was

---

[1] W. Heisenberg, *Z. f. Phys.* **33**, 879 (1925).

[2] M. Born and P. Jordan, *ibid.* **34**, 858 (1925).

making use of quantities called *matrices* which had already been discussed by mathematicians, and that his newly invented operations were those of *matrix algebra.* The Heisenberg quantum mechanics, usually called *matrix mechanics,* was rapidly developed[1] and applied to various problems.

When Schrödinger discovered his wave mechanics the question arose as to the relation between it and matrix mechanics. The answer was soon given by Schrödinger[2] and Eckart,[3] who showed independently that the two are mathematically equivalent. The arguments used by Heisenberg in formulating his quantum mechanics are extremely interesting. We shall not present them, however, nor enter into an extensive discussion of matrix mechanics, but shall give in the following sections a brief treatment of matrices, matrix algebra, the relation of matrices to wave functions, and a few applications of matrix methods to quantum-mechanical problems.

51a. Matrices and Their Relation to Wave Functions. The Rules of Matrix Algebra.—Let us consider a set of orthogonal wave functions[4] $\Psi_0, \Psi_1, \cdots, \Psi_n, \cdots$ and a dynamical quantity $f(q_i, p_i)$, the corresponding operator[5] being $f_{\text{op.}} = f\left(q_i, \dfrac{h}{2\pi i}\dfrac{\partial}{\partial q_i}\right)$. In the foregoing chapters we have often made use of integrals such as

$$f_{mn} = \int \Psi_m^* f_{\text{op.}} \Psi_n d\tau; \qquad (51\text{-}1)$$

for example, we have given f_{nn} the physical interpretation of the average value of the dynamical quantity f when the system is in the nth stationary state. Let us now arrange the numbers f_{mn} (the values of the integrals) in a square array ordered according to m and n, as follows:

---

[1] M. Born, W. Heisenberg, and P. Jordan, *Z. f. Phys.* **35**, 557 (1926); P. A. M. Dirac, *Proc. Roy. Soc.* **A109**, 642 (1925).

[2] E. Schrödinger, *Ann. d. Phys.* **79**, 734 (1926).

[3] C. Eckart, *Phys. Rev.* **28**, 711 (1926).

[4] These functions include the time factor; a similar discussion can be made with use of the functions $\psi_0, \psi_1, \cdots$ not including the time.

[5] In this chapter we shall use the symbol $f_{\text{op.}}$ to represent the operator corresponding to the dynamical function f. The subscript "op." was not used in the earlier chapters because there was no danger of confusion attending its omission. See Secs. 10, 12.

$$
\mathbf{f} = (f_{mn}) = \begin{pmatrix}
f_{00} & f_{01} & f_{02} & f_{03} & \cdots \\
f_{10} & f_{11} & f_{12} & f_{13} & \cdots \\
f_{20} & f_{21} & f_{22} & f_{23} & \cdots \\
f_{30} & f_{31} & f_{32} & f_{33} & \cdots \\
\cdots & \cdots & \cdots & \cdots & \cdots
\end{pmatrix}.
$$

This array we may represent by the symbol $\mathbf{f}$ or (f_{mn}). We enclose it in parentheses to distinguish it from a determinant, with which it should not be confused.

We can construct similar arrays $\mathbf{g}$, $\mathbf{h}$, etc. for other dynamical quantities.

It is found that the symbols $\mathbf{f}$, $\mathbf{g}$, $\mathbf{h}$, etc. representing such arrays can be manipulated by an algebra closely related to ordinary algebra, differing from it mainly in the process of multiplication. The rules of this algebra can be easily derived from the properties of wave functions, which we already know.

It must be borne in mind that the symbol $\mathbf{f}$ does not represent a single number. (In particular the array $\mathbf{f}$ must not be confused with a determinant, which is equal to a single number. There is, to be sure, a determinant corresponding to each array, namely, the determinant whose elements are those of the array. We have set up such determinants in the secular equations of the preceding chapters.) The symbol $\mathbf{f}$ instead represents many numbers—as many as there are elements in the array. *The sign of equality in the equation* $\mathbf{f} = \mathbf{g}$ *means that every element in the array* $\mathbf{f}$ *is equal to the corresponding element in the array* $\mathbf{g}$.

Now let us derive some rules of the new algebra. For example, the sum of two such arrays is an array each of whose elements is the sum of the corresponding elements of the two arrays; that is,

$$
\mathbf{f} + \mathbf{g} = \begin{pmatrix}
f_{00} + g_{00} & f_{01} + g_{01} & f_{02} + g_{02} & \cdots \\
f_{10} + g_{10} & f_{11} + g_{11} & f_{12} + g_{12} & \cdots \\
\cdots & \cdots & \cdots & \cdots
\end{pmatrix}. \tag{51--2}
$$

It is seen that the arrays add in the same way as ordinary algebraic quantities, with $\mathbf{f} + \mathbf{g} = \mathbf{g} + \mathbf{f}$. *Addition is commutative.*

On the other hand, *multiplication is not commutative: the product* $\mathbf{fg}$ *is not necessarily equal to the product* $\mathbf{gf}$. Let us evaluate the mnth element of the array $\mathbf{fg}$. It is

$$
\{fg\}_{mn} = \int \Psi_m^* f_{\text{op}}. g_{\text{op}}. \Psi_n \, d\tau.
$$

Now we can express the quantity $g_{\text{op}}.\Psi_n$ in terms of the functions Ψ_k with constant coefficients (Sec. 22), obtaining

$$g_{\text{op}}.\Psi_n = \sum_k g_{kn}\Psi_k.$$

That the coefficients are the quantities g_{kn} is seen on multiplying by Ψ_k^* and integrating. Introducing this in the integral for $\{fg\}_{mn}$ we obtain

$$\{fg\}_{mn} = \sum_k \int\Psi_m^* f_{\text{op}}.\Psi_k d\tau g_{kn};$$

since $\int\Psi_m^* f_{\text{op}}.\Psi_k d\tau$ is equal to f_{mk}, this becomes

$$\{fg\}_{mn} = \sum_k f_{mk}g_{kn}. \tag{51-3}$$

This is the rule for calculating the elements of the array obtained on multiplying two arrays.

We may continue to develop the algebra of our arrays in this way; or we can instead make use of work already done by mathematicians. The arrays which we have been discussing are called *matrices*, and their properties have been thoroughly investigated by mathematicians, who have developed an extensive *matrix algebra*,[1] some parts of which we have just derived.

Problem 51-1. Show that the laws of ordinary algebra hold for the addition and subtraction of matrices and their multiplication by scalars; for example,

$$\mathbf{f} + (\mathbf{g} + \mathbf{h}) = (\mathbf{f} + \mathbf{g}) + \mathbf{h},$$
$$a\mathbf{f} + a\mathbf{g} = a(\mathbf{f} + \mathbf{g}),$$
$$a\mathbf{f} + b\mathbf{f} = (a + b)\mathbf{f}.$$

Matrix methods, especially matrix multiplication, are often very useful in solving problems. Thus we have applied Equation 51-3 in Section 27e, after deriving the equation in order to use it. Another example of the use of this equation is provided by Problem 51-2.

In quantum-mechanical discussions the matrix $\mathbf{f}$ corresponding to the dynamical quantity $f(q_i, p_i)$ is sometimes defined with the use of the wave functions Ψ_n, which include the time (Eq. 51-1), and sometimes with the wave functions ψ_n, with the time factor

---

[1] See, for example, M. Bôcher, "Introduction to Higher Algebra," The Macmillan Company, New York, 1924.

omitted, in which case the matrix elements are given by the integrals

$$f_{mn} = \int \psi_m^* f_{op.} \psi_n d\tau.$$ (51-4)

The matrix elements f_{mn} in the two cases differ only by the time factor $e^{\frac{2\pi i(W_m - W_n)t}{h}}$, and as there is no danger of confusion the same symbol can be used for the matrix containing the time as for that not containing the time.

Problem 51-2. The elements x_{mn} of the matrix $\mathbf{x}$ for the harmonic oscillator are given by Equation 11-25. Using the rule for matrix multiplication, set up the matrices $\mathbf{x}^2(= \mathbf{xx})$, $\mathbf{x}^3$, and $\mathbf{x}^4$, and compare the values of the diagonal elements with those found in Section 23a.

The non-commutative nature of the multiplication of matrices is of great importance in matrix mechanics. The difference of the product of the matrix $\mathbf{q}_j$ representing the coordinate q_j and the matrix $\mathbf{p}_j$ representing the canonically conjugate momentum p_j and the reverse product is not zero, but $\frac{h}{2\pi i}\mathbf{1}$, where $\mathbf{1}$ is the *unit matrix*, discussed in the following section; that is, these matrices do not commute. On the other hand, $\mathbf{q}_j$ and $\mathbf{p}_k$ (with $k \neq j$), etc., do commute, the complete *commutation rules* for coordinates and momenta being

$$\left.\begin{array}{l} \mathbf{p}_j\mathbf{q}_j - \mathbf{q}_j\mathbf{p}_j = \dfrac{h}{2\pi i}\mathbf{1}, \\[2mm] \mathbf{p}_j\mathbf{q}_k - \mathbf{q}_k\mathbf{p}_j = 0, \qquad k \neq j, \\[2mm] \mathbf{q}_j\mathbf{q}_k - \mathbf{q}_k\mathbf{q}_j = 0, \\[2mm] \mathbf{p}_j\mathbf{p}_k - \mathbf{p}_k\mathbf{p}_j = 0. \end{array}\right\}$$ (51-5)

These commutation rules together with the rules for converting the Hamiltonian equations of motion into matrix form constitute matrix mechanics, which is a way of stating the laws of quantum mechanics which is entirely different from that which we have used in this book, although completely equivalent. The latter rules require a discussion of differentiation with respect to a matrix, into which we shall not enter.[1]

Problem 51-3. Verify the commutation rules 51-5 by evaluating the matrix elements $(p_j q_j)_{mn}$, etc.

[1] For a discussion of matrix mechanics see, for example, Ruark and Urey, "Atoms, Molecules and Quanta," Chap. XVII.

51b. Diagonal Matrices and Their Physical Interpretation.—A *diagonal matrix* is a matrix whose elements f_{mn} are all zero except those with $m = n$; for example,

$$\begin{pmatrix} f_{00} & 0 & 0 & 0 & \cdots \\ 0 & f_{11} & 0 & 0 & \cdots \\ 0 & 0 & f_{22} & 0 & \cdots \\ 0 & 0 & 0 & f_{33} & \cdots \\ \cdots & \cdots & \cdots & \cdots & \cdots \end{pmatrix}.$$

The unit matrix, **1**, is a special kind of diagonal matrix, all the diagonal elements being equal to unity:

$$\mathbf{1} = \begin{pmatrix} 1 & 0 & 0 & 0 & \cdots \\ 0 & 1 & 0 & 0 & \cdots \\ 0 & 0 & 1 & 0 & \cdots \\ 0 & 0 & 0 & 1 & \cdots \\ \cdots & \cdots & \cdots & \cdots & \cdots \end{pmatrix}.$$

A constant matrix, **a**, is equal to a constant times the unit matrix:

$$\mathbf{a} = a\mathbf{1} = \begin{pmatrix} a & 0 & 0 & 0 & \cdots \\ 0 & a & 0 & 0 & \cdots \\ 0 & 0 & a & 0 & \cdots \\ 0 & 0 & 0 & a & \cdots \\ \cdots & \cdots & \cdots & \cdots & \cdots \end{pmatrix}.$$

Application of the rule for matrix multiplication shows that the square (or any power) of a diagonal matrix is also a diagonal matrix, its diagonal elements being the squares (or other powers) of the corresponding elements of the original matrix.

In Section 10c, in discussing the physical interpretation of the wave equation, we saw that our fundamental postulate regarding physical interpretation requires a dynamical quantity f to have a definite value for a system in the state represented by the wave function Ψ_n only when f_{nn}^r is equal to $(f_{nn})^r$, for all values of r. We can now express this in terms of matrices: *If the dynamical quantity f is represented by a diagonal matrix **f** then this dynamical quantity has the definite value f_{nn} for the state corresponding to the wave function Ψ_n of the set Ψ_0, Ψ_1, $\cdots$.*

For illustration, let us discuss some of the wave functions which we have met in previous chapters. The solutions

$$\Psi_0\left(= \psi_0 e^{-\frac{2\pi i W_0 t}{h}}\right), \Psi_1, \cdots$$

of the wave equation for any system correspond to a diagonal energy matrix

$$\mathbf{H} = \begin{vmatrix} W_0 & 0 & 0 & 0 & \cdots \\ 0 & W_1 & 0 & 0 & \cdots \\ 0 & 0 & W_2 & 0 & \cdots \\ 0 & 0 & 0 & W_3 & \cdots \\ \cdots & \cdots & \cdots & \cdots & \cdots \end{vmatrix},$$

so that, as mentioned in Section 10c, the system in a physical condition represented by one of these wave functions has a fixed value of the total energy.

In the case of a system with one degree of freedom no other dynamical quantity (except functions of H only, such as H^2) is represented by a diagonal matrix; with more degrees of freedom there are other diagonal matrices. For example, the surface-harmonic wave functions $\Theta_{lm}(\vartheta)\Phi_m(\varphi)$ for the hydrogen atom and other two-particle systems separated in polar coordinates (Secs. 19, 21) make the matrices for the square of the total angular momentum and the component of angular momentum along the z axis diagonal, these dynamical quantities thus having definite values for these wave functions. The properties of angular momentum matrices are discussed in Section 52.

The dynamical quantities corresponding to diagonal matrices relative to the stationary-state wave functions $\Psi_0, \Psi_1, \cdots$ are sometimes called *constants of the motion* of the system. The corresponding constants of the motion of a system in classical mechanics are the constants of integration of the classical equation of motion.

Let us now consider a system whose Schrödinger time functions corresponding to the stationary states of the system are Ψ_0, $\Psi_1, \cdots, \Psi_n, \cdots$ Suppose that we carry out an experiment (the measurement of the values of some dynamical quantities) such as to determine the wave function uniquely. Such an experiment is called a *maximal measurement*. A maximal measurement for a system with one degree of freedom, such as the one-dimensional harmonic oscillator, might consist in the accurate measurement of the energy; the result of the measurement would be one of the characteristic energy values W_n; and the corresponding wave function Ψ_n would then represent the system so long as it remain undisturbed and could be used for predicting average values for later measurements (Secs. 10, 12).

A maximal measurement for a system of three degrees of freedom, such as the three-dimensional isotropic harmonic oscillator or the hydrogen atom with fixed nucleus and without spin, might consist in the accurate determination of the energy, the square of the total angular momentum, and the component of the angular momentum along the z axis. The wave function corresponding to such a maximal measurement would be one of those obtained by separating the wave equation in polar coordinates, as was done in Chapter V.

It is found that the accurate measurement of the values of N independent[1] dynamical quantities constitutes a maximal measurement for a system with N degrees of freedom. In classical mechanics a maximal measurement involves the accurate determination of the values of $2N$ dynamical quantities, such as the N coordinates and the N momenta, or for a one-dimensional system the energy and the coordinate, etc. A discussion of the significance of this fact will be given in connection with the uncertainty principle in Section 53.

Now let us consider a complete set of orthogonal normalized wave functions $\chi_0, \chi_1, \cdots, \chi_{n'}, \cdots$, each function $\chi_{n'}$ being a solution of the Schrödinger time equation for the system under discussion. These wave functions are linear combinations of the stationary-state wave functions Ψ_n, being obtained from them by the linear transformation

$$\chi_{n'} = \sum_n a_{n'n} \Psi_n, \qquad (51-6)$$

in which the coefficients $a_{n'n}$ are constants restricted only in that they are to make the χ's mutually orthogonal and normalized. A set of wave functions $\chi_{n'}$ is said to form a *representation* of the system. Corresponding to each representation matrices[2] f', g', etc. can be constructed for the dynamical quantities f, g, etc., the elements being calculated by equations such as

$$f_{m'n'} = \int \chi_m^* f_{op.} \chi_{n'} d\tau \qquad (51-7)$$

---

[1] The meaning of independent will be discussed later in this section.

[2] We use primed symbols to indicate that the matrices correspond to the representation $\chi_{n'}$.

or obtained from the matrices **f**, **g**, etc. (corresponding to the stationary-state representation Ψ_n), by the use of the coefficients $a_{n'n}$ of Equation 51–6.

So far we have discussed the measurement of constants of the motion of the system only; that is, of quantities which are represented by diagonal matrices relative to the Schrödinger wave functions Ψ_0, Ψ_1, Ψ_2, $\cdots$, and which are hence independent of the time. But in general we might make a maximal measurement consisting in the accurate measurement of N dynamical quantities f, g, etc., whose matrices **f**, **g**, etc., relative to Ψ_0, Ψ_1, $\cdots$, are not all diagonal matrices. In the case of such a maximal measurement we must specify the time $t = t'$ at which the measurement is made. An accurate measurement of the quantities f, g, etc. at the time $t = t'$ requires that at the time $t = t'$ the matrices **f′**, **g′**, etc. be diagonal matrices. In order to find the wave function representing the system at times subsequent to $t = t'$ (so long as the system remain undisturbed), we must find the representation $\chi_{n'}$ which makes these matrices diagonal at the time $t = t'$. The accurate values of f, g, etc. obtained by measurement will be identical with the numbers $f_{n'n'}$, $g_{n'n'}$, etc., occurring as a certain diagonal element of the diagonal matrices **f′**, **g′**, etc., and the wave function representing the system will be the corresponding $\chi_{n'}$.

It is interesting to notice that the condition that the dynamical quantity f be represented by a diagonal matrix **f′** in the χ representation can be expressed as a differential equation. In order for **f′** to be a diagonal matrix, $f'_{m'n'}$ must equal 0 for m' not equal to n' and a constant value, $f_{n'n'}$, say, for $m' = n'$. This means that on expanding $f_{\text{op.}}\chi_{n'}$ in terms of the complete set of functions χ only the one term $f_{n'n'}\chi_{n'}$ will occur; that is, that

$$f_{\text{op.}}\chi_{n'} = f_{n'n'}\chi_{n'}, \qquad (51\text{–}8)$$

in which $f_{n'n'}$ is a number, the n'th diagonal element of the diagonal matrix **f′**. For example, the stationary-state wave functions Ψ_{nlm} for a hydrogen atom as given in Chapter **V** satisfy three differential equations,

$$H_{\text{op.}}\Psi_{nlm} = W_n\Psi_{nlm},$$
$$M_{\text{op.}}^2\Psi_{nlm} = \frac{l(l+1)h^2}{4\pi^2}\Psi_{nlm},$$

and

$$M_{z\,\mathrm{op.}}\Psi_{nlm} = \frac{mh}{2\pi}\Psi_{nlm},$$

corresponding to the three dynamical functions whose matrices are diagonal in this representation; namely, the energy, the square of the total angular momentum, and the z component of the angular momentum. For a discussion of this question from a different viewpoint see the next section.

52. THE PROPERTIES OF ANGULAR MOMENTUM

As pointed out in the previous section, systems whose wave equations separate in spherical polar coordinates (such as the hydrogen atom) possess wave functions corresponding not only to definite values of the energy but also to definite values of the total angular momentum and the component of angular momentum along a given axis (say, the z axis). In order to prove this for one particle[1] let us construct the operators corresponding to M_x, M_y, and M_z, the components of angular momentum along the x, y, and z axes. Since classically

$$M_x = yp_z - zp_y, \qquad (52\text{–}1)$$

with similar expressions for M_y and M_z, the methods of Section 10c for constructing the operator corresponding to any physical quantity yield the expressions

$$\left.\begin{aligned}
M_{x\,\mathrm{op.}} &= \frac{h}{2\pi i}\left(y\frac{\partial}{\partial z} - z\frac{\partial}{\partial y}\right), \\
M_{y\,\mathrm{op.}} &= \frac{h}{2\pi i}\left(z\frac{\partial}{\partial x} - x\frac{\partial}{\partial z}\right), \\
M_{z\,\mathrm{op.}} &= \frac{h}{2\pi i}\left(x\frac{\partial}{\partial y} - y\frac{\partial}{\partial x}\right).
\end{aligned}\right\} \qquad (52\text{–}2)$$

In order to calculate the average values of these quantities it is convenient to express them in terms of polar coordinates. By the standard methods (see Sec. 1b) we obtain

$$M_{x\,\mathrm{op.}} \pm iM_{y\,\mathrm{op.}} = \frac{h}{2\pi i}e^{\pm i\varphi}\left(\pm i\frac{\partial}{\partial \vartheta} - \cot\vartheta\frac{\partial}{\partial \varphi}\right) \qquad (52\text{–}3)$$

[1] The total angular momentum and its z component also have definite values for a system of n particles in field-free space; see, for example, Born and Jordan, "Elementare Quantenmechanik," Chap. IV, Julius Springer, Berlin, 1930.

and

$$M_{z\,\text{op.}} = \frac{h}{2\pi i}\,\frac{\partial}{\partial\varphi}. \qquad (52\text{–}4)$$

We have postulated that the wave equation is separable in polar coordinates; if we also restrict the potential energy to be a function of r alone, the dependence of the wave functions on the angles will be given by

$$\psi_{lmn}(\vartheta,\,\varphi,\,r) = \Theta_{lm}(\vartheta)\Phi_m(\varphi)R_{nl}(r), \qquad (52\text{–}5)$$

where $\Theta_{lm}(\vartheta)\Phi_m(\varphi)$ are the surface-harmonic wave functions obtained in Section 18. Using these and the expressions in Equations 52–3 and 52–4, we can evaluate the integrals of the type

$$M_x(l'm';\,lm) = \int\psi_{l'm'n}^{*}M_{z\,\text{op.}}\psi_{lmn}d\tau. \qquad (52\text{–}6)$$

In order to prove that the square of the total angular momentum M^2 has a definite value for a given stationary state described by ψ_{lmn}, it is necessary for us to show that the average value of any power of M^2 is identical with the same power of the average value (Sec. 10c). By using the properties of matrices given in the previous section we can considerably simplify this proof. As stated there, we need only show that $M_{\text{op.}}^2$ is represented by a diagonal matrix. Furthermore we can obtain the matrix for M^2 from the matrices for M_x, M_y, and M_z by using the relation defining M^2 in classical mechanics,

$$M^2 = M_x^2 + M_y^2 + M_z^2, \qquad (52\text{–}7)$$

and applying the rules for matrix multiplication and addition.

If we carry out this procedure, we first find on evaluation of the proper integrals that

$$M_x(l'm';\,lm) = -\frac{h}{4\pi}[\{l(l+1) - m(m+1)\}^{1/2}\delta_{m',m+1} +$$
$$\{l(l+1) - m(m-1)\}^{1/2}\delta_{m',m-1}]\delta_{l',l}, \qquad (52\text{–}8)$$

$$M_y(l'm';\,lm) = \frac{ih}{4\pi}[\{l(l+1) - m(m+1)\}^{1/2}\delta_{m',m+1} -$$
$$\{l(l+1) - m(m-1)\}^{1/2}\delta_{m',m-1}]\delta_{l',l}, \qquad (52\text{–}9)$$

$$M_z(l'm';\,lm) = \frac{h}{2\pi}m\delta_{l',l}\delta_{m',m}, \qquad (52\text{–}10)$$

in which $\delta_{m',m+1} = 1$ for $m' = m + 1$ and 0 otherwise, etc.

The next step is to obtain the elements of the matrices $\mathbf{M}_x^2$, $\mathbf{M}_y^2$, and $\mathbf{M}_z^2$ from these by using matrix multiplication, and then the elements of $\mathbf{M}^2$ by using matrix addition. The final result is that $\mathbf{M}^2$ is a diagonal matrix with diagonal elements $\dfrac{l(l+1)h^2}{4\pi^2}$. It is therefore true that M^2 has a definite value in the state ψ_{lmn}; in other words, it is a constant of the motion with the value $\dfrac{l(l+1)h^2}{4\pi^2}$.

The proof that M_z is also a constant of the motion is contained in Equation 52–10, which shows that $\mathbf{M}_z$ is a diagonal matrix with diagonal elements $mh/2\pi$ so that its value is $mh/2\pi$ for the state with quantum number m.

Problem 52–1. Carry out the transformation of Equations 52–2 into polar coordinates.

Problem 52–2. Derive Equations 52–8, 52–9, and 52–10.

Problem 52–3. Obtain the matrices for M_x^2, M_y^2, M_z^2 by matrix multiplication and from them obtain the matrix for M^2.

There is a close connection between the coordinate system in which a given wave equation is separated and the dynamical quantities which are the constants of the motion for the resulting wave functions. Thus for a single particle in a spherically symmetric field the factor $S(\vartheta, \varphi)$ of the wave function which depends only on the angles satisfies the equation (see Sec. 18a)

$$\frac{1}{\sin \vartheta} \frac{\partial}{\partial \vartheta}\left(\sin \vartheta \frac{\partial S}{\partial \vartheta}\right) + \frac{1}{\sin^2 \vartheta} \frac{\partial^2 S}{\partial \varphi^2} = -l(l+1)S. \quad (52\text{–}11)$$

It can be shown that the operator for M^2 in polar coordinates has the form

$$M_{\text{op.}}^2 = -\frac{h^2}{4\pi^2}\left\{\frac{1}{\sin \vartheta} \frac{\partial}{\partial \vartheta}\left(\sin \vartheta \frac{\partial}{\partial \vartheta}\right) + \frac{1}{\sin^2 \vartheta} \frac{\partial^2}{\partial \varphi^2}\right\}, \quad (52\text{–}12)$$

so that Equation 52–11 may be written

$$M_{\text{op.}}^2 \psi_{nlm} = l(l+1)\frac{h^2}{4\pi^2}\psi_{nlm}, \quad (52\text{–}13)$$

since $\psi = S(\vartheta, \varphi)R(r)$ and $M_{\text{op.}}^2$ does not affect $R(r)$.

Furthermore the equation for $\Phi_m(\varphi)$, the φ part of ψ, is (Sec. 18a)

$$\frac{d^2\Phi}{d\varphi^2} = -m^2\Phi, \quad (52\text{–}14)$$

whereas from Equation 52–4 we find that

$$M_{z\,\text{op.}}^2 = -\frac{h^2}{4\pi^2}\frac{\partial^2}{\partial\varphi^2}, \qquad (52\text{–}15)$$

so that Equation 52–13 may be written in the form

$$M_{z\,\text{op.}}^2\psi_{nlm} = m^2\frac{h^2}{4\pi^2}\psi_{nlm}. \qquad (52\text{–}16)$$

The formal similarity of Equations 52–13 and 52–16 with the wave equation

$$H_{\text{op.}}\psi_{nlm} = W_n\psi_{nlm}$$

is quite evident. All three equations consist of an operator acting upon the wave function equated with the wave function multiplied by the quantized value of the physical quantity represented by the operator. Furthermore, the operators $H_{\text{op.}}$, $M_{\text{op.}}^2$, and $M_{z\,\text{op.}}^2$ will be found to commute with each other; that is,

$$H_{\text{op.}}(M_{\text{op.}}^2\chi) = M_{\text{op.}}^2(H_{\text{op.}}\chi),$$

etc., where χ is any function of ϑ, φ, and r.

It is beyond the scope of this book to discuss this question more thoroughly, but the considerations which we have given above for this special case can be generalized to other systems and other sets of coordinates. Whenever the wave equation can be separated it will be found that the separated parts can be thrown into the form discussed above, involving the operators of several physical quantities. These physical quantities will be constants of the motion for the resulting wave functions, and their operators will commute with each other.

53. THE UNCERTAINTY PRINCIPLE

The *Heisenberg uncertainty principle*[1] may be stated in the following way:

The values of two dynamical quantities f and g of a system can be accurately measured at the same time only if their commutator is zero; otherwise these measurements can be made only with an uncertainty $\Delta f \Delta g$ whose magnitude is dependent on the value of the commutator. In particular, for a canonically conjugate coordinate

[1] W. HEISENBERG, *Z. f. Phys.* **43**, 172 (1927).

q and momentum p the uncertainty $\Delta q \Delta p$ is of the order of magnitude of Planck's constant h, as is $\Delta W \Delta t$ for the energy and time.

To prove the first part of this principle, we investigate the conditions under which two dynamical quantities f and g can be simultaneously represented by diagonal matrices. Let these matrices be $\mathbf{f}'$ and $\mathbf{g}'$, $\chi_{n'}$ being the corresponding representation. The product $\mathbf{f}'\mathbf{g}'$ of these two diagonal matrices is found on evaluation to be itself a diagonal matrix, its n'th element being the product of the n'th diagonal elements $f_{n'}$ and $g_{n'}$ of the diagonal matrices $\mathbf{f}'$ and $\mathbf{g}'$. Similarly $\mathbf{g}'\mathbf{f}'$ is a diagonal matrix, its diagonal elements being identical with those of $\mathbf{f}'\mathbf{g}'$. Hence the *commutator* of $\mathbf{f}'$ and $\mathbf{g}'$ vanishes: $\mathbf{f}'\mathbf{g}' - \mathbf{g}'\mathbf{f}' = 0$. The value of the right side of this equation remains zero for any transformation of the set of wave functions, and consequently the commutator of $\mathbf{f}$ and $\mathbf{g}$ vanishes for any set of wave functions; it is invariant to all linear orthogonal transformations. We accordingly state that, *in order for two dynamical quantities f and g of a system to be accurately measurable at the same time, their commutator must vanish; that is, the equation*

$$\mathbf{fg} - \mathbf{gf} = 0 \tag{53-1}$$

must hold.

A proof of the second part of the uncertainty principle is difficult; indeed, the statement itself is vague (the exact meaning of Δf, etc., not being given). We shall content ourselves with the discussion of a simple case which lends itself to exact treatment, namely, the translational motion in one dimension of a free particle.

The wave functions for a free particle with coordinate x are $Ne^{\pm\frac{2\pi i \sqrt{2mW}(x-x_0)}{h}}e^{-\frac{2\pi i Wt}{h}}$ (Sec. 13), the positive sign in the first exponential corresponding to motion in the x direction and the negative sign in the $-x$ direction. On replacing W by $p_x^2/2m$ this expression becomes $Ne^{\frac{2\pi i p_x(x-x_0)}{h}}e^{-\frac{2\pi i p_x^2 t}{2mh}}$, in which positive and negative values of the momentum p_x refer to motion in the x direction and the $-x$ direction, respectively. A single wave function of this type corresponds to the physical condition in which the momentum and the energy are exactly known, that is, to a stationary state of the system. We have then no knowledge of the position of the particle, the uncertainty Δx in the coordinate x being infinite, as is seen from the probability distribution function $\Psi^*\Psi$, which is constant for all values of x between $-\infty$ and $+\infty$. When Δp_x is zero Δx is infinite.

Now let us suppose that at the time $t = 0$ we measure the momentum p_x and the coordinate x simultaneously, obtaining the values p_0 and x_0, with the uncertainties Δp_x and Δx, respectively. Our problem is to set up a wave function χ which represents this physical condition of the system. One way of doing this is the following. The wave function

$$\chi = A \int_{-\infty}^{\infty} e^{-\frac{(p_x - p_0)^2}{2(\Delta p_x)^2}} e^{\frac{2\pi i p_x (x - x_0)}{h}} e^{-\frac{2\pi i p_x^2 t}{2mh}} dp_x \qquad (53\text{-}2)$$

corresponds to a Gaussian-error-curve distribution $e^{-\frac{(p_x - p_0)^2}{(\Delta p_x)^2}}$ of the values of the momentum p_x about the average value p_0, with the uncertainty[1] Δp_x. (The factor $\frac{1}{2}$ in the exponent in Equation 53–2 results from the fact that the coefficients of the wave functions are to be squared to obtain probability values.) A is a normalization constant. On evaluating the integral we find for χ at the time $t = 0$ the expression

$$\chi(0) = B e^{-\frac{2\pi^2 (\Delta p_x)^2 (x - x_0)^2}{h^2} + \frac{2\pi i p_0 (x - x_0)}{h}}, \qquad (53\text{-}3)$$

which corresponds to the probability distribution function for x

$$\chi^*(0)\chi(0) = B^2 e^{-\frac{(x - x_0)^2}{(\Delta x)^2}} \qquad (53\text{-}4)$$

with

$$\Delta x = \frac{h}{2\pi \Delta p_x}. \qquad (53\text{-}5)$$

This is also a Gaussian error function, with its maximum at $x = x_0$ and with uncertainty Δx given by Equation 53–5. It is seen that the wave function χ corresponds to the value $h/2\pi$ for the product of the uncertainties Δx and Δp_x at the time $t = 0$, this value being of the order of magnitude h, as stated at the beginning of the section.

Problem 53–1. Evaluate the normalization constants A and B^2 by carrying out the integration over p_x and then over x.

[1] The quantity Δp_x is the reciprocal of the so-called *precision index* of the Gaussian error curve and is larger than the probable error by the factor 2.10; see R. T. Birge, *Phys. Rev.* **40**, 207 (1932).

Problem 53-2. Carry through the above treatment, retaining the time factors. Show that the center of the wave packet moves with velocity p_0/m, and that the wave packet becomes more diffuse with the passage of time.

A general discussion by the use of the methods of transformation theory (Sec. 54), which we shall not reproduce, leads to the conclusion that the product of the uncertainties $\Delta f \Delta g$ accompanying the simultaneous measurement of two dynamical quantities f and g is at least of the order of magnitude of the absolute value of the corresponding diagonal element in their commutator **fg** − **gf**. (The commutator of **x** and $\mathbf{p}_x$ is $\frac{h}{2\pi i}\mathbf{1}$ (Eq. 51–5), the absolute value of the diagonal elements being $h/2\pi$, in agreement with the foregoing discussion.) This leads to the conclusion that the energy W and time t are related regarding accuracy of measurement in the same way as a coordinate and the conjugate momentum, the product of the uncertainties ΔW and Δt being of the order of magnitude of h (or $h/2\pi$). In order to measure the energy of a system with accuracy ΔW, the measurement must be extended over a period of time of order of magnitude $h/\Delta W$.

Problem 53-3. Show that the commutator **W**t − t**W** has the value $-\frac{h}{2\pi i}\mathbf{1}$ by evaluating matrix elements, recalling that $W_{\text{op.}} = -\frac{h}{2\pi i}\frac{\partial}{\partial t}$ and $t_{\text{op.}} = t$.

It is natural for us to inquire into the significance of the uncertainty principle by analyzing an experiment designed to measure x and p_x. Many "thought experiments" have been discussed in the effort to find a contradiction or to clarify the theory; in every case these have led to results similar to the following. Suppose that we send a beam of light of frequency ν_0 along the axis AO of Figure 53–1, and observe along the direction OB to see whether or not the particle, restricted to motion along the x axis, is at the point O or not. If a light quantum is scattered into our microscope at B, we know that the particle is in the neighborhood of O, and by analyzing the scattered light by a spectroscope to determine its frequency ν, we can calculate the momentum of the particle by use of the equations of the Compton effect. But for light of finite frequency the resolving power of the microscope is limited, and our measurement of x will show a corresponding uncertainty Δx, which decreases as the frequency increases. Similarly the measurement of the momentum by the Compton effect will show an uncertainty Δp_x, increasing as the frequency increases. The detailed analysis of the experiment shows that under the most favorable conditions imaginable the product $\Delta x \Delta p_x$ is of the order of magnitude of h.[1]

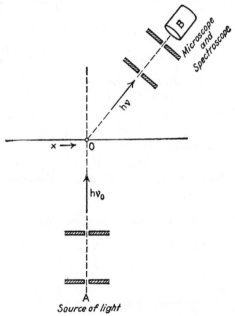

Fɪɢ. 53-1.—Diagram of experiment for measuring x and p_x of particle.

54. TRANSFORMATION THEORY

In discussing the behavior of a system the following question might arise. If at the time $t = t'$ the dynamical property f is

[1] For the further discussion of the uncertainty principle see W. Heisenberg, "The Physical Principles of the Quantum Theory," University of Chicago Press, Chicago, 1930; N. Bohr, *Nature* **121**, 580 (1928); C. G. Darwin, *Proc. Roy. Soc.* **A117**, 258 (1927); A. E. Ruark, *Phys. Rev.* **31**, 311, 709 (1928); E. H. Kennard, *Phys. Rev.* **31**, 344 (1928); H. P. Robertson, *Phys. Rev.* **34**, 163 (1929); **35**, 667 (1930); **46**, 794 (1934); and also Ruark and Urey, "Atoms, Molecules and Quanta," Chap. XVIII; and other references listed at the end of the chapter.

found on measurement to have the value f', what is the probability that the immediately subsequent measurement of the dynamical property g will yield the value g'? We know one way to answer this question, namely, to find the wave function[1] χ (one of the representation which makes $\mathbf{f}'$ a diagonal matrix) corresponding to the value f' of f, to use it to calculate the average value of all powers of g, and from these to construct a probability distribution function for g. This is not a very simple or direct procedure, however; it is of interest that an alternative method has been found by means of which these probability distribution functions can be calculated directly. This method, called the *transformation theory*,[2] is a general quantum mechanics within which wave mechanics is included, the Schrödinger wave equation being one of a large number of equations of the theory and the Schrödinger wave functions a particular type of transformation functions. We shall not enter into an extensive discussion of transformation theory but shall give only a brief description of it.

Let us represent by $(g'|f')$ a *probability amplitude function* or *transformation function* such that $(g'|f')^*(g'|f')$ is the probability under discussion, $(g'|f')^*$ being the complex conjugate of $(g'|f')$. [In case that g' can be any one of a continuum of values, $(g'|f')^*(g'|f')$ is interpreted as a probability distribution function, the probability that g have a value between g' and $g' + dg'$ being $(g'|f')^*(g'|f')dg'$.]

The Schrödinger stationary-state wave functions are probability amplitude functions between the energy and the coordinates of the system. For a system with one degree of freedom, such as a harmonic oscillator, the wave functions ψ_n are the transformation functions $(x'|W')$ between the coordinate x and the characteristic energy values, and for the hydrogen atom the wave functions $\psi_{nlm}(r, \vartheta, \varphi)$, discussed in Chapter V, are the transformation functions $(r'\vartheta'\varphi'|nlm)$ between the coordinates r, ϑ, and φ of the electron relative to the nucleus and the characteristic energy values W_n, the square of angular momentum values

---

[1] In case that the measurement of f is not a maximal measurement many wave functions might have to be considered.

[2] The transformation theory was developed mainly by P. A. M. Dirac. *Proc. Roy. Soc.* **A113**, 621 (1927), and P. Jordan, *Z. f. Phys.* **40**, 809 (1927); **44**, 1 (1927).

$\dfrac{l(l+1)h^2}{4\pi^2}$, and the angular momentum component values $mh/2\pi$,

represented by the symbols n, l, and m, respectively.

Two important properties of transformation functions are the following:

The transformation function between f and g is equal to that between g and f:

$$(f'|g') = (g'|f')^*. \qquad (54\text{-}1)$$

The transformation function between f and h is related to that between f and g and that between g and h by the equation

$$(f'|h') = \int (f'|g')^*(g'|h')dg'. \qquad (54\text{-}2)$$

In this equation the integration includes all possible values g' which can be obtained by measurement of g; in case that g' represents a set of discrete values, the sum over these is to be taken.

We have often written the Schrödinger wave equation in the form

$$H_{\text{op.}}\psi_n = W_n\psi_n.$$

In the nomenclature of transformation theory this is

$$H_{\text{op.}}(q_j'|W') = W'(q_j'|W'),$$

W' representing a characteristic value W_n of the energy and $(q_j'|W')$ the corresponding transformation function to the coordinates q_j. In transformation theory it is postulated that a similar equation

$$g_{\text{op.}_{(f)}}(f'|g') = g'(f'|g') \qquad (54\text{-}3)$$

is satisfied by every transformation function $(f'|g')$. In this equation $g_{\text{op.}_{(f)}}$ is the operator in the f scheme representing the dynamical quantity g. We shall not discuss the methods by means of which the f scheme of operators is found but shall restrict our attention to the q scheme, in which the operators are obtained by the familiar method of replacing p_k by $\dfrac{h}{2\pi i}\cdot\dfrac{\partial}{\partial q_k}$.

The transformation functions are normalized and mutually orthogonal, satisfying the equation

$$\int (f'|g')(g'|f'')dg' = \delta_{f'f''}. \qquad (54\text{-}4)$$

It is interesting to note that this equation signifies that, if the dynamical quantity f has been found on measurement to have the value f', immediate repetition of the measurement will give the same value f' with probability unity, inasmuch as the integral of Equation 54-4 is the transformation function $(f'|f'')$ (see Eq. 54-2) and Equation 54-4 requires it to vanish except when f'' is equal to f', in which case it has the value 1.

From the above equations we can find any transformation function $(f'|g')$, using the q system of operators only, in the following way: we find the transformation functions $(q'|f')$ and $(q'|g')$ by solving the corresponding differential equations 54-3, and then obtain $(f'|g')$ by integrating over the coordinates (Eq. 54-4). As an example, let us obtain the transformation function $(p_x'|W')$ between the energy W and the linear momentum p_x of a one-dimensional system. The function $(x'|W')$ is the Schrödinger wave function, obtained by solving the wave equation

$$H_{op.}(x'|W') = W'(x'|W')$$

as described in the preceding chapters of the book. The transformation function $(x'|p_x')$ between a Cartesian coordinate and its canonically conjugate momentum is the solution of the equation

$$p_{x_{op.}}(x'|p_x') = p_x'(x'|p_x')$$

or

$$\frac{h}{2\pi i}\frac{\partial}{\partial x'}(x'|p_x') = p_x'(x'|p_x'),$$

and hence is the function

$$(x'|p_x') = Ce^{\frac{2\pi i x'p_x'}{h}}, \tag{54-5}$$

C being a normalizing factor. The transformation function $(p_x'|W')$, the momentum probability amplitude function for a stationary state of the system, is accordingly given by the equation

$$(p_x'|W') = \int Ce^{-\frac{i}{h}\frac{ix'p_x'}{}}(x'|W')dx' \tag{54-6}$$

or

$$(p_x'|W_n) = \int Ce^{-\frac{2\pi i x'p_x'}{h}}\psi_n(x')dx'. \tag{54-7}$$

On application of this equation it is found that the *momentum wave functions* for the harmonic oscillator have the same form (Hermite orthogonal functions) as the coordinate wave functions (Prob. 54–1), whereas those for the hydrogen atom are quite different.[1]

Problem 54–1. Evaluate the momentum wave functions for the harmonic oscillator. Show that the average value of p_x^r for the nth state given by the equation

$$\int_{-\infty}^{\infty} (p_x'|W_n)^*(p_x'|W_n)p_x'^r dp_x'$$

is the same as given by the equation

$$\int_{-\infty}^{\infty} \psi_n^* \left(\frac{h}{2\pi i}\right)^r \frac{\partial^r}{\partial x^r} \psi_n dx.$$

Problem 54–2. Evaluate the momentum wave function for the normal hydrogen atom,

$$(p_x'p_y'p_z'|nlm) = \int\int\int Ce^{-\frac{2\pi i(x'p_x'+y'p_y'+z'p_z')}{h}} (x'y'z'|nlm)dx'dy'dz'.$$

It is convenient to change to polar coordinates in momentum space as well as in coordinate space.

The further developments of quantum mechanics, including the discussion of maximal measurements consisting not of the accurate determination of the values of a minimum number of independent dynamical functions but of the approximate measurement of a larger number, the use of the theory of groups, the formulation of a relativistically invariant theory, the quantization of the electromagnetic field, etc., are beyond the scope of this book.

General References on Quantum Mechanics

Matrix mechanics:

M. Born and P. Jordan: "Elementare Quantenmechanik," Julius Springer, Berlin, 1930.

Transformation theory and general quantum mechanics:

P. A. M. Dirac: "Quantum Mechanics," Oxford University Press, New York, 1935.

---

[1] The hydrogen-atom momentum wave functions are discussed by B. Podolsky and L. Pauling, *Phys. Rev.* **34**, 109 (1929), and by F. A. Hylleraas, *Z. f. Phys.* **74**, 216 (1932).

J. v. Neumann: "Mathematische Grundlagen der Quantenmechanik," Julius Springer, Berlin, 1932.

Questions of physical interpretation:

W. Heisenberg: "The Physical Principles of the Quantum Theory," University of Chicago Press, Chicago, 1930.

General references:

A. E. Ruark and H. C. Urey: "Atoms, Molecules and Quanta," McGraw-Hill Book Company, Inc., New York, 1930.

E. U. Condon and P. M. Morse: "Quantum Mechanics," McGraw-Hill Book Company, Inc., New York, 1929.

A. Sommerfeld: "Wave Mechanics," Methuen & Company, Ltd., London, 1930.

H. Weyl: "The Theory of Groups and Quantum Mechanics," E. P. Dutton & Co., Inc., New York, 1931.

J. Frenkel: "Wave Mechanics," Oxford University Press, New York 1933.

APPENDIX I

VALUES OF PHYSICAL CONSTANTS[1]

Velocity of light..................... $c = 2.99796 \times 10^{10}$ cm sec^{-1}

Electronic charge.................... $e = 4.770 \times 10^{-10}$ abs. e.s.u.

Electronic mass...................... $m_0 = 9.035 \times 10^{-28}$ g

Planck's constant.................... $h = 6.547 \times 10^{-27}$ erg sec

Avogadro's number.................. $N = 0.6064 \times 10^{24}$ mole^{-1}

Boltzmann's constant................ $k = 1.3709 \times 10^{-16}$ erg deg^{-1}

Fine-structure constant.............. $\alpha = \dfrac{2\pi e^2}{hc} = 7.284 \times 10^{-3}$

Radius of Bohr orbit in normal hydro-
gen, referred to center of mass....... $a_0 = 0.5282 \times 10^{-8}$ cm

Rydberg constant for hydrogen........ $R_{\mathrm{H}} = 109677.759$ cm^{-1}

Rydberg constant for helium.......... $R_{\mathrm{He}} = 109722.403$ cm^{-1}

Rydberg constant for infinite mass..... $R_\infty = 109737.42$ cm^{-1}

Bohr unit of angular momentum....... $\dfrac{h}{2\pi} = 1.0420 \times 10^{-27}$ erg sec

Magnetic moment of 1 Bohr magneton $\mu_0 = 0.9175 \times 10^{-20}$ erg gauss^{-1}

Relations among Energy Quantities

1 erg $= 0.6285 \times 10^{12}$ v.e. $= 0.5095 \times 10^{16}$ cm^{-1} $= 1.440 \times 10^{16}$ cal/mole

1.591×10^{-12} erg $= 1$ v.e. $= 8106$ cm^{-1} $= 23055$ cal/mole

1.963×10^{-16} erg $= 1.234 \times 10^{-4}$ v.e. $= 1$ cm^{-1} $= 2.844$ cal/mole

0.6901×10^{-16} erg $= 4.338 \times 10^{-5}$ v.e. $= 0.3516$ cm^{-1} $= 1$ cal/mole

[1] These values are taken from the compilation of R. T. Birge, *Rev. Mod. Phys.* **1,** 1 (1929), as recommended by Birge, *Phys. Rev.* **40,** 228 (1932). For probable errors see these references.

APPENDIX II

PROOF THAT THE ORBIT OF A PARTICLE MOVING IN A CENTRAL FIELD LIES IN A PLANE

The force acting on the particle at any instant is in the direction of the attracting center O (see **F**, Fig. 1). Let the arrow marked **v** in the figure represent the direction of the motion at any instant. Set up a system of Cartesian axes $x\,y\,z$ with origin at the point P and oriented so that the z axis points along **v** and the y axis points perpendicular to the plane of **F** and **v**, being directed up from the plane of the paper in the figure.

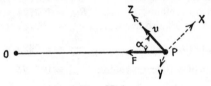

FIG. II–1.

Then the equation of motion (in Newton's form) in y is

$$m\frac{d^2y}{dt^2} = 0,$$

since there is no component of the force F in the y direction. Therefore the acceleration in the y direction is zero and the velocity in the y direction, being initially zero, will remain zero, so that the particle will have no tendency to move out of the plane determined by **F** and **v**.

APPENDIX III

PROOF OF ORTHOGONALITY OF WAVE FUNCTIONS CORRESPONDING TO DIFFERENT ENERGY LEVELS

We shall prove that, if $W_n \neq W_k$, the solution ψ_n of the wave equation

$$\sum_{i=1}^{N} \frac{1}{m_i} \nabla_i^2 \psi_n + \frac{8\pi^2}{h^2}(W_n - V)\psi_n = 0 \tag{1}$$

and the solution ψ_k^* of the equation

$$\sum_{i=1}^{N} \frac{1}{m_i} \nabla_i^2 \psi_k^* + \frac{8\pi^2}{h^2}(W_k - V)\psi_k^* = 0 \tag{2}$$

satisfy the relation

$$\int \psi_k^* \psi_n d\tau = 0; \tag{3}$$

i.e., that ψ_k is orthogonal to ψ_n.

Multiply Equation 1 by ψ_k^*, Equation 2 by ψ_n, and subtract the second from the first. Since V is real, the result is that

$$\sum_{i=1}^{N} \frac{1}{m_i} (\psi_k^* \nabla_i^2 \psi_n - \psi_n \nabla_i^2 \psi_k^*) + \frac{8\pi^2}{h^2}(W_n - W_k)\psi_k^* \psi_n = 0. \tag{4}$$

If we now integrate the terms of this equation over configuration space, we obtain

$$\frac{8\pi^2}{h^2}(W_n - W_k) \int \psi_k^* \psi_n d\tau = -\sum_{i=1}^{N} \frac{1}{m_i} \int (\psi_k^* \nabla_i^2 \psi_n - \psi_n \nabla_i^2 \psi_k^*) d\tau. \tag{5}$$

If we introduce the expression for ∇_i^2 in terms of Cartesian coordinates into the integral on the right, it becomes

$$\sum_{j=1}^{3N} \frac{1}{m_j} \int_{-\infty}^{\infty} \cdots \int_{-\infty}^{\infty} \left(\psi_k^* \frac{\partial^2 \psi_n}{\partial q_j^2} - \psi_n \frac{\partial^2 \psi_k^*}{\partial q_j^2} \right) dq_1 dq_2 \cdots dq_{3N}, \tag{6}$$

441

in which we have written $q_1, q_2, \cdots, q_{3N}$ in place of $x_1, y_1, z_1,$ $x_2, \cdots, z_N$. We next make use of the identity

$$\frac{\partial}{\partial q_j}\left(\psi_k^*\frac{\partial \psi_n}{\partial q_j} - \psi_n\frac{\partial \psi_k^*}{\partial q_j}\right) = \psi_k^*\frac{\partial^2 \psi_n}{\partial q_j^2} - \psi_n\frac{\partial^2 \psi_k^*}{\partial q_j^2}, \qquad (7)$$

from which we see that

$$\int_{-\infty}^{\infty}\left(\psi_k^*\frac{\partial^2 \psi_n}{\partial q_j^2} - \psi_n\frac{\partial^2 \psi_k^*}{\partial q_j^2}\right)dq_j = \left[\psi_k^*\frac{\partial \psi_n}{\partial q_j} - \psi_n\frac{\partial \psi_k^*}{\partial q_j}\right]_{-\infty}^{\infty} = 0,$$

because of the boundary conditions on ψ.

Since every term of the sum can be treated similarly, the expression 6 is equal to zero and therefore

$$\frac{8\pi^2}{h^2}(W_n - W_k)\int \psi_k^*\psi_n d\tau = 0,$$

from which Equation 3 follows, since $W_n - W_k \neq 0$.

If $W_n = W_k$, so that ψ_k and ψ_n are two linearly independent wave functions belonging to the same energy level, ψ_k and ψ_n are not necessarily orthogonal, but it is always possible to construct two wave functions $\psi_{k'}'$ and $\psi_{n'}'$ belonging to this level which are mutually orthogonal. This can be done in an infinite number of ways by forming the combinations

$$\psi_{k'}' = \alpha\psi_k + \beta\psi_n \quad \text{and} \quad \psi_{n'}' = \alpha'\psi_k + \beta'\psi_n,$$

with coefficients $\alpha, \beta, \alpha', \beta'$ satisfying the relation

$$\int\psi_{k'}'^*\psi_{n'}'d\tau = \alpha^*\alpha'\int\psi_k^*\psi_k d\tau + \alpha^*\beta'\int\psi_k^*\psi_n d\tau + \alpha'\beta^*\int\psi_n^*\psi_k d\tau +$$
$$\beta^*\beta'\int\psi_n^*\psi_n d\tau = 0. \qquad (8)$$

APPENDIX IV

ORTHOGONAL CURVILINEAR COORDINATE SYSTEMS

In Section 16 the general formulas for the Laplace operator ∇^2 and for the volume element $d\tau$ were given in terms of the quantities q_u, q_v, and q_w defined by Equation 16–4. In this appendix there are given the equations of transformation (in terms of Cartesian coordinates) and the expressions for the q's for the 11 sets of orthogonal coordinate systems listed by Eisenhart[1] as the only such systems in which the three-dimensional Schrödinger wave equation can be separable. In addition the explicit expressions[2] for ∇^2 and $d\tau$ are given for a few of the more important systems. These quantities may be obtained for the other systems by the use of Equations 16–3 and 16–5.

Cylindrical Polar Coordinates

$x = \rho \cos \varphi,$

$y = \rho \sin \varphi,$

$z = z.$

$q_\rho = 1, q_z = 1, q_\varphi = \rho.$

$d\tau = \rho d\rho dz d\varphi.$

$$\nabla^2 = \frac{1}{\rho} \frac{\partial}{\partial \rho}\left(\rho \frac{\partial}{\partial \rho}\right) + \frac{1}{\rho^2} \frac{\partial^2}{\partial \varphi^2} + \frac{\partial^2}{\partial z^2}.$$

Spherical Polar Coordinates

$x = r \sin \vartheta \cos \varphi,$

$y = r \sin \vartheta \sin \varphi,$ (Fig. 1–1),

$z = r \cos \vartheta.$

$q_r = 1, q_\vartheta = r, q_\varphi = r \sin \vartheta.$

$d\tau = r^2 \sin \vartheta dr d\vartheta d\varphi.$

$$\nabla^2 = \frac{1}{r^2} \frac{\partial}{\partial r}\left(r^2 \frac{\partial}{\partial r}\right) + \frac{1}{r^2 \sin \vartheta} \frac{\partial}{\partial \vartheta}\left(\sin \vartheta \frac{\partial}{\partial \vartheta}\right) + \frac{1}{r^2 \sin^2 \vartheta} \frac{\partial^2}{\partial \varphi^2}.$$

[1] L. P. EISENHART, *Phys. Rev.* **45**, 428 (1934).

[2] E. P ADAMS, "Smithsonian Mathematical Formulae," Washington, 1922. This book contains extensive material on curvilinear coordinates as well as other very useful formulas.

Parabolic Coordinates

$$x = \sqrt{\xi\eta}\cos\varphi,$$
$$y = \sqrt{\xi\eta}\sin\varphi,$$
$$z = \tfrac{1}{2}(\xi - \eta).$$

$$q_\xi = \frac{1}{2}\sqrt{\frac{\xi+\eta}{\xi}},\ q_\eta = \frac{1}{2}\sqrt{\frac{\xi+\eta}{\eta}},\ q_\varphi = \sqrt{\xi\eta}.$$

$$d\tau = \tfrac{1}{4}(\xi + \eta)d\xi d\eta d\varphi.$$

$$\nabla^2 = \frac{4}{\xi+\eta}\frac{\partial}{\partial\xi}\left(\xi\frac{\partial}{\partial\xi}\right) + \frac{4}{\xi+\eta}\frac{\partial}{\partial\eta}\left(\eta\frac{\partial}{\partial\eta}\right) + \frac{1}{\xi\eta}\frac{\partial^2}{\partial\varphi^2}.$$

Confocal Elliptic Coordinates (Prolate Spheroids)

$$x = a\sqrt{\xi^2-1}\sqrt{1-\eta^2}\cos\varphi,$$
$$y = a\sqrt{\xi^2-1}\sqrt{1-\eta^2}\sin\varphi,$$
$$z = a\xi\eta.$$

In terms of the distances r_A and r_B from the points $(0, 0, -a)$ and $(0, 0, a)$, respectively, ξ and η are given by the expressions

$$\xi = \frac{r_A + r_B}{2a};\ \eta = \frac{r_A - r_B}{2a}.$$

$$q_\xi = a\sqrt{\frac{\xi^2-\eta^2}{\xi^2-1}},\ q_\eta = a\sqrt{\frac{\xi^2-\eta^2}{1-\eta^2}},\ q_\varphi = a\sqrt{(\xi^2-1)(1-\eta^2)}.$$

$$d\tau = a^3(\xi^2 - \eta^2)d\xi d\eta d\varphi.$$

$$\nabla^2 = \frac{1}{a^2(\xi^2-\eta^2)}\left[\frac{\partial}{\partial\xi}\left\{(\xi^2-1)\frac{\partial}{\partial\xi}\right\} + \frac{\partial}{\partial\eta}\left\{(1-\eta^2)\frac{\partial}{\partial\eta}\right\} + \frac{\xi^2-\eta^2}{(\xi^2-1)(1-\eta^2)}\frac{\partial^2}{\partial\varphi^2}\right].$$

Spheroidal Coordinates (Oblate Spheroids)

$$x = a\xi\eta\cos\varphi,\ y = a\xi\eta\sin\varphi,\ z = a\sqrt{(\xi^2-1)(1-\eta^2)}.$$

$$q_\xi = a\sqrt{\frac{\xi^2-\eta^2}{\xi^2-1}},\ q_\eta = a\sqrt{\frac{\xi^2-\eta^2}{1-\eta^2}},\ q_\varphi = a\xi\eta.$$

Parabolic Cylinder Coordinates

$$x = \tfrac{1}{2}(u - v),\ y = \sqrt{uv},\ z = z.$$

$$q_u = \frac{1}{2}\sqrt{\frac{u+v}{u}},\ q_v = \frac{1}{2}\sqrt{\frac{u+v}{v}},\ q_z = 1.$$

Elliptic Cylinder Coordinates

$$x = a\sqrt{(u^2 - 1)(1 - v^2)},\ y = auv,\ z = z.$$

$$q_u = a\sqrt{\frac{u^2 - v^2}{u^2 - 1}},\ q_v = a\sqrt{\frac{u^2 - v^2}{1 - v^2}},\ q_z = 1$$

Ellipsoidal Coordinates

$$x^2 = \frac{(a^2 + u)(a^2 + v)(a^2 + w)}{(a^2 - b^2)(a^2 - c^2)},\ y^2 = \frac{(b^2 + u)(b^2 + v)(b^2 + w)}{(b^2 - c^2)(b^2 - a^2)},$$

$$z^2 = \frac{(c^2 + u)(c^2 + v)(c^2 + w)}{(c^2 - a^2)(c^2 - b^2)}.$$

$$q_u^2 = \frac{(u - v)(u - w)}{4(a^2 + u)(b^2 + u)(c^2 + u)},\ q_v^2 = \frac{(v - w)(v - u)}{4(a^2 + v)(b^2 + v)(c^2 + v)},$$

$$q_w^2 = \frac{(w - u)(w - v)}{4(a^2 + w)(b^2 + w)(c^2 + w)}.$$

Confocal Parabolic Coordinates

$$x = \frac{1}{2}(u + v + w - a - b),\ y^2 = \frac{(a - u)(a - v)(a - w)}{b - a},$$

$$z^2 = \frac{(b - u)(b - v)(b - w)}{a - b},\qquad u > b > v > a > w.$$

$$q_u^2 = \frac{(u - v)(u - w)}{4(a - u)(b - u)},\ q_v^2 = \frac{(v - u)(v - w)}{4(a - v)(b - v)},$$

$$q_w^2 = \frac{(w - u)(w - v)}{4(a - w)(b - w)}.$$

A Coordinate System Involving Elliptic Functions

$$x = u\,\text{dn}(v, k)\,\text{sn}(w, k'),\ y = u\,\text{sn}(v, k)\,\text{dn}(w, k'),$$
$$z = u\,\text{cn}(v, k)\,\text{cn}(w, k'),\ k^2 + k'^2 = 1.$$
$$q_u^2 = 1,\ q_v^2 = q_w^2 = u^2\{k^2\,\text{cn}^2(v, k) + k'^2\,\text{cn}^2(w, k')\}.$$

For a discussion of the elliptic functions dn, sn, and cn see W. F. Osgood, "Advanced Calculus," Chap. IX, or E. P. Adams, "Smithsonian Mathematical Formulae." p. 245.

APPENDIX V

THE EVALUATION OF THE MUTUAL ELECTROSTATIC ENERGY OF TWO SPHERICALLY SYMMETRICAL DISTRIBUTIONS OF ELECTRICITY WITH EXPONENTIAL DENSITY FUNCTIONS

In Section 23b there occurs the integral

$$I = \frac{Ze^2}{32\pi^2 a_0} \int \int \frac{e^{-\rho_1}e^{-\rho_2}}{\rho_{12}} d\tau_1 d\tau_2,$$

in which $\rho_1 = 2Zr_1/a_0$ and $d\tau_1 = \rho_1^2 d\rho_1 \sin \vartheta_1 d\vartheta_1 d\varphi_1$, with similar expressions for ρ_2 and $d\tau_2$, r_1, ϑ_1, φ_1 and r_2, ϑ_2, φ_2 being polar coordinates for the same system of axes. The quantity ρ_{12} represents $2Zr_{12}/a_0$, in which r_{12} is the distance between the points r_1, ϑ_1, φ_1 and r_2, ϑ_2, φ_2.

This integral (aside from the factor $Ze^2/32\pi^2 a_0$) represents the mutual electrostatic energy of two spherically symmetrical distributions of electricity, with density functions $e^{-\rho_1}$ and $e^{-\rho_2}$, respectively. It can be evaluated by calculating the potential due to the first distribution, by integrating over $d\tau_1$, and then evaluating the energy of the second distribution in the field of the first.

The potential of a spherical shell of radius ρ_1 and total charge $4\pi\rho_1^2 e^{-\rho_1} d\rho_1$ is,[1] at a point r,

$$4\pi\rho_1^2 e^{-\rho_1} d\rho_1 \cdot \frac{1}{\rho_1} \qquad \text{for } r < \rho_1$$

and

$$4\pi\rho_1^2 e^{-\rho_1} d\rho_1 \cdot \frac{1}{r} \qquad \text{for } r > \rho_1;$$

that is, the potential is constant within the shell and has the same value outside of the shell as if the entire charge were located at the origin.

---

[1] See, for example, Jeans, "Electricity and Magnetism," Cambridge University Press, Cambridge, 1925, Sec. 74.

The potential of the complete distribution is hence

$$\Phi(r) = \frac{4\pi}{r} \int_0^r e^{-\rho_1}\rho_1^2 d\rho_1 + 4\pi \int_r^\infty e^{-\rho_1}\rho_1 d\rho_1,$$

which is found on evaluation to be

$$\Phi(r) = \frac{4\pi}{r}\{2 - e^{-r}(r + 2)\}.$$

The integral I then has the value

$$I = \frac{Ze^2}{32\pi^2 a_0} \int \Phi(\rho_2) e^{-\rho_2} d\tau_2$$

$$= \frac{Ze^2}{2a_0} \int_0^\infty \{2 - e^{-\rho_2}(\rho_2 + 2)\} e^{-\rho_2}\rho_2 d\rho_2,$$

which gives on integration

$$I = \frac{Ze^2}{2a_0} \cdot \frac{5}{4} = \frac{5}{4}ZW_\text{H}.$$

APPENDIX VI

NORMALIZATION OF THE ASSOCIATED LEGENDRE FUNCTIONS

We can obtain the orthogonality property of the functions $P_l^{|m|}(z)$ and $P_{l'}^{|m|}(z)$ as follows: Multiply the differential equation 19–9 satisfied by $P_l^{|m|}(z)$ by $P_{l'}^{|m|}(z)$ and subtract from this the differential equation satisfied by $P_{l'}^{|m|}(z)$ multiplied by $P_l^{|m|}(z)$. The result is the relation

$$P_{l'}^{|m|}\frac{d}{dz}\left\{(1-z^2)\frac{dP_l^{|m|}}{dz}\right\} - P_l^{|m|}\frac{d}{dz}\left\{(1-z^2)\frac{dP_{l'}^{|m|}}{dz}\right\}$$
$$= \frac{d}{dz}\left[(1-z^2)\left\{P_{l'}^{|m|}\frac{dP_l^{|m|}}{dz} - P_l^{|m|}\frac{dP_{l'}^{|m|}}{dz}\right\}\right]$$
$$= \{l'(l'+1) - l(l+1)\}P_{l'}^{|m|}P_l^{|m|}.$$

If we integrate this between the limits -1 and 1, we obtain the result

$$\{l'(l'+1) - l(l+1)\}\int_{-1}^{+1}P_{l'}^{|m|}(z)P_l^{|m|}(z)\,dz$$
$$= \left[(1-z^2)\left\{P_{l'}^{|m|}\frac{dP_l^{|m|}}{dz} - P_l^{|m|}\frac{dP_{l'}^{|m|}}{dz}\right\}\right]_{-1}^{+1} = 0.$$

Therefore, if $l' \neq l$,

$$\int_{-1}^{+1}P_{l'}^{|m|}(z)P_l^{|m|}(z)\,dz = 0. \tag{1}$$

This result is true for any value of m, so it is also true for the Legendre functions $P_l(z)$, since $P_l(z) = P_l^0(z)$.

We can now obtain the normalization integral for the Legendre polynomials. Replacing l by $l-1$ in Equation 19–2 gives the equation

$$P_l(z) = \frac{1}{l}\{(2l-1)zP_{l-1}(z) - (l-1)P_{l-2}(z)\}.$$

Using this and the orthogonality property just proved, we obtain the relation

448

$$\int_{-1}^{+1} \{P_l(z)\}^2 dz = \frac{2l-1}{l}\int_{-1}^{+1} P_{l-1}(z)zP_l(z)dz.$$

Equation 19–2 can be written in the form

$$zP_l(z) = \frac{1}{2l+1}\{(l+1)P_{l+1}(z) + lP_{l-1}(z)\},$$

so that, again employing the orthogonality property, we get

$$\int_{-1}^{+1} \{P_l(z)\}^2 dz = \frac{2l-1}{2l+1}\int_{-1}^{+1} \{P_{l-1}(z)\}^2 dz.$$

This process can be repeated until the relation

$$\int_{-1}^{+1} \{P_l(z)\}^2 dz =$$
$$\frac{(2l-1)(2l-3)(2l-5)\cdots 3\cdot 1}{(2l+1)(2l-1)(2l-3)\cdots 5\cdot 3}\int_{-1}^{+1} \{P_0(z)\}^2 dz$$
$$= \frac{1}{2l+1}\int_{-1}^{+1} \{P_0(z)\}^2 dz$$

is obtained. $P_0(z)$ is by definition (Eq. 19–1) the coefficient of t^0 in the expansion of $(1 - 2tz + t^2)^{-\frac{1}{2}}$ in powers of t. It is therefore equal to unity, so that

$$\int_{-1}^{+1} \{P_l(z)\}^2 dz = \frac{1}{2l+1}\int_{-1}^{+1} dz = \frac{2}{2l+1}. \tag{2}$$

To obtain the normalization integral for the associated Legendre functions we proceed as follows.[1] By differentiating Equation 19–7 and multiplying by $(1 - z^2)^{\frac{1}{2}}$ we obtain

$$(1 - z^2)^{\frac{1}{2}}\frac{dP_l^{|m|}(z)}{dz} = (1 - z^2)^{\frac{|m|+1}{2}}\frac{d^{|m|+1}}{dz^{|m|+1}}P_l(z) -$$
$$|m|z(1 - z^2)^{\frac{|m|-1}{2}}\frac{d^{|m|}}{dz^{|m|}}P_l(z) = P_l^{|m|+1}(z) - |m|z(1 - z^2)^{-\frac{1}{2}}P_l^{|m|}(z).$$

Transposing, squaring, and integrating gives

$$\int_{-1}^{+1} \{P_l^{|m|+1}(z)\}^2 dz = \int_{-1}^{+1}\left[(1 - z^2)\left\{\frac{dP_l^{|m|}(z)}{dz}\right\}^2 + \right.$$
$$\left. 2|m|zP_l^{|m|}\frac{dP_l^{|m|}(z)}{dz} + \frac{m^2z^2}{1 - z^2}\{P_l^{|m|}(z)\}^2\right]dz$$

[1] Whittaker and Watson, "Modern Analysis," Sec. 15·51.

$$= -\int_{-1}^{+1} P_l^{|m|}(z)\frac{d}{dz}\left\{(1 - z^2)\frac{dP_l^{|m|}(z)}{dz}\right\}dz - |m|\int_{-1}^{+1}\{P_l^{|m|}(z)\}^2dz$$

$$+ \int_{-1}^{+1}\frac{m^2z^2}{1 - z^2}\{P_l^{|m|}(z)\}^2dz,$$

where integration by parts[1] has been employed to obtain the first two terms of the last line.

If we now use the differential equation 19–9 for $P_l^{|m|}(z)$ to reduce the first term of the last line, we obtain, after combining terms, the result

$$\int_{-1}^{+1}\{P_l^{|m|+1}(z)\}^2dz = (l - |m|)(l + |m| + 1)\int_{-1}^{+1}\{P_l^{|m|}(z)\}^2dz.$$

We can continue this process and thus obtain

$$\int_{-1}^{+1}\{P_l^{|m|}(z)\}^2dz = (l - |m| + 1)(l - |m| + 2) \cdots l$$
$$(l + |m|)(l + |m| - 1) \cdots (l + 1)\int_{-1}^{+1}\{P_l(z)\}^2dz,$$

so that

$$\int_{-1}^{+1}\{P_l^{|m|}(z)\}^2dz = \frac{2}{2l + 1}\frac{(l + |m|)!}{(l - |m|)!},$$

where we have used the result of Equation 2.

[1] In the general equation $\int u\,dv = uv - \int v\,du$, we set $u = (1 - z^2)\dfrac{dP_l^{|m|}}{dz}$,

$dv = \dfrac{dP_l^{|m|}}{dz}$ in order to reduce the first term, and $u = z$,

$dv = 2P_l^{|m|}\dfrac{dP_l^{|m|}}{dz}dz = d\{P_l^{|m|}\}^2$

to reduce the second term. The term in uv vanishes, in the first case because $(1 - z^2)$ is zero at the limits, and in the second case because $P_l^{|m|}(z)$ is zero at the limits, if $m \neq 0$.

APPENDIX VII

NORMALIZATION OF THE ASSOCIATED LAGUERRE FUNCTIONS

In order to obtain Equation 20–10, we make use of the generating function given in Equation 20–8, namely

$$U_s(\rho, u) \equiv \sum_{r=s}^{\infty} \frac{L_r^s(\rho)}{r!} u^r \equiv (-1)^s \frac{e^{-\frac{\rho u}{1-u}}}{(1-u)^{s+1}} u^s.$$

Similarly let

$$V_s(\rho, v) \equiv \sum_{t=s}^{\infty} \frac{L_t^s(\rho)}{t!} v^t \equiv (-1)^s \frac{e^{-\frac{\rho v}{1-v}}}{(1-v)^{s+1}} v^s.$$

Multiplying these together, introducing the factor $e^{-\rho}\rho^{s+1}$, and integrating, we obtain the equation

$$\int_0^{\infty} e^{-\rho}\rho^{s+1} U_s(\rho, u) V_s(\rho, v) d\rho = \sum_{r,t=s}^{\infty} \frac{u^r v^t}{r! t!} \int_0^{\infty} e^{-\rho}\rho^{s+1} L_r^s(\rho) L_t^s(\rho) d\rho$$

$$= \frac{(uv)^s}{(1-u)^{s+1}(1-v)^{s+1}} \int_0^{\infty} \rho^{s+1} e^{-\rho\left(1+\frac{u}{1-u}+\frac{v}{1-v}\right)} d\rho$$

$$= \frac{(s+1)!(uv)^s(1-u)(1-v)}{(1-uv)^{s+2}} = (s+1)!(1-u-v+uv)$$

$$\sum_{k=0}^{\infty} \frac{(s+k+1)!}{k!(s+1)!}(uv)^{s+k},$$

where we have expanded $(1-uv)^{-s-2}$ by the binomial theorem.[1]

The integral we are seeking is $(r!)^2$ times the coefficient of $(uv)^r$ in the expansion, which is

[1] For the value of the integral $\int_0^{\infty} \rho^{s+1} e^{-a\rho} d\rho$ see Peirce's "Table of Integrals."

$$(r!)^2(s+1)!\left\{\frac{(r+1)!}{(r-s)!(s+1)!} + \frac{r!}{(r-s-1)!(s+1)!}\right\}$$

$$= \frac{(r!)^3(2r-s+1)}{(r-s)!}.$$

In order to obtain the integral of Equation 20–10 we must put $r = n + l$ and $s = 2l + 1$, yielding the final result

$$\int_0^\infty e^{-\rho}\rho^{2l+2}\{L_{n+l}^{2l+1}(\rho)\}^2 d\rho = \frac{2n[(n+l)!]^3}{(n-l-1)!}.$$

APPENDIX VIII

THE GREEK ALPHABET

A, α	. . .	Alpha	N, ν	. . .	Nu
B, β	. . .	Beta	Ξ, ξ	. . .	Xi
Γ, γ	. . .	Gamma	O, o	. . .	Omicron
Δ, δ	. . .	Delta	Π, π	. . .	Pi
E, ϵ	. . .	Epsilon	P, ρ	. . .	Rho
Z, ζ	. . .	Zeta	Σ, σ	. . .	Sigma
H, η	. . .	Eta	T, τ	. . .	Tau
Θ, ϑ, θ	. .	Theta	Υ, υ	. . .	Upsilon
I, ι	. . .	Iota	Φ, φ, ϕ	. .	Phi
K, κ	. . .	Kappa	X, χ	. . .	Chi
Λ, λ	. . .	Lambda	Ψ, ψ	. . .	Psi
M, μ	. . .	Mu	Ω, ω	. . .	Omega

INDEX

A

Absorption of radiation, 21, 299
Accessible wave functions, 396, 397
Action, 25
Action integrals, 29
Activation energy, 412
Adams, E. P., 201
Adsorption, unactivated, 388
Alkali atom spectra, 207
Alternating intensities in band spectra, 356
Amplitude equation, 56
in three dimensions, 86
Amplitude functions, definitions of, 58
Amplitudes of motion, 286
Anderson, C., 209
Angular momentum, of atoms, 237
conservation of, 11
of diatomic molecules, 265
of electron spin, 208
of hydrogen atom, 147
properties of, 425
of symmetrical top molecule, 280
Antisymmetric wave function, definition of, 214
Approximate solution of wave equation, methods of, 191
(*See also* Wave functions.)
Approximation by difference equations, 202
Aromatic carbon compounds, energies of, 379
Associated Laguerre functions, normalization, 451
polynomials, 131
table of, 135
Associated Legendre functions, 127
table of, 134

Asymptotic solution of wave equation, 68
Atanasoff, J. V., 228
Atomic energy states, semi-empirical treatment, 244
Atomic terms, Hund's rules for, 246
Atomic wave functions, 250
Atoms, with many electrons, 230*ff.*
variation treatments for, 246
Average values, in quantum mechanics, 89
of dynamical quantities, 65
of r^s for hydrogen atom, 144
Azimuthal quantum number, 120

B

Bacher, R. F., 258
Balmer formula, 27
Balmer series, 43
Bartholomé, E., 310
Bartlett, J. H., Jr., 254
Beach, J. Y., 385
Bear, R. S., 415
Beardsley, N. F., 249
Benzene, plane structure of, 411
structure of, 378
Beryllium atom, wave functions for, 249
Bichowsky, R., 208
Birge, R. T., 41, 336, 439
Black, M. M., 254
Black body, 25
Bôcher, M., 419
Bohr, N., 26, 36, 112
Bohr frequency rule, 27
Bohr magneton, 47
Bohr postulates, 26
Boltzmann distribution law, 399
Boltzmann statistics, 219

455

458 *INDEX*

Electron-pairing approximation, 374

Electron-spin functions for helium, 214

Electron-spin quantum number, 208

Electronic configuration, definition of, 213

Electronic energy function for diatomic molecules, 266

Electronic energy of molecules, 259

Electronic states, even and odd, 313

Electronic wave function for molecule, 261

Elliptic orbit, equation of, 38

El-Sherbini, M. A., 179

Emde, 343

Emission of radiation, 21, 299

Empirical energy integrals for atoms, 244

Energy, of activation, 412
  of classical harmonic oscillator, 5
  correction to, first-order, 159
    second-order, 176
  and the Hamiltonian function, 16
  of hydrogen molecule-ion, 336
  kinetic, definition of, 2
  of molecules, separation of, 259
  potential, definition of, 2
  of resonance in molecules, 378
  of two-electron ions, 225
  values of, for atoms, 246

Energy level, lower limit for, 189
  lowest, upper limit to 181

Energy levels, 58
  approximate, 180
  for diatomic molecule, 271, 274
  for harmonic oscillator, 72
  for plane rotator, 177
  for symmetrical top molecule, 280
  vibrational, of polyatomic molecule, 288

Epstein, P. S., 36, 179, 191

Equation, homogeneous, 60

Equations of motion, in Hamiltonian form, 14
  in Lagrangian form, 8
  Newton's, 2

Ericson, A., 225

Ethane molecule, free rotation in, 280

Eucken, A., 26

Eulerian angles, 276

Even and odd electronic states, 313

Even and odd states of molecules, 354

Even and odd wave functions for molecules, 390

Exchange degeneracy, 230
  integral, 212, 372

Excited states, of helium atom, 225
  of hydrogen molecule, 353
  of hydrogen molecule-ion, 340
  and the variation method, 186

Exclusion principle, 214

Expansion, of $1/r_{ij}$, 241
  in powers of h, 199
  in series of orthogonal functions, 151

Eyring, H., 374, 376, 414

F

Factorization of secular equation for an atom, 235

Farkas, A., 358, 414

Fermi, E., 257, 403

Fermi-Dirac distribution law, 403

Fermi-Dirac statistics, 219, 402

Field, self-consistent, 250ff.

Fine structure, of hydrogen spectrum, 207
  of rotational bands, alternating intensities in, 356

Finkelstein, B. N., 331

Fock, V., 252, 255

Force, generalized, 7

Force constant, definition of, 4

Forces between molecules, 383

Formaldehyde, rotational fine structure for, 282

Formulas, chemical, meaning of, 380

Fourier series, 153

Fowler, R. H., 412

Franck, J., 310

Franck-Condon principle, 309

Frank, N. H., 275